TUBE
AMP
TALK
FOR THE
GUITARIST
AND
TECH

Kendrick Books
531 County Road 3300
Kempner, Texas
76539-5755
gerald@kendrick-amplifiers.com

ISBN: 0-9641060-1-9

Editor: Jeanne Schmitt Watkins
Art Direction and production: R. K. Watkins
Scans and prepress: Spit'n Image
Printed in the United States of America

Disclaimer: Tube amplifiers contain high voltages which may be lethal, even if the amplifier has been off for some time. We do not recommend that you open your amplifier or try to perform any repair operations unless you are properly trained in electronic servicing. Gerald Weber and Kendrick Books accept no responsibility for accidents resulting in personal injury or destruction of property. Again, there are large voltages present in your amplifier that can kill, even with your amplifier unplugged from the wall.

DEDICATION

I dedicate this book to my mother, Gloria Mae Weaver Weber, without whom my life would not have been possible.

ACKNOWLEDGEMENT

In no particular order:

Jill Kendrick Weber, my devoted wife and soulmate, for her constant unselfish support.

Ken Fischer, my friend and mentor, for contributing so much to my life.

Billy "F" Gibbons, my tone hero, for defining that good "Texas tone" back in 1970 at the Town House in Groves, Texas.

Terry Oubre, the worlds best guitarist, for his support and valuable insight.

Robert and Jeanne Watkins for their support and enthusiasm.

Alan and Cleo Greenwood, for publishing the articles from which this book was compiled, in *Vintage Guitar* magazine.

All of the employees at Kendrick for providing me the freedom to spend time on this project.

Everyone else who participated in this book.

Anyone else I forgot to acknowledge.

PREFACE

In 1993, when I released my first book, *A Desktop Reference of Hip Vintage Guitar Amps*, I couldn't imagine ever writing another book. That book, like this one, took three years to write. This book, that you hold in your hands, is a compilation of articles that were written for *Vintage Guitar* magazine from 1993 through 1996, along with my advice columns ("Ask Gerald") in that same magazine. As a special bonus, Ken Fischer contributed three years' worth of articles that he originally wrote for *Vintage Guitar* magazine. These are the "Trainwreck Pages" in Section lll of this book

Many things have changed in the last few years. Guitarists nowadays are much more knowledgeable about tubes and their tube guitar amps. For instance, guitarists who would not have even thought about changing fuses in their amps a few years ago, now feel confident to re-tube and re-bias their own amps. Many guitarists are learning to do their own cap jobs and even tonal modifications. In the spirit of a sports car driver who likes to "tune up" his own car, many players feel it is time for them to "tune up" their own amps.

Whereas my first book assumed the reader knew nothing, *Tube Amp Talk for the Guitarist and Tech* assumes that the reader has at least read my first book, *A Desktop Reference of Hip Vintage Guitar Amps* or has more knowledge of a tube guitar amplifier than the average drummer.

Whether or not you intend to perform your own amp servicing, *Tube Amp Talk for the Guitarist and Tech* is my conversation with you about one of my favorite subjects—*tube guitar amps*!

FOREWORD

Tone and Loudness, Yeah, Tone and Loud, make you wanna jump and shout. And...inside is fine work, Mr. Gerald Weber contains most of the classic and many examples of how to get there. There's a single element which may be found as the highlight of all of it, too...quality. Make that "Quality." Each example studied is an illustration of inside and out this unlikely combination of circuit, power, componentry and arrangement. The remarkable conclusion allows a mystery factor of Voo Doo to remain amidst this neatly laid architectural diagram of tone-making. "Turn it Up!" is a fitting tribute to the needed collection of "The Best of the Bestest" and, who knows?? Volume 2 may well be a double-sized, quantum-prized gift of Xtra stuff goin' for more Tone and, of course, more volume. Tube yo' cube and CRANK.

If you're willing to get under-the-hood, there ain't no shortage to wrench 'n rock on in here. Squeezin' out more horsepower from your fave-o-rite mule is all at hand in these pages of rage. Everything from a quick-change in bias, to elaborate rewire schemes goin' top to bottom and bottom to top.

So, from the glitz and glimmer of Hollywood, to the Texas rude, crude and super-blue'd, New York City 'cross to loudness-in-London, and easin' through Euro-town to ampin' up in Africa, turn to your favorite section here and let it rip.

Red Beans and Ricely Yours,

Billy F Gibbons

Billy F. Gibbons

SERVICING YOUR GUITAR AMPLIFIER

GETTING RID
OF NOISE AND HUM

At one time or another, all of us have experienced the sputter, coughing, hissing, rattling, popping, humming, crackling, spitting, and yes, the intermittent static of an old vintage tube guitar amplifier. If your amplifier sounds like a popular breakfast cereal, then it looks like we are just in time. Since every noise is related to some malfunction in the amplifier, the subject will be how to identify these different noises, isolate the malfunctions through troubleshooting techniques and ultimately correct the malfunctions. Would you like your vintage amplifier to be quiet and noise-free, just like a brand new amplifier? That's what I thought, so let's get started.

First, we will identify the types of noises that could possibly come from an amplifier and then I will introduce you to a few very simple, yet effective, troubleshooting concepts to help you locate where noise is coming from in your amp.

TYPES OF NOISES

Knowing exactly what type of noise you have will give you a clue about the malfunction and where to go from there. Here are some of the most poplar noises found in vintage amplifiers:

HISSING—Hissing is a constant "sssssssssss" type of sound that occurs whether signal is being put through the amp or not. It will usually increase in volume as the volume control is turned up. This type of noise can be troubleshot using the stage isolation technique, described later in this chapter. It is usually related to a preamp tube or plate load resistor.

STATIC—Static is intermittent and has more of a pop to it than hissing. This is also troubleshot using the stage isolation technique and the chopstick procedure. Possible causes of static include: loose electrical connection, faulty solder joint, arcing coupling capacitor, bad tube, bad resistor, bad ground connection, dirty standby switch and arcing between adjacent components. There are probably a few other causes besides these.

HUMMING—Humming falls into two categories—60 cycle wall hum and 120 cycle power supply hum. The 60 cycle hum is related to either the filament supply or hum originating outside the amplifier being picked up by the guitar pickups. Possible causes of 60 cycle hum include cross-wiring of output tube filament connections (this makes a push-pull amp hum inducing instead of hum canceling), a burnt hum balance pot or resistor, or proximity of the filament wires to a grid wire. Another possible cause on a fixed bias amp would be a bad bias supply filter cap. If the humming stops when the guitar is unplugged, it is the guitar that is humming and not the amp.

A 120 cycle hum is always related to the D.C. ripple current. When 60 cycle wall A.C. is rectified to D.C. it becomes 120 cycle pulsating D.C. This type of hum is generally caused by bad filter caps but it can also be caused by a coupling cap becoming conductive and draining so much current from the power supply that the filter caps become ineffective by comparison. A voltage check on the plates of the tubes will reveal coupling cap failure because the plate voltage of the corresponding tube will be noticeably low.

CRACKLING—Crackling is similar to static but it has more of a "pop" sound to it and it is more severe. It is trouble shot in the same manner as static. Possible causes are the same as static but would also include humidity being absorbed by resistors (this would make small arcing occur inside the resistor which accounts for this noise), arcing in a transformer or socket, internally broken lead of a resistor or coupling cap, or a faulty pot.

RATTLING—Rattling can be of two different types—electronic rattling and acoustic rattling. Both types of rattling can be sympathetic to certain frequencies. The electronic rattling can include speaker problems such as voice coil rub, loose internal components in a power tube or rectifier tube, loose speaker connections, intermittent grounding, loose sockets, among other things. Acoustic rattling can be caused by the cabinet itself—especially a loose back-panel, nameplate, speaker mounting screw, or baffleboard. I have even experienced a hairline fracture in the baffleboard that caused severe rattling yet was virtually not noticeable to the eye. Believe it or not, I have actually experienced grille cloth rattling against the baffleboard.

MOTORBOATING—Motorboating is a oscillating sound that gets its name from the "putt putt putt" sound of a motorboat. The exact cause is sometimes very hard to find because it is related to some type of positive feedback happening in the amp that will not be obvious. This positive feedback could be related to improper lead dress, improper grounding, bad filtering (inadequate de-coupling), or a conductive component board.

UNNATURAL HARMONICS OR GHOST NOTES—Ring modulator type sounds that include an unnatural harmonic on top of the note being played are always related to the filtering of the D.C. power supply or a subharmonic being generated in the amp. Filter cap replacement will generally take care of the problem. In the case of the filter cap malfunction, the problem will be most noticeable when playing a B flat on the "G" string (15th fret). In the case of the subharmonic, the problem is most noticeable in the higher register, but not necessarily on the B flat. In this case, changing the coupling caps feeding the output stage to a smaller value will usually stop the problem. For instance, changing the .1uf coupling caps feeding the grids of the 6L6's of a Super Reverb to .047 uf or .02uf will stop the subharmonic without adversely affecting the tone.

POPPING—Popping is similar to crackling except it is more definite and more intermittent. Possible causes include rectifier arcing, tube socket arcing, intermittent component connection, bad tube socket, arcing transformer, arcing capacitor. Troubleshoot this using the stage isolation technique.

SPUTTERING—Sputtering is like crackling but with a little hiss and pop mixed in. Some causes would include bad filter caps, dirty standby switch, noisy output tubes or rectifier.

BUZZING AND FIZZING—Buzzing or fizzing is similar to rattling, but is a higher pitch sound. All of the malfunctions associated with rattling are also common with buzzing and fizzing. In addition to those types of malfunctions, buzzing and fizzing could also be the result of parasitic oscillations. Improper lead dress, too long grid wires, and improper power tube grid resistor mounting are some of the other causes of such a malfunction.

SQUEALING—Squealing is a high pitched ringing that is usually associated with microphonic preamp tubes feeding back. This problem is troubleshot by stage isolation. Other causes could be acoustic feedback originating from the reverb circuit. Usually this problem is remedied by installing weather stripping on the reverb pan to dampen the acoustic ringing and making sure there is a Tolex bag for the pan.

POSSIBLE SOURCES OF NOISE AND HUM

A.C. LINE CORD—Vintage amps never used three-prong plugs and therefore do not have an earth ground. Change the power cord to a three-prong power cord. (Be sure to save old parts.) The green wire (ground), on the power cord, goes to the chassis of the amp. If you are not absolutely certain of what you are doing, bring the amp to a competent technician.

POWER SUPPLY FILTERING—When capacitors in the power supply get old, they sometimes loose their effectiveness in reducing power supply hum. A simple cap job (all electrolytic capacitors in the amp replaced and charged properly) can solve this problem. Also, the capacitor on the ground switch (usually a .047 or .05 mfd. @ 630 volts) might be in need of replacement. You must use a 630 volt or higher-rated capacitor for this replacement.

PREAMP TUBES—There are times when a preamp tube will hum. This can be checked and corrected by replacing the tube with several preamp tubes that are known to be good and pick the one with the least hum. Be careful that the tube is not simply reducing hum because it has weak gain. A preamp tube should have good tone, good gain and not hum.

UNBALANCED OUTPUT TUBES—Push-pull amps (these usually have two or four output tubes) will cancel hum if they are matched. This is another reason why matched tubes are better. Each tube hums a little, but since they are out of phase, they cancel each other's hum. Matching within 5 mA of current is close enough to cancel the hum coming from the output tubes. Also when filaments are hooked up in parallel (this includes almost every tube amp ever made for guitar), it is important that they be wired in phase. For example, the same wire that goes to pin 7 on one tube must go to pin 7 on the other tube. The wire that goes to pin 2 on one tube, goes to pin 2 on the other tube. We are talking about output tubes here. This keeps them hum-canceling.

Improper Ground Reference on Filament Heaters—The 6.3 volt filament heater supply should be grounded on the chassis. It is best if the amp has a center-tap going from the filament heater transformer winding to ground. If it doesn't have a center-tap that is grounded, you can make an artificial center-tap by hooking a 100 ohm ½ watt resistor from one side of the filament supply to ground and then put another 100 ohm from the other side of the filament heater supply to ground. Some early amps used a ground connection for one side of the filament heater and only ran a single filament heater hook-up wire to the other side. This works, but it would probably work better if each side had its own filament supply wire. If you decide to convert such an amp, remember to twist the pairs as you are running them from tube to tube to reduce hum. Of course, the artificial ground mentioned earlier would apply for this set-up.

Shielding Inadequacies—Many vintage amps are not shielded properly and therefore are subject to interference. For instance, all Fender Tweed amps had a wooden back-panel with no shielding. This allows all sorts of hum and noisy interference. They had asbestos on the back, but only to keep the tweed from catching on fire. A simple remedy is to cover the inside back panel with heavy-duty aluminum foil. An even coat of Elmer's glue on the inside back-panel should make that aluminum foil stick nicely. Make sure that the foil touches the chassis (you may have to re-adjust the chassis mounting screws to assure that this happens.) If that is not possible, staple the stripped end of an insulated wire to the foil and then solder the other end (stripped also ¼") to the chassis ground. This will reduce hum and noise.

Another spot that is inadequately shielded on most vintage amps is the wire going from the input jack to the grids of the preamp tube (pin 2 or pin 7 on both the 12AX7 and/or 12AY7). Replacing this wire with a shielded wire will usually reduce hum. I usually just ground the shielding on the input end and do not ground the shielding at the grid end.

External Hum Causes—Sometimes the hum is coming from something outside the amp, in which case working on the amp is futile. For instance, the guitar may be humming. Another example might be that an otherwise quiet guitar is picking up hum from the fluorescent lights in the room. You could be standing too close to your amp. To check for external hum, unplug your guitar. If it goes away, it's not the amp.

BAD CONNECTIONS BETWEEN TUBE PINS AND SOCKET CONNECTORS—
Have you ever had a battery cable not making good connection in your car? Even though the cable looked connected, it was not connected well enough? Just because the tube is in the socket does not necessarily mean that all of the pins of the tube are adequately connected to the rest of the circuit. Loose sockets can be troubleshot by simply rocking the tube in a circular motion while the amp is "on" and in the "play" mode. With the volume up, you rock the tube gently and listen for static and noise. If you hear any static or noise, the socket needs service. The small preamp tubes are usually not very hot and can be handled with your bare hands. Big tubes (rectifier, output tubes or any other big tubes) will generally be **very hot** and should not be handled with your bare hands.

Possible cures are to clean and possibly re-tension the socket, or replace the socket. To clean and re-tension, turn the amp off and unplug it. Use alcohol or suitable tech spray (there are differing opinions about which is better; try both and judge for yourself) on both the socket and the tube pins. Insert the tube in the socket with a gentle circular rocking or scrubbing motion as the tube pins are going in. Now with the same gentle circular rocking, withdraw the tube almost all the way, but not quite. Now re-insert in a scrubbing motion. You want to rid the tube of any oxide or film that would otherwise corrode the connection. This alone may be enough to repair the problem. You may have to do this more than one time to really get the inside connector clean. Sometimes the connector inside the socket is "sprung" and is not making good contact because it is "sprung." In this case you will need to re-tension the socket's internal connectors.

To re-tension the tube sockets, unplug the amp, turn it off. If it has a standby switch, take it off the "standby" mode and put it in the "play" mode. You must drain the power supply to avoid accidental electric shock. You may already be familiar with draining the power supply by grounding any plate lead in the socket to the chassis with a jumper-wire. The plate lead is either pin 1 or 6 on a 12AX7, 12AT7, or 12AY7. If you are performing this procedure with the amp chassis still in the cabinet, remember you are looking at the side of the socket that the pins are numbered counter-clockwise. If the chassis is out of the amp and you are looking at the soldered side of the socket, the pins are numbered clockwise. I like draining the power supply from the preamp tubes because all of the power must go through the typical 100K ohm plate load resistor. This resistor will limit the current and therefore discharge the power supply capacitors slowly. It is for this reason that you must leave the jumper connected for 20 to 30 seconds before the capacitors will be totally drained.

Find a tool to re-tension the socket. A safety pin, dental pick, paper clip or some other small tool is suggested. The idea is to re-tension the internal connectors of the socket, so that they will make tighter contact against the tube pins upon re-installing the tube. Be careful not to chip the insulators on the socket.

Time to re-test the amp. If the problem persists, you may need to replace the socket.

LOOSE CONNECTIONS ON JOINTS AND COMPONENTS—Loose connections can cause static, crackling, electronic rattle and many other noises. These are easy to locate and repair; however, do not attempt this procedure unless you are certain that you know what you are doing. It involves working on a "live" amp with the chassis removed from the cabinet and therefore is not recommended for the novice.

Remove the chassis from the cabinet and hook up a suitable speaker. Turn the amp "on" and put the standby switch in the "play" mode. Set the volume and tone controls up all the way. Find a wooden chopstick and use it for a probing device. Starting at the input jack, and using the chopstick as an insulated probe, begin gently poking at the connections on the input jack. Follow the wire to related components and poke around at joints, wire and component connections. You are looking for noise when a joint is disturbed by a gentle poke or tap from the chopstick. Be sure and check every connection and every component in the entire amp.

This procedure can get tricky sometimes because the mechanical shock produced by tapping or poking one component or joint can easily disturb some adjacent joint. The noise may be coming from the adjacent component or connection and not the one you are tapping. **Therefore I recommend that you double check** surrounding components and then re-evaluate before believing you know where the noise is coming from. Also, the lead for a component will sometimes be broken right at the point where the lead enters the component. This condition is sometimes unnoticeable upon visual inspection but will be revealed in this "chopstick" test.

If you find a bad joint, it is important to remove the old solder and then resolder using new solder. Be careful not to overheat associated components when soldering near them.

PREAMP TUBES—Certainly the easiest thing to troubleshoot for noise is the preamp tube. Starting with the amp "on" and in the "play" mode, remove the phase inverter tube. (This is the preamp tube next to the power tubes). If the noise stops, then you will know that it is coming from either the phase inverter tube or any other preamp tube as well. You would know this because removing the phase inverter tube disconnects the preamp from the output stage and if the noise stops, you know it is not coming from the output stage. This leaves, by process of elimination, only the phase inverter and preamp.

Proceed by replacing the phase inverter tube and removing the preamp tube next to it. If the noise persists, it is coming from the phase inverter tube or related circuit because that tube and circuit are now connected to the output stage. If the noise stops, it is not the phase inverter tube.

Proceed in a similar manner by replacing the preamp tube and removing the one next to it. The idea is to find which tube or related circuit is noisy and therefore in need of repair.

Sometimes, and especially in high-gain amps, the tube noise can present itself as a squealing microphonic feedback sound. Sometimes the noise will present itself as hiss or static. I have even heard intermittent popping sounds.

Test by replacing the suspected noisy tubes with good quiet tubes. If replacing with a known good tube does not fix the problem, the tube could actually be OK and the noise could simply be coming from the circuit associated with that particular preamp tube. More on this later.

BAD SPEAKER CONNECTIONS—A source for loud static, intermittent sound or sympathetic electronic rattle, bad speaker connections are easily troubleshot. Probing connections on speaker leads and jack/plug connections and speaker wire could reveal if this condition exist. Substituting a different speaker with its own speaker cable could let you know if the noise is coming from the amp or the speaker and connections.

When re-soldering old joints, always de-solder and repair with new solder.

ARCING COMPONENTS—Although sometimes more difficult to troubleshoot because of their intermittent nature, arcing transformers, arcing resistors and capacitors can be a source of intolerable noise.

An arcing resistor will sometimes look charred or burnt. The "chopstick" testing recommended earlier in this article is an excellent way to find arcing resistors. Another suggestion is to troubleshoot as if you are troubleshooting noisy preamp tubes described above. This will isolate the source of the noise. Tapping an arcing resistor with a chopstick will make it arc which in turn will usually make an awful sound come from the speaker.

Arcing capacitors and arcing transformers can be troubleshot by isolating the noise and then either substituting a good part for the suspected bad part or....

Use a non-conductive hose-type stethoscope with the diaphragm assembly removed. You will be wearing the stethoscope as normal, except the end with the diaphragm will be removed and you will place the end of the hose (that was originally connected to the diaphragm) on components that you suspect are arcing. You are listening for arcing in the suspect component. Again, don't try this unless you are certain you know what you are doing. This procedure involves working directly in a working "live" chassis and is not for the novice to attempt.

INPUT JACKS—Especially a problem on dual inputs where a shorting jack is used, input jacks can cause loud static noises or loud hissing and other objectionable noises. This shorting jack is supposed to short out a resistor when the jack is not being used. If the contacts on the jack are not making an adequate connection, loud noises can occur. The jack can be loose or corroded so that it is not making a good ground connection to the chassis.

Visual inspection of the jack while inserting a plug or turning the amp on and plugging into the suspected jack (listening for noise changes) are good ways to troubleshoot this problem. I have sometimes even forced the connection with a jumper to see if this would cure the noise.

A jack can rarely be re-tensioned, once it has sprung. Sometimes they are not sprung but in need of just a minor adjustment. Best to replace the jack.

But if you want to try a re-tension, assuming that the jack is not sprung, here's the procedure. Remove the chassis and drain all electricity as described earlier. Insert a plug into the input jack so that the contacts on the jack switch open. Clean the contact surface with tech spray. Then, using needle-nose pliers, gently bend the switching contacts closer to each other. Be careful to bend evenly. A little dab will do you. Be careful not to bend too far. Now remove the plug and look at the jack. If necessary, repeat this procedure. Test the re-tensioned jack for noise. You may still have to replace the jack, but there are times when re-tensioning is all that is needed.

EXTERNAL NOISE—Maybe we should have talked about this first. What if the noise is coming from your guitar or cord? Ouch! I hope you read the entire article before troubleshooting every way described above. Oh, well. Turn on the amp and unplug the guitar cable from the amp input jack. Did the noise disappear? If the noise disappeared, the noise is not coming from the amp. Faulty connections on the guitar cord or guitar input jack or bad pots or connections in the guitar could be causing the noise.

FIVE SIMPLE TROUBLESHOOTING TECHNIQUES

Although there are many troubleshooting techniques available, the five most common that I feel will yield the fastest results with the least effort are: visual inspection, stage isolation, voltage checking, chopstick testing, and substitution.

VISUAL INSPECTION—You may be surprised to learn that visual inspection is probably the most valuable troubleshooting technique. Time spent looking very carefully at each component and each connection in an amp will almost certainly help you locate malfunctions quickly. Look for burnt resistors, burnt or frayed wires, bubbles forming on electrolytic capacitors, and any other thing that looks like it may have been malfunctioning. Carbon deposits on tube sockets are a dead giveaway that arcing has been occurring. Look for leads that are to close to other connections and tell-tale signs of cold solder joints or otherwise questionable connections. Once, I had an apprentice who knew almost nothing about amps, but he could easily find a malfunction more quickly than a trained technician because he was so good at using his eyes and looking in the amp.

STAGE ISOLATION—The technique of stage isolation is simply to remove a tube from the circuit and check to see if the noise goes away. If it does, the malfunction was either originating from that stage or an earlier stage. Actually the quickest way is to remove the phase inverter tube (usually the preamp tube closest to the power tubes) first. If the problem persists, it is in the output stage or speaker or cabinet. If the problem stops, replace the phase inverter tube and remove the preamp tube next to it. If the problem stops, replace the tube and remove yet another preamp tube. Continue until you find which tube both stops the problem and starts the problem. This narrows down the malfunction to that particular stage of the circuit.

VOLTAGE CHECKING—Once you have located the stage in which the malfunction occurs, a voltage check may reveal what is going on. For instance, if the plate voltage reads low, you could have power supply problems or a leaky coupling cap to that stage. High voltage on the cathode would indicate a bad ground on the cathode resistor circuit. D.C. voltage on the grid could indicate, in some cases, that the coupling cap from the previous stage or grid emission from that stage is the problem.

CHOPSTICK TESTING—Perhaps one of the best ways to troubleshoot intermittent noise is with a technique I call chopstick testing. Periodically, we eat lunch at a Chinese buffet. Since I always eat Chinese food with chopsticks, I always have plenty of chopsticks laying around the office. When an amp makes an intermittent noise such as crackling, I turn the amp on with a speaker load hooked up and the volume control full up. By carefully physically tapping each and every connection, lead, socket, component, transformer, tube, etc., I will most certainly discover any broken leads, bad joints, faulty grounds, or other intermittent connections.

SUBSTITUTION—When a part is suspected of malfunctioning, substituting a known good part can sometimes be the fastest way to confirm the malfunction's source. For instance, if you have a squealing preamp tube, putting a known good tube in the socket could confirm the malfunction if the squealing stops. Resistors and coupling caps are relatively inexpensive and simple substitution can save time and let you know if your troubleshooting is going well. Speakers and cabinets can rattle, so hooking the amp chassis to a known good speaker cabinet may save time in chasing down that rattle.

EPILOGUE

Making a vintage amp noise-free can sometimes become a major headache because, more often than not, there will be multiple problems occurring intermittently. This could make you feel like you're working in a maze because coincidences can make you think one component is malfunctioning when the noise is really coming from somewhere else. When my daughters were little, I use to play a game where I would push on the rear-view mirror in my car with my right hand while blowing the horn with my left hand. This would create the illusion to my young daughters that the rear-view mirror was the controller for the horn. Of course, when they tried to "blow the rear-view mirror," they couldn't do it and were completely amazed to see me perform this magical feat. Troubleshooting can be that way at times. You may be performing a chopstick test on a lead and at that moment the transformer will intermittently arc, making you think the lead is the problem. Make sure and check and double check. It may take patience and time, but a noise-free vintage amplifier is worth the effort.

PERFORMING A CAP JOB ON YOUR AMP

So you got that cool amp you wanted, and it may have sounded great for a while, but lately when you crank it up you get some weird harmonics that follow the note in an unusual way. Maybe the amp just sounds sluggish—as if it is straining just to get a sound out. Maybe the volume is weak and the gain seems a little too low. All of these symptoms could be the tip-off that it's time for a cap job.

WHAT'S A CAP JOB?

Let's backtrack and keep it simple. All tube amplifiers use high voltage that is D.C. (all the current moves in one direction), but your wall voltage is 120 volts A.C. (the current moves in both directions); therefore the amplifier has a power supply to convert 120 volts A.C. to, let's say, 400 volts D.C. This power supply consists of a step-up transformer that steps-up the voltage, a rectifier (either solid-state or tube) that converts the A.C. into pulsating D.C., and some filtering devices to smooth out the pulsating D.C. to smooth D.C. These filtering devices include filter capacitors and sometimes a smoothing choke. The capacitors smooth out the voltage and the choke smoothies out the current. The type of filter capacitor typically used to stabilize voltage is called an electrolytic capacitor. This type of capacitor uses an electro-chemical reaction to store an electron charge.

An electrolytic capacitor is almost exactly like a high voltage battery. Like the battery, it has a positive and negative lead. Both use an electro-chemical reaction to store electricity, and both **do not last forever**. The big difference is that the power supply is constantly charging the electrolytic capacitor when the amplifier is "on."

A capacitor manufacturer told me that a typical electrolytic capacitor will last approximately 6 to 10 years. My experience is that the voltage rating of the cap as used in a particular circuit and the heat at which the cap is operated also play a big part in how long a capacitor will last.

And finally, a cap job is: **the act of changing all electrolytic capacitors in an amp**. If the amp is a fixed-bias amp, there will be a cap or two in the negative voltage supply. Electrolytic caps are also used as bypass capacitors in the preamp tube circuits and power tube circuits on cathode-biased amps. These are used to bypass A.C. signal voltage around the cathode resistors in the tubes. When these are bad, the sound is lacking gain/volume and sometimes mushy. Dynamic response will also suffer.

It's important to know that filter caps do not all go bad or fail in the same way. Sometimes filter caps leak current, in which case the power transformer is burdened with having to supply a lot more current than it was designed to supply. This can cause a transformer to overheat and possibly blow. If the problem is severe enough, the amp may blow fuses. Sometimes the positive ends of the cap will rupture and chemicals will ooze out of the rupture. Sometimes the cap will not leak current, but will just not filter, in which case there will be severe hum (this is called D.C. ripple current and it will be 120 Hz— exactly twice the wall hum of 60 Hz).

Ruptured capacitors (courtesy Cari Duty)

FIRST LOCATE THE CAPS

The best way to find the main filter caps is to locate the rectifier circuit. If the amp has a rectifier tube (unless modified, almost all vintage amps do), the first of these filter caps would start from pin 8 of the rectifier tube socket, assuming that the amp uses a standard rectifier tube. Standard rectifier tubes include 5Y3, 5V4, 5AR4, 5U4, 5R4, GZ32, and GZ34. If the amp uses a non-standard rectifier tube or a solidstate rectifier, the first of these filter caps would start at the cathode of the rectifier device. Capacitor values are typically 8 to 50 microfarads at 450 to 600 volts. Generally there are three to five of these caps in an amp. They start from the rectifier and are separated by power resistors. Sometimes the rectifier goes to a standby switch with a small value cap on the rectifier side of the switch and the filter caps start on the other side of the switch. In most Fender amps, there is a pan on the chassis with the filter caps inside the pan. Some vintage amps have a multi-cap which is actually three or four caps in one. These look like a metal cylinder on the chassis that resembles a tube in size.

The filter cap/caps in the negative voltage supply will be connected to the anode of the solidstate rectifier. There may be another one slightly down circuit from there. Typical values will be 8 to 60 microfarads at 100 to 200 volts.

To find the bypass capacitors simply locate the cathode resistors of the preamp tubes and the cathode resistor for the power tubes (in a cathode-biased amp). Each bypass cap will be connected across a cathode resistor with the positive lead towards the tube and the negative lead to ground.

NOTE POLARITIES

Of utmost importance is getting the polarity correct. This means that the electrolytic caps only go in one way. There is a plus or positive lead and a minus or negative lead on a filter cap. One side of the cap goes to power and the other side goes to ground. Before removing the bad caps, notice how they are situated in the circuit. Unless the cap is used to filter negative bias supply voltage, the plus will go to power and the minus will go to ground. On a multi-cap, one terminal will be minus (ground) and will be internally connected to all minus leads of each cap while the plus terminal will each be individual. Again the best way for the novice is to notice which way they go before removing any caps. Maybe a drawing would be useful, so you do not have to remember anything.

USE CORRECT VALUES

At this point you would want to make a list of all capacitors needed to perform a cap job. Correct value on microfarad rating is essential to maintain the original tone of the amp; however, you may go more on the voltage rating without affecting the tone of the amp. For instance, if you need a 8 microfarad 160 volt cap and you replace it with an 8 microfarad 175 volt cap, don't expect to hear any difference.

A common difficulty is finding a 600 volt capacitor that will fit into an amp. I've seen techs replace a 600 volt cap with a 500 volt cap because they couldn't find a 600 volt cap. This works in some cases, but not all. For instance, vintage 4x10 Bassmans use 20 microfarad 600 volt caps which are simply not available. If you use 500 volt caps on the main filter supply (these are the two 20 mfd. 600 volt caps connected to the standby switch), they will soon fail. Using 500 volt caps for the other filter caps in the amp will work just fine, so what's a guy to do?

Use two 100 mfd. 350 volt caps in series instead of the two 20 mfd. 600 volt caps in parallel (stock) for the main filter supply. Putting the two 100 mfd. 350 volt caps in series will give a total of 50 mfd. 700 volt; stock set-up is 40 mfd. 600 volt. This will give you the advantage of a little extra voltage rating (caps will last longer) and a 20% upgrade in terms of microfarads. Capacitors are rated plus or minus 20%, so by upgrading the main filters by 20%, you are taking no chances. Also, heat affects cap performance, so having that little bit extra, the amp will have an advantage when it does get hot.

When placing two capacitors in series, the positive of one cap goes to the power and the negative of that same cap goes to the positive of the other. The remaining negative lead would then go to ground. I would recommend installing two 220K -1 watt resistors; one across each cap. This will keep the voltage equal on each cap so that you will get the full 700 volt rating. You wouldn't want one cap to be taking 75% of the voltage and the other 25%!

Another recommendation: when replacing the main filters in a blackface Fender Pro, Super Reverb, etc., use 100 mfd. caps instead of the stock 70 mfd. caps. This is for the same reason stated above. These amps already have the two 220K -1 watt resistors because they are in series.

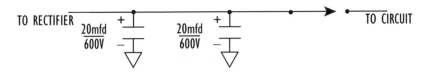

TWO 20MFD IN PARALLEL

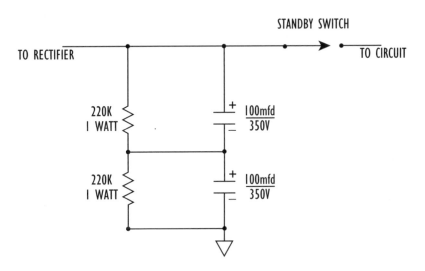

TWO 100UF IN SERIES

CHANGE ALL ELECTROLYTICS

Even though the filter caps generally last longer than bypass caps, I like to go ahead and change them all at once. They are about the same age and consequently subject to failure at any time. This way, you can be certain that all the caps are good, and you may prevent the situation where the amp fails because a cap that was good two weeks ago finally failed. I like to think that it's like changing tires on a car. If you have a couple of tires on your automobile with the steel belt showing, you would probably want to go ahead and change the entire set rather than chance a future blow-out on the freeway!

As always, unplug the amp and drain all voltages properly before changing anything. I like to use a solder sucker and remove all of the old solder in a joint, and then install the capacitors using new solder. New solder is better.

If you don't do cap jobs every day, you might want to change one capacitor at a time so that you don't get confused about what value goes where or correct polarity, etc. When you have changed each capacitor, it is time to charge it up.

CHARGING IT UP

After changing all of the electrolytics in the amp, I would recommend checking your work and then bringing the amp up slowly on a Variac. I usually set the Variac on 40 volts and leave it on overnight. The next day I bump up the Variac voltage about 10 volts per hour and at the end of the day, the amp is being given the full 120 volts. Electrolytic capacitors need to have D.C. across the terminals so that the dielectric has a chance to form. Do not evaluate the tone of the amp until the capacitors have been on for several hours and the dielectric is fully formed.

When replacing caps, be aware of the physical location of the capacitors because they can fail prematurely if they are located too close to a hot resistor or other component. Particularly for a cathode biased amp, make sure that the cap is not anywhere near the cathode resistor or screen resistor of a power tube. Heat will make a 50 volt cap think it's a 25 volt cap and it could rupture prematurely. Heat can also make a 500 volt cap think its a 400 volt cap. Be aware.

There is one last thing that I would like to clear up. I've heard people remark that they sure wish they could get some of the old style paper filter caps. There is no such thing as a paper filter cap. The paper outer covering was used to insulate the aluminum can electrically and also to label the value of the cap, etc. If you have an electrolytic that has a paper outside, tear off the paper and there is an aluminum can inside. The caps made today are also aluminum can type, except shrink wrap is used to label the cap and insulate the can instead of paper.

After fully forming the caps, the amp is ready to play. Listen for improved dynamic response, gain, and overall aliveness of sound.

FINE TUNE THE REVERB ON YOUR BLACKFACE REVERB AMP

Although most players prefer the reverb sound of a stand-alone reverb unit such as the Fender 6G15 or the Kendrick Model 1000, there are many players who have blackface amps that have on-board reverb already. These reverb circuits are very limited because they cannot control the actual tone of the reverb and one must settle for however it happens to sound, regardless of its actual tone. The reverb control on these amplifiers is just a mix control that will only allow the player to get more or less of the reverb in the mix. I am going to offer some suggestions for how to fine tune your blackface reverb to sound more like an original Fender 6G15 or a Kendrick Model 1000. Depending on how far you would like to take these modifications, you could amaze yourself for less than $10, provided your blackface has a normal channel that you do not use. The modification doesn't actually change anything on the amp itself, so your amp remains dead stock.

YOU DON'T HAVE TO BE A TECH

Here's the most basic and simple mod to dramatically improve your reverb tone. My good friend from Finland, Sampo Kolkki, invented this mod and I must say it works very well. You will only need a 6-foot shielded patchcord with a quarter-inch phone plug on one end and an R.C.A. phono plug on the other end.

On the back of your amp, you will notice an R.C.A. jack marked "Reverb Output." There will be a patchcord running from this jack to the jack on the reverb pan. Unplug this patchcord at both ends and set it aside. Now take your new patchcord and plug the R.C.A. connector into the pan and plug the quarter-inch plug into the normal channel of the amp. (There will be nothing plugged into the jack marked "Reverb Output".) Now that wasn't so hard, was it?

LET'S PLAY THIS AMP

The normal channel volume control is now the mix control for the reverb; the bass and treble controls will allow you to get the reverb tone you've been missing. That bass control can really thicken things up a bit and the treble control will add more presence. If you like this modification and would like to make it permanent, it can be wired internally by simply doing away with the new patchcord. Use the original patchcord (hooked up stock), and run a wire internally from the hot lead of the Reverb Output jack to pin #2 of the first preamp tube. The wire that was already on pin #2 must be removed. I suggest you tape this wire up, in case you ever want to put it back to stock.

ANYONE WANT TO CARRY THIS EVEN FURTHER?

Even though this modification is very cool, you still only have tone controls and mix controls, and you cannot control how hard the reverb pan is being driven. (Incidentally, the dwell control on the Fender 6G15 and the Kendrick Model 1000 are the controls that allow the user to adjust this parameter.) You can carry this mod to the next level by adding a dwell control. Your stock Reverb knob is not being used at this point.

LET'S HOOK UP THE DWELL CONTROL

You'll need to have a little expertise to do this modification. Remove the center lead from the stock reverb pot and ground it. Remove the right lead and tape it up. Now remove the pot and replace it with a 1 meg ohm pot, but don't solder anything to it yet. There are three leads on the back of your new 1 meg pot. Looking at the back of your 1 meg ohm pot, the left lead gets grounded. Now look at your reverb driver tube. (The 12AT7 is the third preamp tube.) Pins #7 and #2 are jumped together and there is a wire that runs from one of these leads to the circuit board. Leave pins #7 and #2 jumped, but disconnect the lead that runs from them to the board. (You might want to tape this lead up.) Now run a wire from either

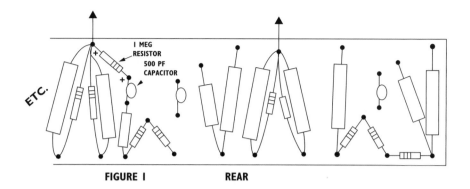

FIGURE I   **REAR**

pin #7 or #2 to the middle lead of your new 1 meg pot. On the board you will notice a 1 meg resistor that we will remove (see Figure 1). Originally this resistor was grounded at one end and connected to a 500 picofarad cap at the other end. From the 500 picofarad cap lead, run a wire to the right lead of the 1 meg pot and you are done. (This wire connects to the lead of the cap that originally went to the 1 meg resistor that you just removed).

LET'S PLAY THIS BABY

Your knob marked Reverb is now actually a dwell control, so you can alter how hard the reverb pan is being driven. If you turn it up all the way, you are dead stock; however, as you turn it down, it will make the pan less driven, thus altering the sustain characteristics of your reverb pan. A little experimentation on your new "four knob tube reverb for under $10" can be as rewarding tonally as it is fun.

GETTING RID OF THOSE SVT BLUES

Perhaps the most popular bass amplifier of all time, and yet the most difficult to set up correctly, is the Ampeg SVT. When set up correctly with the proper tubes, these amps can provide hundreds of hours of fabulous bass tone. They rival a blacksmith's anvil in both size and weight and when you finish working on one, you are likely to look like you have just finished rebuilding an internal combustion engine, because of all the internal dirt and grease deposits that build up from the cooling fan. They have two chassis, one for the preamp and the other for the power amp. I have talked to so many owners who have had numerous problems with these amps, but they still love them for that killer bass tone. The problems come from improper set-up. I will reprint one such typical letter I received from an SVT owner and I'll describe the proper setup of an SVT.

Dear Gerald,

I have an Ampeg SVT, purchased new in 1978. After about 30 hours of use, V9 blew (6550). I replaced it with a GE 6550 and after another 30 hours V9 blew again. I replaced with a Sylvania 6550 and after 30 hours V9 and V8 blew. I replaced them with GE 6550 tubes and after another 30 hours V9 and V8 blew again. After opening the amp and pulling the chassis, the V9 and V8 plate resistors were cooked. I replaced all plate resistors and, for good measure, all the screen resistors and cathode resistors. I also did the 12AX7 conversion recommended in the Groove Tubes book and replaced the output tubes with a matched set of Groove tubes. I also replaced all of the electrolytic caps. After about 25 hours of use, V9 and V8 blew again. I put the amp in a closet for about 3 years so I could save up for some more output tubes.

June of 1992—This time I took the amp to a well known repair tech. He replaced all of the preamp tubes and driver tubes with Groove tubes. He also replaced the output tubes with a matched set of Groove tubes 6550 (small bottle). After 24 hours of certified use, the amp started making a loud popping sound which escalated to a continuous sound of pops and screeches. I opened up the amp and found a cooked resistor (R33) in the preamp section and replaced it. I put the amp back together, turned it on, plugged in and it sounded great, for about 2 minutes. You guessed it—V8 blew.

I've always operated the amp with a load, correct bias and never spilled any beer in it. I've never played the amp for more than three one-hour sets. Gerald, what gives? Is 30 hours normal tube life for an SVT? Any suggestions or tips would be greatly appreciated.

Kindest regards,
Puzzled

The truth is that 30 hours is certainly not the normal operating time for a SVT. In fact, those amps, when set up correctly, should give many hundreds of hours of playing time and almost never go down at a gig. First things first. The SVT is a very powerful device that is capable of tearing itself up if it is not properly calibrated. We are dealing with a push-pull circuit that is so powerful that if one side is unbalanced with the other side, it will virtually self-destruct.

Absolutely critical to these adjustments are two 1 ohm 1 watt resistors (R35 and R36). These resistors should be checked and matched to exactly 1 ohm; otherwise your setup and adjustments will be inaccurate and self-destruction will be imminent.

These three adjustments can be made without actually removing the chassis, but you may have to remove the chassis to replace those two 1 ohm resistors we talked about earlier. I feel I must re-state that none of these adjustments will be valid unless those two 1 ohm 1 watt resistors are exactly 1 ohm each!

Underneath the head's front baffle panel you will find three variable resistors and three test points. The three variable resistors are marked VR1, VR2, and VR3 and the three test points are marked K1, K2, and G (for ground). Come to think of it, you can check those 1 ohm resistors we talked about earlier from these points. With the

SVT (courtesy of Bill Welker, South Austin Music, Austin, Texas)

amp off and unplugged, check the resistance from K1 to ground and see if it is 1 ohm. K2 to ground should also read exactly 1 ohm. If these readings are not exact, I recommend opening the amp and changing R35 and R36 to exactly matched 1 ohm 1 watt resistors. (Bring your meter to an electronics store and go through a box of resistors and find two that are exactly 1 ohm!) K1 to K2 should read exactly 2 ohms. This is where that expensive American-made Fluke or Beckman meter is a must.

Now that we are sure our control resistances are correct, we will do three adjustments. First, we will set the idle current of one bank of 6550s; second, we will set the idle current of the other bank of 6550s; third, we will balance the driver section of the phase inverter so that each bank of 6550s are driven equally. This is not really as hard as it sounds, so stay with me.

Also critical to these adjustments are the use of a matched sextet of quality (not Chinese) 6550s. These should be matched for quiescent cathode current. This is critical, because if you are measuring a bank of three 6550s and they are not matched for cathode current, one may be drawing most of the current while the others are resting. This condition could cause the one tube to fry prematurely.

Now we are going to set the idle current of one bank of 6550s by using a very high quality digital meter (I use a Fluke 8060A). With the amp on, we want to measure from K1 to G (ground) and adjust

VR1 until we see .072 volts D.C. on the meter. You can see now why we must use a very high quality meter—a variance of even a tenth of a volt could prove disastrous! It takes 72 milliamps of current going through that 1 ohm resistor to create a .072 D.C. voltage reading on the meter. That 72 milliamps is the idle current of the one bank of three 6550s (24 milliamps per tube provided the tubes were matched for cathode current).

Next we want to adjust the idle current of the other bank by measuring the volts from K2 to G (ground) and adjusting VR2 until we see .072 volts D.C. on the voltmeter. When this is achieved, you should have a reading of zero volts between K1 and K2, and you can use this measurement as a double check.

At this point, the only thing left to do is adjust the drive of the phase inverter to assure equal drive per side. This can be accomplished by running a 1000 hz signal through the amp with the amp turned up about three quarters of the way up and adjusting VR3 so that you get zero volts between K1 and K2.

We are not finished yet. Tubes have a tendency to drift as they are played, so I would leave the amp on for a few hours, play it a little, and when it is nice and hot, repeat the entire balancing and set-up procedure. This step is very important to assure proper balance. Do not assume that only one balancing will be correct. It is vital to repeat this procedure and make sure the amp is adjusted under actual heated playing conditions.

There are a few other things you should know about SVT amps prior to any set-up and balancing. For one thing, there is a 99.999% chance that the hum balance control of any SVT is fried. All it takes is one tube to short plate to heater and the hum balance control will be history. Very few techs even think of that, and with six output tubes, the chances are that one of them shorted plate to heater at one time or another. Change the pot to a 100 ohm 3 watt "J" style military pot and adjust the pot for the least amount of hum. Also, all 10 ohm 5 watt plate resistors should be changed and all 22 ohm 1 watt screen resistors should be changed. Don't even bother checking them, just change them. These are in the circuit as fuses so that if a tube shorts while the amp is being played, the resistor burns, takes the tube out of the circuit and the amp continues to play. If any of these resistors were ever strained from a previous bad tube or bad set-up, they could go open on their own, which would take that particular tube out of the circuit. This is really a great design, because if you are playing a gig, the amp will continue to play and not go down; however, it is best to use all new ones when doing a set-up so that premature and unnecessary failure will not occur.

Another tidbit of information about SVTs: the internal impedance switch of the output transformer is deliberately wired backwards. This was done because, when the amps were first sold, some players did not want to use two speaker cabinets and would blow speakers which were under warranty. To stop this, Ampeg wired the impedance switch on the extra speaker jack so that, when using one 4 ohm cabinet, the transformer was actually set at 2 ohms. This reduced the power somewhat and kept the speakers from blowing. You could actually increase the power of an SVT simply by plugging a dummy ¼" phone plug (not connected to anything) into the external speaker jack. This will trip the impedance switch to 4 ohms and increase power when using a 4 ohm speaker load.

SEMI-PRO RECOVERING TECHNIQUES— "UNCOVERED"

Have you ever owned a ratty looking vintage amp that you wish looked like brand-new old stock but at the same time wondered, "What effect will recovering have on the value of my vintage amp?" The answer is, "It all depends." I would never suggest recovering a vintage amplifier that is original and in fair or good condition. In fact, minor defects add character to a vintage piece. However, an amplifier in poor to ratty condition can be an eyesore. Let's use a vintage car analogy. Would you rather have a vintage automobile with a ripped up original interior, rusted and rotted original paint with lots of rust and peeling on the bumper? Or would you rather it have a re-upholstered interior done right, using original style materials and color, a fabulous new paint job with correct colors and re-chromed bumpers? I'm sure you'll agree that a perfect restoration can enhance the value of an otherwise ratty vintage piece. On the other hand, use the wrong materials and colors and you could actually take away some of the value. I have clients that have bought vintage amps, had them restored to original spec and sold them for hundreds and, in certain cases, thousands of dollars more than their buying price. On the other hand, there are people that tried to recover their own amp, screwed it up a little, and lost what money they thought they saved by doing it themselves, when they sold it!

Unless you are absolutely certain that you know what you are doing, don't attempt to recover a vintage amplifier unless you are willing to take a loss if and when you sell it. A recovering will only enhance value if it is done perfectly!

OH WELL, IF YOU MUST KNOW

I know there are some of you who are going to try recovering your amp anyway, so maybe I can save you a screw-up by sharing my experience. There are four steps to recovering with tweed and three steps for Tolex.

STEP ONE—PROPER PLANNING

Take some close-up pictures of the amp from various angles for starters. You might want to make a few drawings of distinguishing characteristics, such as which way the tweed stripes are going, what the corners look like, which pieces were applied first, what size each piece of covering material must be, where will the handle and other hardware mount, what is the spacing of the feet, etc.

Remove hardware, speakers, chassis, etc., and look at how the inside of the cabinet should look. Make drawings or take photos if needed.

Next you will need to order your materials. To determine how much covering material you will need, try this little trick. After figuring the size of each piece of covering material and considering the way pieces are wrapped around internal batten boards, use graph paper to cut out miniature pieces using a scale of one square of the graph paper equal to one inch. After finding out how wide the covering material comes, make a scale of that width on another piece of graph paper. Now take your scale model pieces that have been cut out and lay them on top of this graph paper. Place them where you will end up with the least amount of waste and each piece is positioned so the the tweed stripes are correct. From this you will be able to determine how each piece will be cut out and how many yards of covering material you will need. When using tweed, remember to draw the direction of the strip on each graph paper scale model so that the stripes will be running in the correct direction when they are actually cut out.

You will also need to purchase some glue. Although I use animal glue, a semi-pro recoverer probably will not have access to a hot glue machine. A hot glue spreader machine costs thousands of dollars and requires an experienced operator. This only leaves a few options. I would recommend using Kraft Bond Glue, you will need 5 or 6 bottles in most cases. There are other glues that have been used; however, the contact cement is difficult to use and toxic, and aerosol spray glues are very expensive and don't work as well. The Kraft Bond is non-toxic, water based and sticks better than the spray glues.

You will also need a few small paint rollers for applying the glue. The 3" size works very well.

A heat gun or hair dryer is essential as well as a utility knife with extra blades and a few lint-free clean white rags.

STEP TWO—PREPARE THE CABINET

Just like body work prepares a car for its paintjob, cabinet preparation is essential before any covering can be applied. All previous covering and glue should be removed. Most vintage amps used animal glue originally. It can be removed with heat and it is water based (water dissolves it). A heat gun can be very useful in removing the outer covering. If you don't have a heat gun, use a hair dryer set on high.

All stripped mounting holes should be filled in with epoxy (Bond-o) and re-drilled. This would include both chassis mount holes, back panel screw holes and, sometimes, speaker screw holes.

Finger joints that are loose should be re-glued with wood glue. Any cracks should be repaired. Look for cracks around the chassis mount holes. These can be repaired with epoxy as well. Fill and sand dings and dents. Sand cabinet until smooth.

STEP THREE—APPLY THE COVERING

Of course, you must cut out the pieces first. Use the carpenter's rule—measure twice, cut once! When you are ready to apply a piece of material to the cabinet, apply glue to both facing surfaces with a small paint roller. You should not use too much glue, but every square inch should have glue and the glue must be even. When you have the glue spread properly on both surfaces, apply some heat to pre-dry the glue. A heat gun or hair dryer is used for this. It is important to pre-dry the glue but don't get it completely dry. Now apply the piece. When using tweed, if you cut the pieces with the stripes going the right way in the first place, you should have no problem duplicating the positioning.

Squeeze out any and all air bubbles that may occur. When using Tolex, a straight-pin can be used to poke a tiny hole to allow air to escape and thus eliminate air bubbles. Use your lint-free rag to wipe up any excess glue.

I would suggest starting with the bottom corners first. That way you can practice cutting a couple of corners before you cut the top corners (that will be clearly visible from across a 20,000 square foot Convention Center!)

When you finish covering the amp, leave it alone for a day to dry. If it is Tolex it may take several days to dry, depending on how much the glue was pre-dried during initial application.

After it's dried, replace the hardware and you are done if you are doing Tolex. If you are doing tweed, you will need to do one more step.

STEP FOUR—SEAL AND AGE THE TWEED

Tweed is a porous fabric that will look awful if it is not sealed. Any little bit of dust or dirt can become permanently embedded in the tweed and water can penetrate it easily, thus cause irreparable damage. Therefore, sealing the tweed will make it look original as well as protecting the tweed by making it non-porous and thus, water resistant.

There are many types of lacquers and polyurethane products that would work for this. I think the best looking is the Polyshades by Minwax that is available at any Sherwin Williams paint store. It comes in different colors and I recommend Honey-Pine Matte finish. Use a paint brush to apply and sand lightly with steel wool or Scotch-brite in between coats. You will need 3 coats total. Make sure that the glue has completely dried before applying any sealant and let each coat of polyurethane thoroughly dry (approximately 24 hours) before sanding between coats.

Reassemble and you are done.

TWEAKING YOUR AMP FOR BASS

I recently received a letter from a musician who said he'd heard of so many ways to mod a bass amp for guitar but could a bass amp be used for bass? And—what qualities are needed to have a good bass amp? Here are the basics on tweaking your tube amp for bass guitar.

WHAT IS NEEDED FOR AMPLIFICATION OF BASS GUITAR?

Consider that a bass guitar is producing low frequency signals that should be relatively clean with a fast attack. This points to a heavy duty power supply and bigger than normal capacitors. Generally, all the capacitors will be a little bigger. Bypass caps should be bigger so that lower frequencies will bypass the cathode resistors. Filter caps should be bigger to provide constant steady power to the tubes and it takes more power to amplify low frequency sounds. Coupling caps should be a little larger as well to aid in coupling those low notes. The tone caps will generally be a little bigger to put them more in the range of the instrument. A solidstate rectifier is preferred over a tube rectifier to provide tightness in the lower frequencies. Certain tubes sound better for low frequencies as well. Generally 6550s are the power tube of choice, but 7581As and 7027As are also very desirable.

LET'S BEEF UP THE POWER SUPPLY

Since an amplifier is basically a modulated power supply, we will start with the power supply. If the rectifier is tube-type, replace it with a solidstate rectifier replacement or better yet, take the rectifier tube out and wire some diodes across the socket. You will need four 1N5408 diodes. Take a pair of them and solder the cathode of one to the anode of the other. Now do that with the other pair. Now take the unsoldered cathode lead from each pair and twist them together.

This goes to the cathode pin of your rectifier tube socket (pin 8 of a 5AR4, 5V4, 5Y3, 5U4 tube socket). Solder it. Now take one of the unsoldered anode leads and wire it to one of the plate leads of the socket. The other unsoldered anode lead goes to the other plate lead on the socket (plate pins are pin 4 and 6 of a 5AR4, 5V4, 5Y3, 5U4). Make sure all connections are properly soldered. Using 4 1N5408s is much better than using the plug-in type rectifier because you will end up with a 3 amp and 2000 volts PIV (peak inverse voltage) rating instead of the normal 1 amp and 1000 volt PIV rating of the standard plug-in variety. This adds a reliability factor and peace of mind of knowing that the rectifier is not going to fail.

Next we will need to double or triple the values of the main filter on the B+ supply. These are the electrolytic capacitors that are ultimately connected to the center-tap of your output transformer, either directly or through a "standby" switch. Often there will be two or more of these either in parallel or series. You need to know what value the circuit "sees" and design for two or three times that amount. For instance, if you are dealing with an original 5F6A Bassman, there will be two 20 uf capacitors in parallel, so the circuit would actually "see" 40 uf of capacitance. (When caps are wired in parallel, the total capacitance is the sum of the two). On the other hand, an AB165 Bassman has two 70 uf capacitors in series so the circuit "sees" 35 uf of capacitance. (When two capacitors of the same value are in series, the total capacitance is the cap value divided by two but the voltages are added). In either of the above examples, I would recommend using two 220 uf at 350 v. caps in series. This would have the circuit "see" 110 uf at 700 volts! Remember to put a 220K 1 watt resistor across each cap so that the voltage is divided evenly across each cap. (These resistors offer the added benefit of bleeding off the voltage in the caps every time the amp is turned off.) Always observe correct polarity when installing electrolytic caps. When series connected in a B+ power supply, the positive of one cap goes to the negative of the other—the remaining positive goes to the circuit and the remaining negative goes to ground.

GETTING THE TUBES RIGHT

If you are modifying an amp that originally used 6L6 type tubes, a 7581A is a direct drop-in that will give you more headroom.

Another good choice would be the 7027A; however, one pin on the socket would need to be rewired. The 7027A has an internal connection connecting the screen to both pin 4 and pin 1. In a 6L6 circuit, since pin 1 is not used on the tube, the pin 1 of the socket is sometimes used as a mounting post for other circuitry and this is where the problem lies. We do not want to accidentally connect the screen voltage to whatever happens to be mounted on pin 1 of the socket. Look at the output tube socket. If it is a 6L6 type amp, remove whatever is connected to pin 1 of the socket and wire it elsewhere. Then you will simply plug the 7027A into the tube socket and you're set. Regardless of what type of tube you change to, rebiasing is a must.

My favorite tube for bass is the 6550. This tube has loads of headroom, tons of bottom-end and is clean as a whistle. Not only that, it has the same pin-out as a 6L6. There is one consideration when changing to this tube. The filament heater draws 1.8 amps of heater current per tube! This is twice as much current as the 6L6, the 7581A or the 7027A. The problem is this: your power transformer may not be able to handle the extra heater current. So what's a person to do? You will most probably need an auxiliary heater transformer. When I convert a 6L6 style amp to 6550s, I usually leave the main heater supply connected to the output tube sockets, but I disconnect the preamp tube heaters from the circuit and give them their own 6.3 volt heater transformer. (A 3 amp 6.3 volt auxiliary heater transformer can handle up to nine 12AX7 type tubes!) As always, biasing is a must and remember the 6550s need to idle at more plate current than the 6L6s.

LET'S START TWEAKING

At this point, it's time to listen to the amp and see where you're at. You may not want to carry this any further. However, you may want to experiment with different coupling cap values. Typical coupling cap values for bass might be somewhere between .01 and .1 uf at 400 to 600 volts. The 600 volt coupling caps generally have better bottom-end, but they are very difficult to obtain. (Orange drop caps work great in a bass amp and they are available in 630 volt ratings). The coupling values should be smaller at the front of the amp and get gradually larger as the signal is amplified through each stage of the amp.

Finally, the tone caps should be tweaked for proper range. Generally the cap values should be about twice of what you would expect to find for guitar amps. For instance, instead of using a 250 pf treble cap, try a 500 pf replacement. Similarly a .02 bass cap might turn into a .047 after tweaked. You want to be careful not to go overboard and end up with a "boomy" sound, so make sure and listen after each individual mod. This will give you the advantage of knowing the direction the tone is going.

Another tweaking area would be the cathode bypass capacitors. These are wired between ground and the cathode of a tube (pin 3 or pin 8 of a 12AX7 style tube). Typical values are usually around 25 uf at 25 volts for guitar amps. If you want a little more bass, change to a 50 uf at 25 volts. If you need more, try a 100 uf, a 220 uf, or even a 330 uf—all at 25 volts. The bigger the value, the easier it is for bass frequencies to get through the amp's circuitry.

OTHER TWEAKING AND GENERAL GUIDELINES

The speaker cabinet is what's making the sound. A closed back cabinet will sound better than an open back cabinet because there will be no phase cancellation of lower frequencies. A low note played through an open back cabinet will have the effect of the air rushing around to the other side and phase canceling some of it—not a good idea for bass!

Generally, bigger magnet speakers can handle bass better. Ten inch speakers will have more punch than 15" speakers because there is less cone inertia. On the other hand, 15" speakers will have better bottom end because the resonant frequency will be lower. Bigger cabinets usually sound better for bass because their resonant frequency is lower. Bigger speaker cables are a must for bass, try some 10 or 12 gauge speaker wire.

Always remember, good tone takes time to tweak—be patient.

TWENTY TRICKS TO STABILIZE THE UNSTABLE AMP

Unwanted oscillations can be a major problem in a vintage tube amp. Sometimes these stability problems will just start happening one day, even though the amp has worked fine for many years. Often a vintage amp will work fine at a low volume, but when cranked, it will begin to oscillate either at a low frequency or have parasitic oscillations on top of the note. Sometimes the problem will come and go inconsistently. These kind of problems are very difficult to troubleshoot and can make you think there is a poltergeist residing in your amp.

OTHER GHOST-LIKE SYMPTOMS

Some unstable amps appear to have a loose tube socket. You are using a known good tube; you rock the tube; the problem seems to go away. You change the socket only to have the problem remain. The capacitance of your hand, near the tube, actually temporarily cured the oscillation but you were thinking you had a bad tube socket!

This reminds me of a game I played with my children when they were small. When driving down the road, I would touch the rear-view mirror with my right hand while simultaneously blowing the horn with my left hand. The children were fascinated and when they tried pushing the rear-view mirror, I would simply blow the horn again with my left hand. They became totally convinced that they were blowing the horn by pushing the rear-view mirror! An unstable amp will sometime plays a similar game with the technician. You poke around and notice that a particular component seems to be related to the problem, only to find out later that this was not the case.

Almost all glass-front Marshalls have a parasitic oscillation/stability problem if the treble and presence and volume are all turned up. Many 50s style tweed amps have a low frequency oscillation when cranked. Sometimes you will hear a mosquito-like buzz trail the note. I have heard amps with oscillations that would cause the notes to sound exactly like a blown speaker or even a blown output transformer, but changing those components offered no help. Some unstable amps will cut-out completely, while others can produce a machine gun sound. The 5E3 Tweed Deluxe will sometimes have a parasitic oscillation in the number one instrument channel when the volume and tone controls are cranked. This manifests itself by sounding like a rattle in the speaker (although the speaker is fine).

Yes, and have I got a story for you. Many years ago, while experimenting with a tube circuit, the B+ voltage would unpredictable surge to 700 volts from a 400v supply. After I spent a couple of days trying to track this down, I finally stumbled on to the fact that two of the preamp tubes were too close together and somehow that would make the B+ jump up and stay there. This would only occur on certain settings and sporadically even then. The physical rotation of the tube in relationship to the adjacent tube affected the problem as well.

VALUABLE TRICKS

I'm going to give you some trade secrets. These took me many years and many hours experimenting to perfect. I am going to share knowledge gained from these experiences.

There are four basic ways or approaches you can use to stabilize an unstable amplifier. They are: improve component grounding, alter layout, reduce high-end response, or reduce gain. Within these basic approaches, there are specific actions you can take. Let us begin.

IMPROVE GROUNDING

1. INSTALL GROUNDING BUSS - Often times the pots in a vintage amp are grounded to the body of the pot which is supposed to be mechanically grounded to the chassis by its mounting nut and star-washer. Sometimes these mechanical grounds are corroded or oxidized or otherwise not good enough. Installing a bare solid wire that goes from each pot's ground lead to the next pot's ground lead, etc., and eventually to the input jack's ground lead will sometimes

cure the problem. You may want to remove the input jack first and clean any corrosion off the chassis near the input jack's mounting hole. Remember to clean the jack itself so that it will make a good mechanical ground to the chassis.

2. CHECK RESISTANCE IN ALL GROUNDS - Here's a good way to make sure your grounds are good. Use an ohmmeter on the lowest register (probable 100 or 200 ohm setting), hook the common lead to the power transformer center-tap ground and leave it connected. Using the positive meter lead as a probe, check every component that is supposed to be chassis grounded and read the resistance. I would be suspect of any reading over 1 or 2 ohms. I have seen amps with a shiny clean ground that have read as much as 65 ohms—ouch, not good!! This could be the source of your amp's instability.

3. CHECK OUTPUT TRANSFORMER GROUND - What is sometimes not so obvious is a bad secondary winding ground connection on your output transformer. This can be checked with an ohmmeter as described above; however, sometimes a problem can occur if the jack and ground are too close to the input jack. Remake the ground here as necessary and remember to clean all surface areas to improve the mechanical ground of the output jack.

4. CHECK SPEAKER GROUND - Believe it or not, I have seen some amps whose parasitic oscillation problem was solved when the speaker's frame was grounded to the ground terminal of the speaker.

ALTER LAYOUT / LEAD DRESS

1. MOVE OUTPUT TRANSFORMER. If the amp's instability is related to the output transformer's phantom coupling to some other component in the amp, moving the transformer could stop the oscillation. After you disconnect the mounting screws holding the transformer near the chassis, and with the amp "on," move the transformer to various chassis locations and notice how it affects the problem. On a high-gain amplifier, it is extremely important to keep the output transformer away from the input jacks. Sometimes changing the transformer mounting position from horizontal to vertical will do the trick. When you find a spot that the transformer likes, mount the transformer there. Be aware that there are other transformers on the chassis and do not mount one transformer next to another, unless the core laminates of the one transformer are mounted perpendicular the the second transformer. If you mistakenly mount a transformer where the laminates are parallel to an adjacent transformer, you will almost certainly have some magnetic coupling between the two transformers. This is undesirable, unless you want major 60 Hz. hum in your output transformer!!!!

2. RE-INSTALL EXISTING SERIES GRID RESISTORS CLOSER TO SOCKET LEAD. Sometimes an amplifier oscillates from positive feedback being fed through the output tubes. The way that the series grid resistor is mounted to the output tube socket could possibly affect the problem. Typical values for this resistor range from 1K to 5.6K and sometimes even higher. For example, in 6L6, 6V6, and EL34 circuits, pin #5 is the grid connection. Desolder and remount the power tube grid resistors so that the body of the resistor is directly against the grid socket. It is imperative that there is zero lead length between the grid socket lug and the resistor's body.

3. SHORTEN WIRES FEEDING THE POWER TUBE GRID CIRCUIT. The wires that come from the phase inverter to the power tube grids can also affect positive feedback (oscillation) in the output tubes. If the amp uses series grid resistors on the output tubes, this is the wire feeding the resistors; however, in some amps the wire goes directly to the grid lead of the socket. This wire should be as short as possible. Sometimes a difference of even a ¼" can be the difference between a stable and an unstable amp.

4. SHORTEN WIRES IN ALL GRID CIRCUITS. Good layout practices dictate short grid wires. Go through the amplifier and shorten every wire that feeds a grid circuit. These are the wires that go to pins #2 and #7 in a 12AX7 or pin #5 in 6L6, 6V6, EL34, etc. A grid wire can sometimes act as an antenna that picks up oscillation. Therefore, grid wires should be as short as possible to minimize this tendency.

5. USE SHIELDED WIRE FOR THE INPUT JACK TO GRID. But connect the shielding to the plate of the same tube (and not to ground, as you would normally). Do not hook up the shielding on the input jack side. In fact, I recommend clipping the shielding very short on the input jack end and put some shrink tubing over it so that it does not connect to anything accidentally. The grid end of the shielded cable has the shielding connected to the plate of the tube.

6. REPOSITION A COUPLING CAPACITOR. Coupling capacitors can sometimes feedback via the input jack. This is sometimes a problem in the 5E3 Tweed Deluxe amp. Specifically, the .1 mfd / 400 volt capacitor that feeds the grid of the power tube can be unsoldered on top and moved 90 degrees counter clockwise. This will necessitate moving the connecting 220K resistor as well. A new wire connecting the grid of the output tube (to the junction of the 220K and the .1 mfd at 400 volt capacitor) should be used, because the wire originally ran under the board and is too long anyway. Although this modification of lead dress can be used to stabilize a tweed Deluxe, there are other amps that could benefit from this procedure.

REDUCE HIGH-END RESPONSE

Many times, parasitic oscillation can occur in the higher than audible frequencies. This inaudible high frequency can make the amp seem to shut down even though the amp is actually working very hard. It's just that it is working hard at producing inaudible frequencies and only seems to be shut down. Here's a few tricks to stop that high frequency oscillation.

1. INSTALL SERIES GRID RESISTORS. If there are no series grid resistors on the output tube sockets, adding them can sometimes stabilize the unstable amp. One resistor is added to the grid circuit of each power tube. Typical suggested values would range from 1 K to 10K. I recommend using the smallest value that stabilizes the amp. For instance, you might try using a 2.2K. If that doesn't do it, try a larger value. If the 2.2 K stops the parasitic, try a 1.5 K. Proceed in a like manner until you get the smallest value that stops the problem. As in item #2 of ALTER LAYOUT/LEAD DRESS above, mount the resistor body directly up against the socket lug and make sure there is zero lead length on the resistor. This trick can also be performed on the grid circuit of a preamp tube.

2. ADD A GRID-TO-GROUND CAPACITOR. This is the same circuit used in the silver face Fenders. I recommend it only as a last resort, because it adversely affects tone. One end of the capacitor goes to ground, the other goes to the grid of the power tube. A capacitor is used for each power tube socket. Typical suggested values range from 500 pf. to 3000 pf. Use the smallest value that stabilizes the amp. Remember: the larger the value, the more it affects tone.

3. SHUNT A PLATE LOAD RESISTOR WITH A CAPACITOR. I've seen this work in plexi-Marshalls. The idea is to bypass high-end around the plate load resistor (pin #1 or pin #6 of the 12AX7). The capacitor goes in parallel with the plate load resistor. Typical values range from 250 pf. to 1000 pf. Again, I would suggest using the smallest value that gets the result. This is a trial and error process, be patient. In the case of the plexi-Marshall, there is a plate load resistor on V2 that goes from pin #1 to pin #6. Placing a 100 pf or a 250 pf. cap across this resistor can sometimes stop parasitic oscillations.

4. INSTALL A CAPACITOR FROM PLATE TO GRID OR PLATE TO CATHODE. This type of modification actually reduces the very high-end response through degenerative feedback. There are two versions of this modification. In one version, a small value capacitor is connected between the plate and the grid of the preamp tube. In the other version, a small value cap is place between the cathode and the plate. Typical suggested values range from 5 pf to 2000 pf. It is always best to use the lowest value that works.

REDUCE GAIN

1. INSTALL POWER TUBE TO INPUT CIRCUIT FEEDBACK LOOP. This is an old ham radio trick that you will see on many early Boogie amps. I would save this one as a last resort because it will make the tone sound nasally, but it will stabilize an amp that is otherwise unstable. Here's how it works.

Take a piece of solid-core, insulated, hookup wire and wrap it about three times around the insulated input wire that goes from your input jack to the grid of the first preamp tube. You do not connect this wire electrically to the input wire, but only wrap the insulated wire around the insulated input (jack to grid) wire. This will form a capacitive coupling somewhere around 5 picofarads. Take the other end and connect it (electrically) to the plate on one of the output tubes. With a two output tube, push-pull amp, connecting to one plate (pin #3 on a 6L6 or EL34) will make the problem better and connecting to the other plate will make it worse. This is because one plate is in phase with the input and the other is out of phase.

The plate that makes it better is the one that is out of phase with the input. What is happening here is this: The capacitive coupling is so small that only very high frequencies can pass. If your oscillation is in this range, it is fedback, out of phase, to the input and actually phase cancels the oscillations. It reduces gain, but only on the extreme top end. This is a clever idea. Too bad that it makes an amp sound nasal.

2. INSTALL INTERSTAGE FEEDBACK LOOP. This trick is similar to the last trick except that instead of going from the output tube plate back to the preamp input, you are going from the output of one preamp stage back to its input. This trick does not alter the tone very much and certainly doesn't sound as nasal. Here's how this one works.

Wrap an insulated solid wire around the wire feeding the grid of any preamp tube. Here again, you should wrap it around about three times. Do not connect this wire electrically on the input side, but only wrap it. Now connect the other end of this wire electrically to the plate of the same tube and see if that cures the instability.

This idea can be expanded in many ways. One way would be to wrap insulated wire around any insulated grid wire and do some trial and error testing by connecting the other end to various parts of the circuit to see how the circuit likes it. For example, you might have a wire wrapped around the second gain stage grid wire and terminate it on the plate of a phase inverter. In other words, it does not necessarily have to be across only one tube but it does have to be from two points in the circuit that are out of phase, in order to stabilize the amp.

I can remember, for example, a Dual Showman that a hobbyist modified for high gain, and had the lead dress all wrong. He sent me the amp requesting that it be ridded of unwanted oscillations. Nothing seemed to work except the power tube to input circuit feedback loop. This didn't make me very happy because it made the amp sound nasal. I ended up using several interstage feedback loops that took out the parasitic, but without the adverse tone associated with power tube to preamp grid feedback.

3. CHANGE PREAMP TUBE BIASING BY INCREASING THE CATHODE RESISTOR VALUE. Occasionally an amplifier has been modified for more gain and becomes unstable as a result. Increasing the value of the cathode resistor on any preamp tube might be just what you need to fine tune the gain so that the amp is stabilized. For instance, if you have a 1500 ohm cathode resistor (pin # 3 or pin #8 on a 12AX7), changing it to a 2000 ohm or 1800 ohm might be just the right amount of gain reduction to stabilize it. This is where experimenting with different values might be appropriate. I have actually used a 1K ohm pot in series with the stock value resistor and fine tuned the circuit just to the point that it is stable. A small trimmer pot would work just fine for this. You could even unsolder the pot once you have determined the correct value, and measure the ohms resistance. Now replace the pot with a fixed resistor of the same or slightly higher value. You are done.

4. LOWER PLATE VOLTAGES ON THE PREAMP TUBES. Lowering the plate voltages on an amp will generally do two things: reduce gain and curb high-end frequency response. This could be all that is needed to stabilize the amp. Usually the oscillations occur in the upper range, so lower voltages not only drop overall gain slightly, but drop off the top-end dramatically.

Lowering plate voltages usually involves changing a power supply resistor to a larger ohm value. A word of caution: Bigger value resistors drop more voltage across them and therefore may need a higher wattage rating. Here's how to determine the wattage rating. Measure the voltage across the power resistor. Square this number and divide by the resistance value in ohms. This will give you the wattage of the circuit, but for safety reasons — double it, and round up to the next highest standard value! For example, let's say you are dropping 100 volts across a 5K ohm power resistor. You would square 100 volts to get 10,000 and divide by 5000 (ohms) which would give you 2 watts. Double this for safety and you have four watts. But wait—4 watts is not a standard value, so you would use a standard value 5 watt resistor in this case.

TWO SPECIAL CASES

I have seen rare cases where the speaker inductance changed the overall phase relationship just enough to cause the feedback loop to be in such a phase that positive oscillations occurred. This is very rare, but it has happened. To check this condition, try substituting a different speaker. Another way of checking is to disconnect the feedback loop and see if the oscillations disappear. If they do, you know that either the output transformer is hooked up reverse polarity (in which case swapping the primary leads or the secondary leads will cure the problem) or the inductance of the speaker is such that the negative feedback loop occasionally drifts positive.

Another special case that is rather common is the conductive circuit board. This is most common in old Fender amps or garage hobbyist amps that use a black circuit board material. I am not sure of the cause of this condition, but I think it might be related to carbon that is used to dye this type of circuit board and the humidity in the air. I have seen many amps that have worked perfectly for a month, a year, even a decade and then one day; for no apparent reason they begin oscillating when turned up. Sometimes this will be intermittent, with no apparent predictability. Nothing could be found to be wrong with the amp or any component in the amp and yet the only thing that would cure the problem was to change the circuit board! It's a lot of work and a big hassle, but when a black circuit board becomes conductive, the only cure is to change it.

MODIFICATIONS

ADJUSTABLE FIXED-BIAS MOD FOR RE-ISSUE BASSMAN

Many players have changed their re-issue Fender Bassman solid-state rectifiers to 5U4GBs in an effort to correct the voltages in the amp and to create a spongier, more bluesy attack, only to find that the bias must be readjusted to avoid the amp's sounding cold and lifeless. When the tube rectifier is inserted, the plate voltage drops 30 volts (or more depending on initial bias). This lower plate voltage causes the tubes to idle at less current, thus making the amp sound very thin and weak. The bias voltage must be decreased in order to raise the idle current of the output tubes.

Adjusting bias actually helps in other ways as well. By having the output tubes draw more current, the transformer drops more voltage due to the internal resistance of the transformer winding (copper loss). This transformer internal voltage drop is subtracted from the plate voltage, thus correcting even further the more modern high voltage design of the re-issue Fender.

Unfortunately, there is no adjustment to bias a re-issue Bassman and printed circuit boards are a major hassle to work on due to inaccessibility of components. After all, you must change a resistor value in order to adjust bias. Who wants to disassemble an amp in order to change a resistor value? What's a fellow supposed to do?

If you can read and make a few solder joints, you can mod the re-issue without disassembling the amp (other than removing a back panel). This mod will give you the opportunity to actually bias in minutes, without any hassle, whenever you change tubes. How much does it cost? You will only need a 50K ohm Cermet element pot (about $5), a few inches of wire, and a dab of silicone glue.

LET'S DO THE MOD

Unplug the amp and place the standby switch in the play mode. Take the back panel off. You will not need to drain the power supply, because there are bleeder resistors in this particular model that drain the filter capacitors when the amp is off.

Look on the board and you will notice the parts are marked with letters and numbers. Locate R41. Clip each lead of this resistor, leaving as much of the lead on the board as possible. Choke up as close to the body of the resistor as possible. This will facilitate soldering later without having to disassemble the entire amp! See Figure 1.

FIGURE 1   CLIP HERE

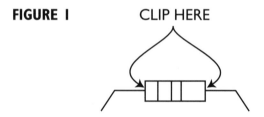

Using a very short piece of hook-up wire (completely stripped 20-gauge solid wire is what I use for this), connect one of the R41 leads to both the middle and one end lead of the Cermet element pot. Now connect the other R41 lead to the remaining end lead of the pot. See Figure 2.

FIGURE 2   ADJUSTMENT
SCREW

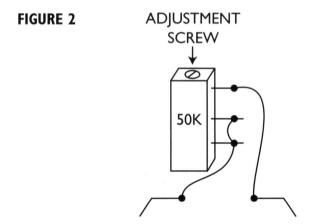

Being very careful not to overheat the R41 leads, solder each of the three connections on the pot and let it cool awhile. Now solder the R41 leads, being very careful again not to overheat the leads, lest you heat the lead so much it becomes unsoldered from the board.

If you do overheat the R41 leads, or you do not trust the connection, simply connect and solder the two leads from the Cermet pot to the plus lead of C23 and solder the other Cermet pot lead to the minus lead of C23. This is the same electronically, you are just connecting to where the R41 leads ultimately terminate.

When you are done soldering, I suggest you secure the pot with a dab of silicone glue and let set for a while. When the glue has set, put the amp in the standby mode and adjust the Cermet element pot's adjustment screw to read approximately minus 52 volts on pin #5 of either output tube. This bias voltage will drop somewhat when the standby switch is put to the play mode; however, it will give you a good starting point to later set your bias.

Bias the amp, using the method you like best, and enjoy the benefits of having your amp biased for the tubes you are using.

Tube Amp Talk

IS YOUR VIBRATO UP TO SPEED?

The many types of vibrato and tremolo circuits that have been incorporated into vintage amps have many different features. However, they all have one thing in common—they all have an oscillator. This is the common denominator. All vibrato/tremolo circuits in tube amps use a "phase shift" type oscillator, a very simple device that uses five or six capacitors, five or six resistors and an adjustment pot (for the speed control).

WHAT IS AN OSCILLATOR?

An oscillator is an electronic circuit whose job is to produce a continuous alternating voltage. This alternating voltage is used to modulate the guitar signal somehow. Depending on the amp involved, the oscillator could:

1. Turn a transistor off and on and ultimately ground out the signal as in the case of the Marshall 1959 T (Tremolo).

2. Change the grid voltage (and therefore the bias) of the output tubes as in the case of the tweed Vibrolux, brown Vibroverb or the blackface Princetons. This will make the output louder or softer and therefore is Tremolo (even though Fender called it vibrato).

3. Change the cathode voltage or grid voltage (and therefore the bias) of a preamp tube as in many Supro, Gretsch, and Rickenbacker amps.

4. Operate a light bulb (neon or LED) which in turn operates a light-dependent resistor that grounds out the signal (Tremolo) as used in most blackface amps.

5. Modulate phase and therefore pitch (vibrato) as in most Magnatones.

Since the oscillator is generating alternating voltage, whatever it is changing/modulating is being done in a cycle.

HOW DOES AN OSCILLATOR WORK?

As we said before, all vibrato/tremolo tube amps use a "phase shift" type oscillator. This tube circuit is actually a feedback circuit where the output is fed back into the input. When sufficient energy is fed back to more than compensate for the loss in the grid circuit, the tube will oscillate. This type of oscillator works on the principle of placing three capacitors in series from the plate to the grid of a vacuum tube (the plate being the output and the grid being the input). Each of these capacitors has a resistor going to ground that gives the resistor/capacitor section a total phase shift of around 60 degrees. Since there are three sections, that adds up to 180 degrees phase shift. When you consider that the output of a tube already shifts 180 degrees from the input, the total phase difference will be 360 degrees and this makes a full cycle. In actuality, the shift is not exactly 360 degrees. The speed of the oscillation will be determined by how close the shift is to 360 degrees.

Every one of these oscillators has a speed control (pot) that alters the shift of one of the R/C feedback sections. Alter one section and you have altered the total phase shift that is being fed back.

Look at Figure 1. C1 and R1 (and the pot) are the first feedback section, C2 and R2 are the second feedback section, and C3 and R3 make up the third feedback section. You will also have the cathode resistor and bypass cap (C4 and R4) for biasing the tube and you will have a plate load resistor (R5) and a coupling cap (C5).

Let's suppose you want to slow down the oscillator. You turn the pot to greater resistance thus increasing the phase shift of the first feedback section and this makes the oscillator slow down. You could stop the oscillator by simply grounding out any section or ungrounding the bypass cap. Grounding a feedback section stops the feedback and therefore the oscillation. Ungrounding the bypass cap reduces gain of the tube so that there will be insufficient gain to compensate for losses in the grid circuit, and therefore the oscillation will cease.

VARIATIONS

There are many variations on this. For instance, in Fender blackface amps, the R2 and R3 normally go to a negative voltage supply which completely stops the oscillation because it turns the tube completely off.

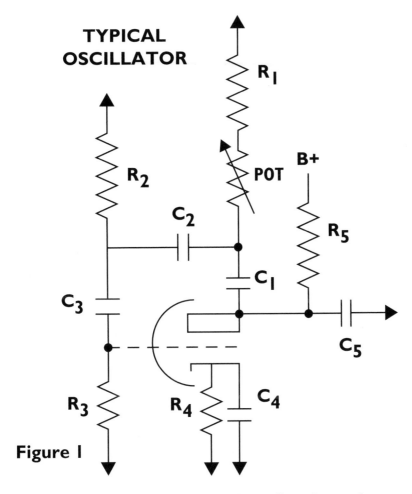

TYPICAL OSCILLATOR

R_1

POT

B+

R_2

C_2

R_5

C_1

C_3

C_5

R_3

R_4

C_4

Figure I

So the footswitch is used to turn on the oscillator by simply grounding out this negative voltage so that the tube can begin working again.

Sometimes you will see the coupling cap hooked to one of the feedback sections instead of the plate and sometimes you will see R2 connected to the cathode as opposed to being grounded, but it is still the same principle.

HOW CAN I SLOW IT DOWN?

Let's suppose you want to slow down your oscillator to get that really slower-than-stock kind of sound. How do you do it? Simply increase one of the feedback caps and the oscillation will get slower. You could even increase a second one to get real slow. Most amps use a .02 cap for C1 and a .01 for C2 and C3, so what we are saying is that changing C2 to a .02 and/or changing C3 to a .02 will really get things slow.

On the other hand, if your oscillator is not going fast enough, reducing the feedback cap values can speed things up a bit.

TROUBLESHOOTING AN OSCILLATOR

If the oscillator is not oscillating, the most common cause of failure is the bypass cap going bad. The second most common problem is one of the feedback caps going bad. Though rare, bad grounds could be a problem. This is where my favorite repair technique, the Texas shotgun approach, works so well. Check the grounds and change all three feedback caps and the bypass cap and you will fix it in minutes without having to figure out which cap is leaking current.

Tube Amp Talk

CONVERTING TO CATHODE-BIAS

So that amp you've been gigging with won't sing until it's cranked and when it's cranked it's a little too loud for the gig. Or maybe you're recording and you'd like a little more natural compression for improved signal to noise ratio. To quote a famous Texas drummer, who had just done an end-over-end in his touring plane, "What we 'posed to do, now?"

You can have more natural output tube compression with a more singing quality with slightly less volume by simply having your fixed-biased amp converted to cathode-bias.

WHAT'S THE DIFFERENCE?

There are two common ways of biasing the output tubes—either fixed-bias or cathode-bias. In a fixed-bias amp, a constant negative voltage is injected to the grids of the output tubes while the cathode is grounded to the chassis. In a cathode-biased amp, the grids are grounded through a grid return resistor and a resistor is placed between the cathodes and ground. This creates a slightly positive charge on the cathodes. Basically, the tube can't tell the difference between having the grid negative with respect to the cathode or the cathode positive with respect to the grids, so either way can be used to bias an amp, however they do sound much different.

LET'S SEE SOME EXAMPLES

Some examples of popular fixed-biased amps would include all Fender Twins, Pros, Supers, Bandmasters, Bassmans made after 1954, all Marshalls that use EL34s or 6550s, the Trainwreck Express Amp, all Kendrick Amplifiers 35 watts and over, all Vibrosonics, and Tremoluxes 6G9 and later.

Some examples of popular cathode-biased amps would include all tweed Deluxes, all tweed, brown and black Champs, all tweed Princetons, Tremolux 5E9 and 5E9A, Vox AC 30, Kendrick Texas Crude series amplifiers, and Trainwreck Liverpool and Rocket amps.

It costs more to build with fixed-bias, because it requires a small negative voltage power supply whereas cathode-biasing only requires a resistor and a capacitor; therefore, many manufacturers used fixed-biasing only on their more expensive models. Also, cathode-biasing is not as efficient in terms of the amp's overall power, making it better suited for lower power amps. Tubes that don't require much bias voltage, such as the EL84 or the 7189A, are very efficient cathode-biased. This could explain why almost all EL84 and 7189A designs are cathode-biased.

HOW IS TONE AFFECTED?

There are two main reasons cathode-bias sounds different from fixed-bias. In a cathode-biased amp, the tubes only see the difference between the cathode and plate voltages. If you have a positive 48 volts on the cathode and another positive 428 volts on the plate, the tube "feels" the same as if the cathode was grounded and the plate had 380 volts on it. If that same tube was fixed-biased, the cathode would be grounded and the tube would "feel" the entire 428 volts from plate to cathode. This is significant, because as plate voltage goes up with respect to cathode voltage, high-end definition is improved. Or you could say as plate to cathode voltage goes down, the high-end is rolled off somewhat creating a browner tone. Everything else being equal, the higher voltage will mean more power. Or you could say the lesser voltage will give less power.

In a cathode-biased amp, the positive cathode-bias is developed by drawing current through the cathode resistor. An electrolytic cap is placed across the resistor to bypass varying alternating current around the resistor in an attempt to stabilize the cathode-bias voltage. However, all of this A.C. does not actually pass through the capacitor. Some passes through the cathode resistor and causes the actual bias relationship to fluctuate with the signal. You hit a note and the output tube draws lots of current. Some of the extra current passes through the cathode resistor causing the positive voltage to rise slightly. This has the effect of cutting the tube current somewhat. As the string begins to decay, less current comes through the tube. This causes the cathode voltage to drop somewhat, thus making the tube current increase. As the string decays, the tube gets louder and as the string is hit harder, the tube gets softer. This is some real tube compression.

In a fixed-bias amp, the cathode is grounded and a constant negative voltage is injected to the grids. There is no current involved in the negative voltage supply, only negative voltage. The tube draws current directly from a grounded cathode and the bias relationship stays stable. Therefore you do not get the compression.

WHAT WE 'POSED TO DO?

Converting to cathode-bias in a push-pull amp requires only three steps:
1. Locate the two grid return resistors (usually 220K ohm) that connect each output tube grid to the negative voltage supply. Disconnect both of them from the voltage supply and ground them to the chassis. (These resistors are not to be confused with the 1500 ohm or 5.6K series resistors that are sometimes on the tube socket!) With 6L6, 5881, and EL34 tubes, it is best to change the 220K resistors to 470K or 500K ½ watt, although 220K will work. With 6V6s, the 220K value will be best.
2. Disconnect the ground wire that connects each cathode to ground and run a wire connecting both cathodes. Now solder a resistor from one of these cathodes to ground. The actual value of the resistor will vary depending on what kind of amp and what kind of tube. For a pair of 6V6s in a Deluxe, anywhere from 250 ohms to 500 ohms at 5 watts would be the range. I would start out at 250 ohms and check the plate current using the transformer shunt method. A larger value resistor would tame the tubes' idle current down.

For a pair of 6L6s, depending on the plate voltage, anywhere from 300 to 600 ohms at 20 watts would be the range. Again, I would monitor plate current and take the educated guesswork approach.

When converting a four output tube amp, you would halve the ohms and double the wattage rating that would be used on only a pair. In fact, when I do a Fender Twin conversion, I use two resistors.

One resistor (600 ohms at 20 watts) is connected from the cathodes of the inside pair of output tubes to ground, and another identical resistor is connected from the outside pair cathodes to ground. This helps keep the output tubes balanced.

3. Connect an electrolytic capacitor in parallel with the cathode resistor, putting the negative side to ground. If you are using two resistors for four output tubes, you will need a capacitor for each resistor. The actual value will vary depending on what value cathode resistor you use. The range will be anywhere from 25 uf to 100 uf at 50 to 100 volts. The higher values will give less compression and more bottom end. This capacitor must be physically located away from the cathode resistors because it can be severely damaged by heat.

4. I know I said there are only three steps, but I almost forgot to tell you the most important step of all. Kick back and enjoy the sweet compression and spongy attack of your "new" amp. Feel it breathe and listen to the blossom of each note. Can you make it "meow?"

GETTING RID OF FILAMENT HUM IN YOUR SINGLE-ENDED AMP

If you play a single-ended amp, the hum level is more than likely unacceptable. This occurs because single-ended amps, unlike push-pull amplifiers, do not have hum cancellation in the output stage. Assuming that the filament heater supply is 6.3 volts A.C. (as is the case with almost all amplifiers), the 60 hertz power supply hum leaks from the filament to the plate of the output tube and ends up as amplified output signal. Smaller single-ended amps such as the tweed Champ sound best cranked, but you are cranking the hum as well. If you are recording, the hum will at best be a major distraction and at worse render a track unusable without considerable gating and EQ to clean up the track. Why not just eliminate the hum in the first place?

POSITIVE BIASING THE FILAMENT SUPPLY

Since the hum is coming from the 60 hertz 6.3 volt A.C. filament voltage in the tube being attracted to the plate of the tube, we could bias the filament supply with positive voltage so that it would not be attracted to the positively charged plate. To perform this simple modification, you will need two resistors and some hook-up wire.

Biasing the filament supply with positive voltage involves only three steps:

1. Rewiring the filament circuit so that it is a "two wire" (twisted pair) circuit that is ungrounded. Some single-ended amps are already a "two wire" circuit, while others such as the tweed Champ use only a "one wire" (daisy chain) circuit. In a "one wire" circuit, the chassis ground is used instead of a second wire. This references one side of each heater to be at ground potential (zero volts). It is not possible to

positively bias a circuit that is grounded to the chassis, so we cannot have any chassis grounds in the filament circuit. Even a "two wire" circuit will have a ground reference which must also be lifted.

2. Create a 40 volt positive bias supply. This is done by using the two resistors mentioned earlier. A voltage divider is made. It draws almost no current so therefore will not burden the existing power supply.

3. Wire one side of the filament supply circuit to the 40 volt positive bias supply.

HOW IT WORKS

Normally a 6.3 volt filament supply is grounded at one end. This means the voltage oscillates between plus 6.3 volts and zero, with respect to ground. If the ground reference is lifted and a 40 volt positive bias applied to the same circuit, the voltage would oscillate between plus 40 volts and plus 46.3 volts, with respect to ground. Since the plate is positive, we can change the charge relationship between the filament and plate and virtually eliminate hum!

LET'S DO THE MODIFICATION!

First things first. Drain the power supply and unplug the amp to avoid a shocking experience. Inspect the filament circuit first. If it is a "one wire system," you will need to rewire it to a "two wire system." On a "one wire system," one lead of the 6.3 volt winding (on the transformer) goes to ground and the other lead will "daisy chain" to one filament connection on each tube. Also, it will go to the 6.3 volt pilot lamp as well. The other filament lead on the tube sockets and pilot lamp terminal will go to ground, thus completing the circuit. (Green is the standardized color most often used on transformer filament winding leads.) To rewire to a "two wire system," you will need a twisted pair of wires connecting the filament leads of one tube in parallel to the filament leads of the next tube and in parallel to the two pilot lamp leads. To refresh your memory, the filament leads on 6V6 and 6L6 tubes are pin 2 and pin 7. On a 12AX7 or 12AY7, pin 9 counts as one lead and pins 4 and 5 are connected together and count as the other lead. A word of caution: Most early single-ended tweed amps had one lead of the pilot lamp grounded in such a way that it is hardly noticeable. If you fail to unground this lead, you will burn up your power transformer!

When you are sure that you have rewired the "one wire system" to a "two wire system," check it with an ohmmeter. If you measure the resistance between the filament winding on the transformer and ground, you should get infinity ohms. If you don't get infinity ohms, go back and find the ground that must be lifted before proceeding.

If you already had a "two wire system," lift the chassis ground and check resistance from filament circuit to ground with an ohmmeter. You must have infinity ohms before proceeding.

Next, you will need to make a 35 to 40 volt positive power supply. This is done by using a 220K ohm one watt resistor (dropping resistor) and a 27K ohm ½ watt resistor (load resistor). Connect the 220K ohm resistor from the B+ to one side of the filament circuit and connect the 27K from the same side of the filament circuit to ground. This will give you approximately 35 to 40 volts positive on the filament circuit at all times. In most amps, the B+ with easiest access will either be the plus side of the main filter cap or the red wire coming from the output transformer.

FINAL ADJUSTMENTS

Take time to double check before proceeding. Measure resistance from the filament circuit to ground. At this point, it should read 27K ohms, the same as your load resistor.

Turn on the amp and measure D.C. voltage from the filament circuit to ground. It should read somewhere between 35 and 40 volts. If it is over 40 volts, you need to change the 27K ohm resistor to a slightly smaller value. Perhaps a 25K or 22K. You do not want more than 40 volts D.C. on the filament circuit. On the other hand, if your meter shows 29 volts, you might want to use a slightly larger resistor in place of the 27K. Perhaps a 30K would be appropriate. The important thing is to see 35 to 40 volts D.C. when measuring D.C. voltage from the filament circuit to ground.

LET'S PLAY

Get out your guitar, plug in and crank out some tunes. Notice how the hum level disappeared! (If there is still hum, unplug the patch cord to the amp and listen—comparing the "amp by itself" hum with the "amp with guitar hooked up" hum. This will show where the hum is really coming from.) If this procedure was performed correctly, your amp's hum level is so quiet that you could make a line level output and drive a power amp without fear of objectionable hum appearing on the output. In fact, this mod evolved from my wanting to put a line level output across the speaker leads of a small single-ended amp (to drive a power amp) and being dissatisfied with the stock hum level appearing on the output of the single-ended amp—especially after hearing it amplified by the power amp!

TOP TEN AMP TRICKS FOR STEEL GUITAR PLAYERS

Setting up an amp for steel guitar is a totally different approach than setting up one for guitar. For one thing, a steel guitar is more of a full range instrument. Its lows dip into the bass guitar range and its highs go up to the fiddle range. Since many steel guitar chords are voiced with close voicings, the amp needs to reproduce those voicings without the dissonant distortion that would normally occur if those voicings were played on an amp set up for regular guitar. (This is especially noticeable when playing major thirds in the upper ranges.) Steel guitar amps need headroom and plenty of solid, tight, bottom-end. To sum it all up: a loud, clean tone with minimal distortion, minimal noise, less envelope, more punch, solid bottom-end with clear note definition is most desirable. Setting up the steel guitar amp can be rather tricky, so here are Ten Tricks for you to try. As with all modifications, try them one at a time and listen to what you have. Either keep the modification or put it back like it was — before proceeding with the next modification.

In general, you need to stiffen and solidify the power supply. Here are three tricks worth trying.

1. THE FILTER CAPACITANCE TRICK - More filtering in the output stage is desirable. This will lower the impedance of your power supply and make it more solid. Actually, the low end will be tighter with more punch. Depending on the type of tubes being used in the output stage, the main power filter (the electrolytic capacitor closest to the rectifier) should be somewhere around 100 uf to 400 uf total capacitance. If you are using a pair of electrolytics in series as your first filter (totem pole), the total capacitance will be half of the single ca-

pacitor's value. Therefore, you will need a pair of 200 uf capacitors to get a total of 100 uf capacitance when using two in series. Why do we use two in series in the first place? To get the voltage rating up! The main power filter needs to have 100 volts or more safety margin above the normal D.C. plate voltage. For one thing, the voltage on the main filter goes up really high when the amp is in standby mode. Also, there is A.C. voltage swing on top of the D.C. voltage when the amp is being used. If you connect two 350 volt capacitors in series, you have the equivalent of a 700 volt cap, but at half the capacitance of the single cap value.

Special note: While increasing the filtering is desirable in the output stages, do not increase filtering in the preamp stages. The preamp stages are RC coupled (resistor and capacitor) and increasing capacitance in the preamp stages will lose tone.

One thing to consider when increasing the filtering of the power supply is this: When large amounts of filtering are used with a solid-state rectifier, the impedance of the filter is so low during "switch on" that much larger amounts of current than normal flow through the rectifier. This could be remedied by either using a larger amperes rated rectifier or installing a "slow start" switch.

A very simple yet effective "slow start" switch can be made with only a simple SPST toggle switch and a 100K ohm 5 watt resistor. Here's how: Find the wire that connects the rectifier to the filtering. Cut this wire and connect the switch in series so that if the switch is "on," the circuit is "dead stock" electrically. Now place the resistor across the switch so that when the switch is "off" the current goes through the resistor, thus limiting the current. In other words, you would place the limit switch in the "limit" mode (off) when you turned on the amp's power. You would then leave it in the "limit" mode when you switch the standby switch to the "play" mode. After fifteen or twenty seconds, you would switch the limit switch to the "on" position. This would short out the 100K ohm resistor and your amp would be ready to play. (By the way, this is also a handy device for forming capacitors that have recently been replaced. If you left the "slow start" switch in the "limit" mode, you could leave your amp on several days and form those caps up really nice. Of course the standby switch must be in the "play" mode to do this).

2. THE SCREEN SUPPLY TRICK - Fluctuations in the screen voltage can affect the gain of the tube, therefore, a more stable screen supply will help make the output stage rock solid. Although there are many complex circuits for achieving this, here is a simple trick.

Having a pi filter between the screen supply and the main filter is helpful. This is the same circuit you see in a Fender Blackface Twin, where there is a choke between the main filter and the screen filter. It is called a pi filter because when drawn in a schematic with the capacitors on each end of the choke going to ground, one is reminded of the greek letter pi (π). Using a larger than stock value inductance choke is better. You could even use two chokes in series to increase the Henries. Better yet, when using two chokes in series, add one more capacitor going to ground (in between them) and you will have two pi filters. This will be very effective at regulating the current from the screen circuit.

3. THE SOLIDSTATE RECTIFIER TRICK - If the amp doesn't already have a solidstate rectifier, installing a solidstate rectifier is a must. The solidstate rectifier offers no envelope or sponginess. This is very helpful when you need that edge on the low end. Steel players use a volume pedal to swell into or out of a note and therefore don't need the amp to breath like it would with a tube rectifier. Also, the solid-state will put out a little higher voltage than a tube rectifier. This helps headroom of the entire amp. It's also good for note definition and clearer highs. I like to use six 1N4007 diodes. These are rated at 1 amp and 1000 peak inverse voltage. I use three in series going to each side of the B plus winding. This gives me a 3000 volt peak inverse voltage rating at 1 amp. There are also 1.5 amp and 3 amp (1000 volt) diodes, respectively the 1N5399 and 1N5408. Use these when you need more current rating than the familiar 1N4007. When installing a solidstate rectifier, always remember: all cathodes face the positive end of the first power filter and all anodes face the B plus winding (usually red wires) on the power transformer.

Choosing the right power tubes and having proper bias plays a big part in both headroom and low-end punch. Here are a couple of tricks.

4. THE POWER TUBE TRICK - Certain types of tubes are better for steel guitar. For instance a 7581A or a 6550 is much better than an ordinary 6L6 or a 5881. The 7581A can handle 35 watts plate dissipation and is very clean at high volumes. A 7581A can simply drop right into a 6L6 socket without any modifications necessary (except setting the bias, of course). It uses the same amount of filament current. These tubes are known for their hard, clean sound.

The 6550 tube might also be an excellent choice for steel guitar. For one thing, it is powerful with loads of clean bottom and it doesn't break up until you drive the living daylights out of it. But, the 6550 may necessitate a couple of small modifications to the amplifier in order to make it work properly. For one thing, the 6550 draws about twice the filament current of a 6L6. An auxiliary filament transformer may be needed to hookup the 6550 filament circuit. A 10 or 15 amp 6.3 volt transformer will work nicely. Also, it is best to change the screen resistors (feeding pin 4) to a higher wattage; possibly a 470 ohm 3 watt or a 1000 ohm 5 watt resistor would be appropriate.

5.THE BIASING TRICK - The amp used for steel guitar should be fixed-bias. If your amp is cathode-biased, you may want to modify it to be fixed-biased.

When adjusting bias, slightly higher (more negative) bias voltage will cause the tubes to be cleaner at louder volume settings. Ultimately, you increase headroom by having the quiescent (idle) plate current set slightly lower than normal. I am not suggesting over-biasing your amp, but only suggesting that you would normally want a steel guitar amp to draw less idle current than if you were setting it up for regular guitar. Actual plate current would really depend on the actual amp you are setting up. Remember this: You can always perform some listening tests and try a few different settings to see what gives you the clarity you need with the tone you want.

About the preamp tubes, certain preamp tubes are better suited for steel guitar. For one thing, the steel has a fairly hot output and a lesser gain structure is generally more desirable. For instance, in the late 50s, steel players generally preferred the two input Bassman instead of the 4 input Bassman. The 2 hole Bassman (5E6) had a lesser gain structure and used two 12AY7s with a 12 AX7 phase inverter whereas the 4 hole Bassman (5F6 and 5F6A) used one 12 AY7 and one 12AX7 and an additional 12 AX7 phase inverter. Six string guitarist, on the other hand, generally preferred the 5F6A amp because of its higher gain structure. In fact, many guitarist actually use all 12AX7s in the 5F6A Bassman when playing regular guitar.

6. The Preamp Tube Trick - This trick will involve a little experimentation. In amps that come stock with 12AX7s (amplification factor = 100), you may want to experiment with 12AY7s (amplification factor = 44), 12AT7s (amplification factor = 60) and 12AU7s (amplification factor = 20). These tubes all have the same pin-out and therefore are direct replacements for each other, but the lesser gain structure may help the amp to stay cleaner longer. For instance, a 12AU7 or a 12AT7 usually makes a great substitute for a 12AX7 phase inverter tube. Likewise, a 12AY7 substitutes nicely for a 12AX7 in the earlier gain stages. Experimentation is the key here. The acid test is: "Does it sound good?" This can only be determined by a little trial and error experimentation.

7. The Voltage Trick - It's good to know that slightly higher voltages in the preamp tubes give more headroom and tighter, clearer bottom. You may want to look at the voltages on the preamp tube plates and make a decision about what needs to be done. If the voltages are down in the 140 to 160 range (on a 12AX7, 12AT7, 12AU7, 12AY7 style tube), you might want to reduce the value of the power resistor feeding that circuit, to increase the plate voltage. I would think that somewhere between 200 and 250 volts (on the plate) would give you what you want. You could do a little experimenting here with some listening tests.

Likewise, if you are using other types of preamp tubes, the rule is the same—slightly higher plate voltages will give you more headroom!

Two very important parts of an amplifier system are the cables connecting the instrument to the amplifier, and the cables connecting the amplifier to the speaker. Here's two more tricks that will give you more stunning bottom-end, note clarity, and more breathtaking highs.

8. The Speaker Cable Trick - Since speakers are drawing high current, you need big wire. You don't want the speaker cabling to "bottle neck" the power that the amp is producing! In fact, the bigger the better. It takes power to reproduce that low end and it takes power

for note integrity. Use the biggest gauge of wire that is practical, perhaps a 10 or 12 gauge. Keep your speaker cable as short as possible. This trick will minimize losses between the amp and the speakers.

9. THE INPUT CABLE TRICK - Input circuits are high impedance and therefore subject to losses of high-end due to capacitance that is present in the input cable itself. In fact, one could say that the input cable itself is a capacitor! A capacitor is two conductors separated by a dielectric (non-conductor). The guitar input cable is an insulated conductor surrounded by shielding. This makes it a capacitor by definition. Likewise, if you take a foot of this wire and measure the capacitance with a meter, you will find that it has a specific amount of capacitance, let's say 20 pf, for example. Capacitance will increase in direct proportion to the number of feet. That is to say that twenty feet of the same type of wire would have 400 pf of capacitance. This is the same as taking a 400 pf capacitor and installing it across your input jack!! You need very low capacitance (measured per foot) input cabling and you need to keep it as short as possible so that the total capacitance is as low as possible. I have used "George L" brand cable (made in the USA—very low capacitance and low microphonics) and "Klotz" brand guitar cable (made in Germany, also very low capacitance and low microphonics). This trick helps keep good note definition and clarity by avoiding the loss of important high frequency signal.

When you consider that the speaker is what actually makes the sound that we hear, you must also look at the range of sounds that a steel guitar produces. We are talking about a range from bass guitar to fiddle! So we need a speaker or speaker system that can make all those sounds.

10. THE SPEAKER TRICK - The trick here is to get the right speaker/speakers for your amp. Generally speaking, full range heavy duty 15" speakers are preferred by most steel players. The JBL D130 is a popular choice. Peavey also makes a very nice 15" steel guitar speaker. Electrovoice makes a 15" full range speaker that is popular. Generally speaking, you need a speaker that has a big magnet, a high power rating, high efficiency, and a full range.

Twelve-inch speakers can also be used. You need something that will produce clear solid bottom and still have a full range on the high-end. The JBL D120, the Kendrick Blackframe 12", and the

Electrovoice 12" full range are all excellent choices. Avoid speakers that bark or break up in the midrange. You want those tight chord voicings to have clarity without dissonance.

Here's a bonus trick. Add a piezo tweeter to the speaker cabinet. Piezo tweeters do not need a crossover, they sound nice, and they are very inexpensive. This trick will help add clarity in the high-end while helping individual note definition. I've even seen players wire a ¼" plug to a piezo tweeter, plug it into the extra speaker jack and just set it on top of the amp!

SAVE THAT PRINCETON— THE SECOND GENERATION

Many moons ago, I wrote an article appearing in *Vintage Guitar* magazine that was titled "Save That Princeton." The article explained how to dramatically improve gain on a Fender Princeton amp. The article was straight-forward and simply showed a "before" and "after" schematic with a few tips regarding mod technique. A good friend of mine from New York's "Unique Recording Studio," Mr. Bobby Nathan, called to discuss this mod. He performed the mod on his amp and came up with some very clever additions to the mod, all of which were tone tested by ear. Here is the updated modification for transforming your modest 'lil 'o Princeton from California to the Princeton Beast of the East.

You either know now or will surely find out later that the blackface and silverface Fender Princeton amps sound nothing like their reverb counterpart, namely the Princeton Reverb amp. The Princeton design actually sacrifices gain in order to have a triode section for use as an oscillator. (This oscillator is necessary in order to make the vibrato function.) The Princeton has a great clean sound but just doesn't have enough gain to really overdrive. This is a real bummer for someone that buys the amp thinking it just needs service, only to find out later that the circuit is weak and will never sound overdriven.

The Princeton Reverb, on the other hand, has two extra tubes for use in the reverb circuit and actually has an extra stage of gain which is used to mix and amplify the dry and wet signals. Of course this extra stage makes a huge difference in the amp's ability to overdrive.

If you can live without vibrato, a fairly simple mod can be performed that will take the vibrato triode and use it to change the phase inverter

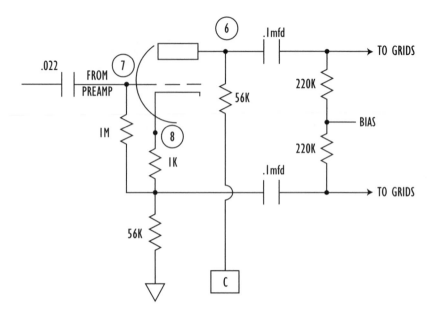

from a distributive load type phase inverter, which supplies no gain and actually loses signal, to a long-tailed pair style phase inverter that supplies considerable gain. In addition to that, you will add a presence control, change the feedback resistor, and improve the power supply voltage to the phase inverter section. This mod is fairly simple, costs less than $15 in parts, is easily reversible if you don't like it, and can be done with six resistors, one or two caps, and a 5K ohm linear potentiometer. This modification will make a simple Princeton amp have as much or more gain than its brother, the Princeton Reverb. This mod can also be done on a Princeton Reverb if you want to make it vicious.

HERE'S HOW IT'S DONE

Look at the "stock" and "modified" schematics and layouts. The "stock" layout corresponds to the "stock" schematic, and the "modified" layout corresponds to the "modified" schematic. Examine these very carefully before you begin. You will find certain parts that remain the same and others that are changed, added, or subtracted. The parts you will need are as follows:

Resistors - one 82K ohm ½ watt

two 100K ohm ½ watt

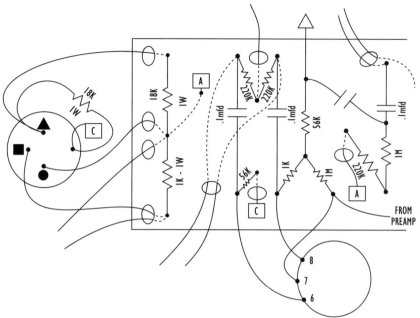

one 470 ohm ½ watt
one 10K ohm ½ watt
one 1 meg ohm ½ watt
one 47 pf. - 400v capacitor
one .02 mfd. 400 volt (if modifying the Princeton Reverb)
one 5K ohm linear taper potentiometer

REWIRE THE PHASE INVERTER

The tube that needs to be rewired is the phase inverter tube (the preamp tube located next to the power tubes). Pins #1, #2, and #3 are used in the stock circuit as the oscillator for the vibrato and these pins are the pins that get rewired. Pins #6, #7, and #8 are used in the stock distributive load phase inverter and do not need to be re-wired. You will need to attach a straight wire from pin #3 to pin #8. Pins #4 and #5 are tied together as one side of the filament and should be left alone while pin #9 is the other filament lead and should also be left alone.

Install the 82K, 100K, 1 meg, 10K and 470 ohm resistors and 47 pf. cap. Double check your work against the "after" layout. Be sure to remove the power wire that goes from point "A" of the power resistors to the 220 K plate load resistor of the oscillator circuit (pin #1 on the phase inverter tube).

MODIFIED PRINCETON PHASE INVERTER
(WITH HIGHER SUPPLY VOLTAGE AND PRESENCE)

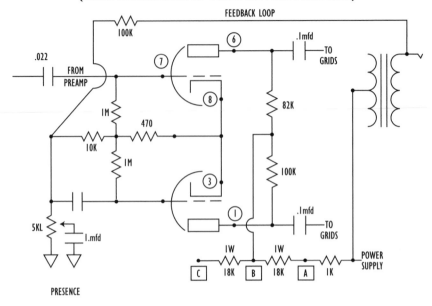

MOVE PHASE INVERTER SUPPLY VOLTAGE

The phase inverter supply voltage is taken off point "C" in the "before" layout/schematic. This wire should be moved to point "B" as shown in the "after" layout/schematic. This will increase your supply voltage resulting in better headroom and slightly more gain.

INSTALL PRESENCE CONTROL

Decide where you want your "presence" control. You can either remove the extra speaker jack and use that chassis hole or remove the speed control and mount it there. Install the .1 mfd on the 5 K linear presence pot. Looking at the back of the pot, the left lead gets grounded.

CHANGE FEEDBACK RESISTOR

This type of circuit will use the 100K ohm feedback resistor. One end connects to the "hot" lead of the speaker jack and the other connects to the presence pot.

OTHER CONSIDERATIONS

In addition to changing the circuit as per the schematic and layout, it is also best to connect a straight wire from the wiper of the inten-

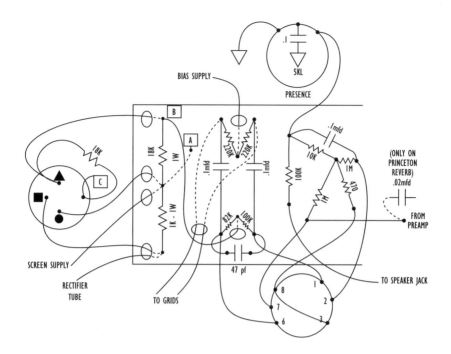

sity pot to the left lead of the pot (looking at the back of the pot.) This will insure stable bias voltage.

Also, if you are performing this modification on a Princeton Reverb, you will need an additional .02 mfd. coupling cap. This goes in series between the preamp and the phase inverter. It is used to isolate the D.C. voltage that is on pin #7 and must be used. You do not need this cap on the Princeton because there is already one on the output of the preamp.

One last modification—change the polarity of the output transformer by swapping the brown wire with the blue wire that go to pin #3 of each output tube socket.

When I do this mod, I usually take it to the next level by using a Kendrick 2112 AHR output transformer. (The AHR stands for Audio Hot Rod). This is an eleven winding interleaved output transformer that is built with minimal losses. Another trick is to go to a 12" speaker by modifying the baffleboard. If you are using a 12" speaker and a 2112AHR, care should be taken to plan exactly where everything will go. These components will barely fit (physically) and a little planning goes a long way. New old stock 6V6s are also recommended for ultimate tone.

Tube Amp Talk

EASY MODS FOR YOUR BLACKFACE CHAMP OR VIBRO-CHAMP

When Fender changed from the tweed series amps to its black-face amps, the tone changed to a more scooped midrange kind of tone. Although it's a signature sound, valid and usable, there are some people that find the tweed sound more desirable for blues. The subject of this chapter is modifying a blackface Champ or Vibrochamp to have that tweed Champ tone. As a matter of fact, the blackface can be converted to virtual tweed specs by performing three basic mods. You will only need two 22K ohm ½ watt resistors and one or two .02 mfd. at 400 volt capacitors and solder.

1. GET RID OF THE TONE CONTROLS - Yes, those tone controls reduce gain and scoop mids. The mod for this involves three steps. Looking at the back of the volume pot, remove the wire from right lead and tape it off. Now remove the wire from the right lead on the treble pot and solder it to the right lead of the volume pot. Looking at the board and coming off pin 1 of the first 12AX7, you will find a 250 pf cap connected to two 100K ohm resistors. Change the 250 pf cap to a .02 mfd. at 400 volts and remove one of the 100K ohm resistors. You have to be sure and remove the right one! The 100K resistor that is mounted longways (horizontally) on the board is the correct one to remove. Actually it is the one that connected the 250 pf cap to the .047 and .1 cap originally.

2. CHANGE THE FEEDBACK LOOP - From the speaker jack, there will be a wire running to the board connected to a 2700 ohm resistor (red violet red). This is the feedback resistor. You are going to remove it and replace it with a 22K resistor but wait—the new resistor is going to be connected differently. One resistor lead goes in the same eyelet

that the wire coming from the speaker goes (like original), but the other end of the resistor goes to pin 8 of the first preamp tube. There is already a wire on pin 8 so make sure and leave that connected.

3. CHANGE A POWER RESISTOR - On the far end of the board near the multi-section can type filter cap, there is a 10K 1 watt resistor (brown black orange). Change this to a 22K ½ watt.

Now you have a virtual tweed Champ (or tweed Champ with vibrato if you converted a Vibro-Champ). Time to play that guitar and let's fine tune the gain.

On the board you will find an electrolytic cap that connects to pin 8 of the first preamp tube. If your amp has too much gain, remove this cap. The tweed Champ did not have this cap, but you might want to listen to it before you remove it because you might like the extra gain. It was put on the blackface to help compensate for loss in the tone circuit.

ADDITIONAL MODS TO TRY

1. CHANGE THE RECTIFIER TUBE - Although all Champs came with a 5Y3 rectifier tube, you might want to experiment with a 5V4 or GZ34 rectifier. These will increase the plate voltage resulting in more power, more headroom and tighter bottom end. A solidstate rectifier will also work with this amp.

2. SLOW DOWN THE VIBRATO ON YOUR VIBRO-CHAMP - If you converted a Vibro-Champ, you may want to slow down the vibrato speed. This can be easily accomplished by changing a cap. On the board you will find three caps in a vertical row whose values are .01, .01, and .02. mfd. Change one of the .01 caps to a .02 mfd for slower speed.

3. CHANGE THE SPEAKER - I should have covered this first because this is the part of the amp that actually makes the sound! Most Champ speakers have either been replaced with the wrong impedance replacement, resulting in very low power, or the original speaker has lost its efficiency, resulting in mushy lows and low volume. Replace with a 4 ohm 8" guitar speaker.

Tube Amp Talk

HEADPHONE HOOKUP FOR YOUR TUBE AMP

So you like the tone of that hip tube amplifier and it really adds meaning to your life, except you can't play it very often. The neighbors on both sides have complained to the police, your wife left with the kids last time you turned it on, and the dog shows you his teeth everytime you make a motion towards the closet where you store your amp. You can't just use a pair of headphones instead of a speaker for fear of blowing up the phones and possibly ruining your amp. Here's an idea that has worked for me. I recommend this for smaller wattage amps — let's say in the 5 to 35 watt range.

WHAT ARE WE GONNA DO?

If you simply used a high wattage resistor in place of the speaker in an amp, the tone would leave much to be desired. Why? Because a pure resistive load lacks an important electrical characteristic that is found in all electro-dynamic speakers. This important characteristic is called "inductance." All electrodynamic speakers are inductive loads. In an actual speaker, the impedance is constantly changing because the inductance of the speaker causes lower resistances to lower frequencies and higher resistances to higher frequencies. This is because of a property that inductors have called "inductive reactance."

You will need a speaker that will be sacrificed/destroyed to complete this project. The speaker must have a good voice coil. The theory is this: A speaker is disassembled such that the voice coil is removed from the speaker and glued into the magnet assembly (which has also been removed from the speaker frame.) An L-pad is installed as a volume control and a headphone jack is added for you to connect your headphones. Now you can crank your amp whenever YOU, THE ARTIST, feel like it. Sound interesting? Read On!

MAKING AN INDUCTIVE DUMMY LOAD

First, determine which amplifier you will be using, and select a speaker that can handle the wattage and impedance of that particular amp. For instance, you may select an 8 ohm 35 watt speaker if using an 8 ohm amp rated for less than 35 watts. If you have such a speaker that has a good voice coil, this will be ideal.

Begin by unsoldering the tinsel wires at the speaker terminals. (These are the two wires that go from the terminals to the voice coil.) Using a razor blade or X-acto knife, carefully remove the dust cap without disturbing the cone. (The dust cap is the small disk that is situated on the inside apex of the cone.) Next, cut a circle around the apex of the cone so that the circumference of the cut is situated just to the outside of the tinsel wires. This can also be done by carefully cutting with a razor knife. At this point, the spider, which is underneath the cone and attaches the apex of the cone to the speaker housing, must be cut away from the housing. Now you should be able to lift the entire voice coil away from the speaker. Using a pair of scissors, carefully remove all of the spider, but without disturbing the coil or what is left of the cone. Set the voice coil aside for now.

Next, we are going to remove the speaker frame from the magnet assembly. (The magnet assembly actually includes the magnet, pole piece, and top-plate.) Different brands of speakers are attached in different ways. Most are staked, in which case a chisel and hammer are the tools of choice. Some are welded, in which case prying or using a hack saw comes to mind. The reason we are removing the speaker frame is because it is not needed and it takes up too much space.

Now, we are going to glue the voice coil into the gap of the magnet assembly. I use a two part epoxy glue that is available in any hardware store. Mix the glue properly, and apply to the gap of the magnet assembly. For good measure, put some glue on the voice coil and place the coil in the gap. Allow to dry completely.

Hint: Get a two-lead terminal strip and glue it to the inside center of this assembly. It is ideal for terminating the tinsel wires and other connections you will be making later. Also a small project box is a nice addition. You could place the dummy load, L-pad, input jack and headphone jack in this enclosure so that the unit would look good and be entirely self-contained.

FINISHING THE HOOK-UPS

Now that you have a magnet assembly with a voice coil glued permanently into it, you must install the L-pad and headphone jack. An L-pad is a high wattage volume control with constant impedance, that will serve as a level control for your headphones. L-pads are available at any place that sells speaker cabinet building supplies. Radio Shack, MCM, and Parts Express are a few sources that come to mind. L-pads come in various different impedances and wattage ratings. Be sure to use one that can handle the wattage rating and impedance of the amp you are using. The L-pad will come with installation instructions. Those instructions will describe how to install the L-pad between the amp and the speaker. You are going to install it this way except instead of using a speaker, you will be using your inductive dummy load. If you are using stereo headphones, you will need a stereo headphone jack, but you will wire it for mono. In other words, you will hook both the tip and ring of the stereo jack together and count this as the "hot" lead. This goes to one end of the dummy load. The sleeve on the jack will count as the "minus" or "ground" lead and must be connected to the other lead of the dummy load. This wiring is essential to make the sound come through both the left and right sides of the headphones.

So to summarize the hookup, the L-pad is connected between the amp and the dummy load. The dummy load goes to the headphone jack (in parallel) and the headphone jack is a stereo jack that is wired for mono.

Headphone impedance could make a difference here. Headphones come in impedances anywhere from 8 to 600 ohms. I would recommend using phones that are 100 ohms are higher. This will insure that the impedance of the phones doesn't significantly alter the overall impedance that the amp "sees."

LET'S START TESTING

When you first begin testing, make sure to start with the amp's level set to low volume. You will need to adjust the L-pad and volume of the amp carefully so as not to blow-out your ears. Once adjusted, make note of the settings so that you can just plug in and go, the next time you feel like playing guitar at 2 o'clock in the morning.

OTHER MODIFICATIONS

You could actually mount the project enclosure inside the amp's cabinet and add a switch to choose between the dummy load/headphone format and the actual speaker of the amp. To do this, you would need a SPDT switch. The switch must be a high current rated switch. Perhaps a 10 amp rating would be fine. A SPDT switch has three leads. One end of the SPDT switch would go the "hot" of the speaker, and the other end would go to the "hot" of the L-pad. The center terminal of the SPDT switch would go to the "hot" of the amp's output. The grounds of the amp, L-pad and speaker are all connected, obviously. (Caution: Do not switch between the two while the amp is "on." Place the amp in the "standby" mode before switching.)

TRAINWRECK PAGES

Tube Amp Talk

INSIDE AMPEG LINDEN

It was the 60s and I had moved back to New Jersey after working for a TV sales and repair chain out of Washington D.C. After hanging around the house for a while, I decided it was time to look for a job, and when I picked up the newspaper I saw a help wanted ad for Ampeg. I drove over to Linden and got a job as the repair tech for Ampeg's final test room. Later, I became a member of the engineering department but still did the electronic repair for final test. So what you are about to read is the true story minus any brain damage I've acquired over the years.

Ampeg was started in New York by Everitt Hull. Everitt was a musician who played bass.

The acoustic bass has an adjustable peg so that the player can raise or lower the instrument to fit their needs. Part of that peg protrudes into the body of the bass. Everitt reasoned that if you installed a microphone on the part of the peg that remains in the body of the instrument you could amplify the volume to whatever level was called for. This was the birth of the Amplified Bass Peg or Ampeg for short. Of course, if you had an Ampeg in your bass, you needed an Ampeg amplifier to make it all work. With the help of a couple of other people, Ampeg was born.

Since Everitt played bass very well indeed and had a good ear, right from square one Ampeg bass amps had the right stuff.

Flash forward to the 60s, and Ampeg's factory is located now in Linden, New Jersey.

Being the second largest guitar and bass amplifier company in the world during the 60s, Ampeg took up the best part of one side of a dead end street in a light industrial zone. Before I left Ampeg, it had expanded to include a building on the opposite side of the street which was converted to warehouse use. Ampeg was different from any factory I had ever worked in because about half the people there were musicians. Also, unlike most factories, everyone had a sense of pride in the product and whenever a production record was broken, a sense of accomplishment was shared by all. Oh, there was some of the typical politics and a few butt kissers like you'll find in any large group, but in general it was more like "The Ampeg Family."

OK, so you have people who take pride in their work and Everitt at the wheel who was big time into quality. What you wind up with is a high quality product. Blue check covering doesn't line up? Recover it! Small scratch on the faceplate? Mark it second and sell it out of our showroom!

Before every amp was boxed, it went through physical, visual, and electronic test bench tests. Each amp was given matched output tubes and tested for noise, hum, and power. After the amps were mated to their cabinets and speakers, they were sent to final test. Each amp was tested with an instrument by a musician for sonic performance, rattles, and buzzes. If it had a problem, it stayed in final test until it could pass every test all over again. It was then sent to packing and shipping where it got one last visual going over.

Hold on! It's payday and I just heard the lunch buzzer! It's the 60s and all our cars at Ampeg have big V-8s. Dennis has a 383 Quad, his brother John a 396, Joe a Pontiac Pony car, and I'm riding an ex-State Police 390 Interceptor. Jump in and head for Walt's! Walt's is a drive-in with a sit down section inside—hot dogs, French fries, and beer. Did you ever see about 3 lbs. of French fries in a huge cardboard barge, grilled—not steamed—hot dogs, and ice cold beer on a 95 degree day? That's Walt's (now gone forever). OK, so the quality control falls off a little for the rest of that day! Lunch over, back at work.

Dennis "383 Quad" is Dennis Kager, who was Ampeg's factory service tech among other roles he had at the factory. Dennis is still the number one Ampeg expert. Dennis was the founder of Sundown Amplifiers. Also in the 60s, his band, the Driftwoods, was the second hottest band in New Jersey—right behind Lowie and The Rhythm Rockers. He played a Strat and a brown Fender concert back then.

Factory service was next to final test, so if things were slow for me and busy for him I'd help out in his section. He did get some strange stuff back. An amp came in from Vegas with a note, "Please fix my amp, it's shot." Sure was—a .38 special right through the chassis. Now that I think of it, that amp was the first ever to lose a shootout! Once he got an amp in that had fallen into a cesspool. Never mind, I can still picture that amp and I've got to eat tonight. Dennis did fix it after a bout with a garden hose on Ampeg's lawn. Dennis also was the Ampeg mouse trapper. He caught them unharmed and returned them to the field out back.

Did you ever notice that most 60s Ampeg guitar amps also had an accordion input? Ampeg sold and installed mike kits that went inside an accordion, so it could be amplified. Dennis was the guy who did the factory installation. Here's the deal—when Everitt Hull was young, everyone in his area of New York took accordion lessons. Like Leo Fender, Everitt didn't like rock music and thought it wouldn't last and accordion would make a comeback. While Everitt owned the company, all Ampeg guitar amps had to work with accordion. This put a major limit on what could be done on the guitar tone side. By then Vox, and later Marshall, were starting to take off and Ampeg was making guitar amps that worked great for jazz but didn't rock.

Ampeg had the talent in-house to design and build a hot contemporary guitar amp during the 60s but, to my mind, while Everitt was the genius behind the spectacular success of Ampeg, by the mid-60s his rather old-fashioned thinking doomed the company. Vox and Marshall became the alternatives to Fender's sound and Ampeg faded away except for the SVT bass amp. The SVT was Ampeg's swan song.

Prozac please—Oh yeah, like I was saying Ampeg was a fun company to work for. We had a demo showroom for the amps and every day one or more famous artists would pop in and play. Look at it this way, a front row seat to a command performance at least once a day and you had a full access pass!

THE AMPEG B-15

The Ampeg B-15 was Ampeg's best selling bass amp. Although it put out only 30 to 40 watts, depending on when it was made, many think it's the best sounding bass amp ever made. With a few minor mods I'd say it just might be!

The B-15 was a Portaflex design; that is, the head was shock mounted to a board that formed the top of the speaker cabinet. This was locked down over a foam rubber gasket by two clamps on each end of the board. This allowed the head to be flipped over and hung via the shock mounts inside the speaker cabinet for transportation. When the head was in the play position it connected to the speaker via an attached speaker cable with a 4 pin XLR connector to a 4 pin XLR port on the side of the cabinet. The use of a four wire XLR connector let Ampeg use a built in safety feature. If the two inner wires in the XLR connector didn't see the shunt inside the cabinet port, it wouldn't allow signal to the power tubes, saving them from being destroyed by playing the amp without a speaker load.

The B-15 was also made in a B-15N version and, after Ampeg was sold, a B-15S version was made. The B-15S was a completely different amp, and I won't cover that model here. The B-15 original version was hand wired and was cathode bias. The B-15N was printed circuit and fixed bias. I like the B-15N best myself. It had a bit more power and a tighter sound. The B-15N used three 6SL7 preamp tubes, a pair of 6L6 tubes, and a 5AR4. The early B-15 used a 7199 in place of one of the 6SL7 tubes.

Both amps had two channels with separate bass and treble controls. The B-15S, which we won't mention, had more controls, more power, and less tone. Ampeg used Jensen and CTS speakers as the stock speakers for the 15 and 15N. The CTS sounded better and was used in most amps. A JBL was an option. Ampeg used three baffle designs in these amps. One design was completely sealed. Another was bass reflex. A few had a double baffle in tunnel port design.

The old Ampeg of Linden, New Jersey should not be confused with the modern Ampeg (St. Louis Music) which makes totally different, contemporary amps—with many models designed by Lee Jackson of Metaltronix fame.

HOT RODDING THE B-15N
1. Check the condition of all power filters.
2. Decide on which channel you want to Hot Rod.
3. On the first pre-amp cathode resistor of this channel install a .1 MFD bypass cap.

4. On the volume control of the same channel install a 500 pf brightness cap.

5. Replace the 5AR4 with a solidstate replacement.

6. Install GE brand 6550-A power tubes in a matched set.

7. Adjust bias to 40 mA per tube.

There you have it, a B-15N with a tight bright tone that will work with round wounds. Don't forget to save the old tubes in case you don't like the sound and want to return to stock.

OTHER MODELS

These amps were also available with a 12" speaker and were exactly the same amp heads with a different speaker box. The models with the 12" speaker were the B12 and B12N. Ampeg also made an entry level Portaflex bass amp, the model SB12. It was a single channel amp with about 25 watts of power and used 7868 power tubes. It also used a 12" speaker. This amp only used one holddown clamp per side of the Portaflex mounting board. To save money, the speaker connections were made through these holddown clamps so when you flipped the head over to the play position and clamped it down you automatically connected the speaker. If you own a SB12 Ampeg, check to see if the holddown clamps are clean and tight. If the clamps are loose or dirty the amp may cut out, lose power, or, worse yet, not play at all.

Ampeg made a higher power Portaflex bass amp known as the B18 and the B18N. The B18 used an 18" Cleveland speaker and pair of 7027A power tubes driven by a 7199 tube to develop 50 watts of power. These were very popular with Motown groups who wanted the "Big Bottom" of the 18" speaker. This is also one of the few Ampegs (and the only American made amps I know of at the time) to offer a British made speaker as an option. For a bunch of extra bucks you could have your choice of a British Fane or British Goodman 18" speaker. Myself, I liked the stock speaker and the Goodman for its musical high end. The Fane sounded harder and a slight bit towards the brittle side to my ears, but of course that's just my personal taste.

Ampeg also made Portaflex guitar amps—some with reverb and true vibrato, but I'm just talking bass amps for now. All tube Portaflex bass amps had a lightup logo. Two light bulbs under the edge of a sheet of clear plexiglass would cause anything engraved into the back face of the sheet to light up. Ampeg engraved their logo and, for a token fee, you could get your name engraved under the Ampeg logo so that it would light up too.

Ampeg also made a transistor Portaflex. It was the BT15 and has 90 to 105 watts of power as it evolved. This amp had a lightup faceplate. It used a Sylvania Electro Luminescent panel behind a faceplate that had the areas that were to light up etched out. Ampeg was always very much an innovator compared to other companies of that time, yet to the public, Ampeg always had an "old fashioned" image. The BT15 even had an up-to-date F.E.T. input stage still used today in some highline solidstate stereo gear.

More inside stuff? O.K. I was in the final test room the day the first short run of BT15 amps were built. They were plugged in on the test bench and allowed to "cook in." After a while, one by one, they "exploded" and went up in smoke. Seems the output would go into runaway and overheat the output cap which, being a 1000 mfd 50 volt electrolytic, would burst, sounding like a firecracker. This condition was fixed but the first few to reach the public still used a single 1000 mfd output cap. If used for bass, things were fine, but if you plugged an organ in, with its continuous tone, the heat would build...then blam! Using two 1000 mfd output caps solved the problem along with a better heat sink for the output transistors. At that point the BT 15 became deadon reliable unless you accidentally shorted its output. It didn't have a workable shorted output protection circuit. If you plugged in a shorted speaker wire or shorted cabinet the amp would fail. I owned a BT15 myself, but instead of the stock cabinet, it was mounted to a 2x12 Portaflex cabinet with two heavy duty CTS 12" bass speakers. That rig really kicked. Transistors done right can beat tubes done wrong. Of course, tubes done right are another thing altogether!

Ampeg not only made amps but also made instruments.

THE BABY BASS

The upright acoustic bass was the standard bass instrument throughout modern musical history. By the 60s, the solidbody electric bass, also known in those days as "The Fender Bass" or

"Horizontal Bass" was gaining acceptance with more and more bassists, particularly for rock music.

Most people think of the Fender style bass as the first mass-produced electric bass. The fact is that years before the Fender-style electric bass was marketed, Ampeg had an electric bass which was produced well into the 60s. This bass had features that are considered modern even today. The Ampeg Baby Bass was a solidbody acoustic-electric bass with a transducer pickup system and a plastic body! While composite petrified spandix is in common use today, building a plastic instrument that was accepted by professional musicians was ground breaking 40 years ago.

The Baby Bass was an upright bass patterned after the acoustic upright bass except for its smaller solid body. The neck, complete with scroll headstock, was all wood as was the traditional wood bridge which gave the instrument the feel and playing characteristics of the acoustic bass but with a smaller, much more durable body that produced less feedback at high volumes. The body was built using two sheets of clear plastic. The front and back halves of the body were vacuum formed from each separate sheet of plastic. The wood grain finish was applied to the inside in the form of a sort of decal. Then the two halves were joined together and trimmed around the edges. To make the body solid and rigid, a two-part chemical was mixed and poured into the body. This mixture formed up and filled the body much like the molded foam ice chests you see in stores.

Since the traditional string of the upright bass was made from cat gut which is really made from twisted lamb intestine and is non-magnetic, a magnetic pickup wouldn't work. Ampeg had the answer, a magnetic transducer. This is how it worked. The bridge was mounted over a metal diaphragm. As the bridge vibrated so did this metal diaphragm. Mounted under this diaphragm was a single pole piece magnetic pickup which responded to the movement of the metal diaphragm.

Later on, Ampeg made a horizontal bass which used this system and let you choose between steel or cat gut strings in the horizontal format. Ampeg also produced the Lucite body Dan Armstrong bass which was designed by electronics whiz Dan Armstrong. There was also a guitar version of this instrument. Ampeg also sold guitars and basses made by Burns of London which came from England.

The guitars and basses used in the final test room were the Burns instruments. For the record, the "Wild Dog" guitars with low impedance pickups weren't well liked by the guitarists at Ampeg, but our British Racing Green Burns Bass was really excellent.

THE EXPLODING BABY BASS

The exploding Ampeg Baby Bass is the rarest of the lot. Only one production run of these terrorist tools was made. A defective batch of the chemical foam was at fault. It turned out that the defective foam would reactivate under heat, such as hot stage lighting. The foam would then re-expand and burst the instrument. The first hint of the problem was when one bass started going very sharp on stage in Las Vegas. It then popped open, oozing foam like a pod from the movie "The Invasion of the Body Snatchers."

THE V4B AND SVT

The V4B and SVT were put into production after I left Ampeg but it would be an incomplete picture not to include them in this article. Keep in mind, I left Ampeg on friendly terms and would often drop by to visit until they moved from Linden. I kept up on their stuff even after I left. In fact, I once owned a V4 guitar amp.

The V4 (also known as a VT22 in combo form), the V4B bass amp, the SVT, and the V 9 guitar amp all shared the same basic preamp design. The V4 and VT22 had reverb and the others didn't. The V4, V4B, and VT22 used four 7027A power tubes and were rated at 100 watts. My V4 put out much more than that and, like a large transformer Marshall or a DR103 Hiwatt, the Ampeg rating was very conservative. The SVT and the V9 used six output tubes. The first amps used six 6146 tubes for a 330 watt rating. Later 6550 tubes were used, the amps still being rated at 330 watts, but in fact the 6550 tubes only put out a "puny" 240 watts. I really don't know why Ampeg switched to 6550's but can tell you the 6146 tubes sound better, last longer, put out more power, and they're not hard to find. Yes, a 6550 SVT can be converted to use 6146 tubes. The V9 was the guitar version of the SVT. I'll bet when you saw the Rolling Stones playing guitar through SVTs they were really the V9 version, although I don't know that for sure.

When the real deal British KT88 tubes were in production, I converted some SVTs to KT88s, upped the voltage on the screen supply and sailed past the true 400 watt R.M.S. mark with ease! Mullard, where are you now?!!

All the V4 and SVT series had a semiparametric midrange E.Q.; a three position switch selected one of three preset midrange frequencies and a control knob would let you cut or boost that frequency if you wished. The SVT used an 8x10 cabinet with eight 32 ohm voice coil speakers wired to 4 ohms. This setup gave a strong midbass punch as opposed to using 15" or 18" speakers which give more of a "feel it rather than hear it" type of bass. Some folks use a SVT with one Ampeg cabinet and another cabinet with larger speakers to get the best of both worlds.

AMPEG QUESTIONS ANSWERED

A common question that keeps coming up is, "My Ampeg's circuit doesn't quite match the schematic glued inside the amp, is it a prototype or rare model?" The answer is that Ampeg made many models of amplifiers that were in a constant state of change. Small changes were made so often the drafting department wasn't asked to draw a new schematic until enough changes were made to justify drawing and printing an updated schematic. Even so, there are enough different Ampeg schematics on there to fill a book. Other changes occurred that wouldn't affect the schematic but would be of interest to a collector. For example, you could find a 100% handwired B15N. You can also find a handwired B15N but with printed circuit boards on the volume and tone controls. Yet again you can find a 100% printed circuit B15N.

Another question I've been getting is, "When did one change or another occur or when did this particular model go into production?" Well, to be honest, I don't know. I didn't keep track of that kind of stuff 'cause back in the 60s, nobody cared. There was no such thing as a "vintage" amp or guitar back then and who would have thought that 30 years into the future anyone would ask about these things? Back when I worked at Ampeg, I could have bought as many late 50's flame top Les Pauls as I wanted for $125! Gibson stopped making them because nobody was buying them. When I was at Ampeg, an Alnico Jensen P12R was a $8 "junk" speaker. Little did we know!

A question I wasn't expecting to get (but nonetheless got asked many times) is, "How can I hotrod my B15 (B12 & B18) for guitar use?" Also, I was asked about putting EL34s into Ampegs. OK then, first I'd recommend getting a copy of *The Tube Amp Book,* 4th Edition, by Aspen Pittman. This book belongs in the hands of anyone who loves tube amps and it covers all the important brands. It has a section on Ampeg history. Its schematic section has a bunch of useful Ampeg schematics, and for the hotrod guys, it describes their "Substitube" device. The "Substitube" allows you to plug a 9 pin tube such as a 12AX7 into an 8 pin tube socket such as a 6SL7. They make three models, so chances are if you have an 8 pin preamp tube amp of any brand, you can use 9 pin tubes without doing any mods to your amp.

Why go to 9 pin tubes? The answer is simple in the case of a B15, for example, the 9 pin 12AX7 will give you more gain than the 6SL7. If you have your tech go in and change some resistor values and add cathode bypass caps the gain boost is impressive. An old Trainwreck B 15 mod trick is to leave the "bass" channel stock and, since the tone circuit is all on the back of the volume and tone controls, you can put in the pot values for a Blackface Fender, add the right caps and resistor values for a Blackface Fender and have both an Ampeg and a Fender channel!!

As for EL34s, they can be used, that's up to your taste. Never put EL34 tubes into a V4, V4B, or VT22. They can not take the power of those amps. The 6550 is the proper substitute for the 7027A in these amps. As for the SVT, if you have 6146 tubes, stick with them. They're easy to find and sound better than 6550s. If you have a SVT with 6550s, don't use the 6550 tube from China. If you have a 7591 output tube equipped Ampeg, the socket can be rewired to take a 5881 tube. Yes, the 7591 tubes sound better, but you'll have to go to Japan to find one these days.

AMPEG GUITAR AMPS

Although they were better known for their bass amps, Ampeg made lots of different guitar amps. When I left Ampeg, there were more Ampeg guitar amps in the world than any other brand except, of course, Fender. I should also point out that at that time my favorite guitar amp was a Vox AC30 followed by various Fender models. I always thought Ampeg guitar amps worked great for jazz and that holds true even today. However, only a few Ampeg models ever did much for blues, and fewer yet could rock and roll.

Ampeg, while owned by Everitt Hull, only made "clean" amps. No Ampeg at that point was ever designed for, tested for, or sold for its distorted sound. In fact, the owners' manual warned players to turn down the volume to avoid distortion. However, as chance would have it, a few rare models had a really good overdrive sound. Before we get into those, some inside Ampeg stuff might be of interest.

Ampeg's production line was not built on Henry Ford's principles. There were no motors pulling the amps from work station to work station. Each amp stayed at its work station until every operation done at that station was done and done right. Ampeg tried a quality control department, I know 'cause they put me in charge of it. After a couple of months, I disbanded it because the production workers took such pride in their work I never found one defect caused by a worker being lazy or careless!

Ampeg also hired an outside time/motion expert who installed a system to improve efficiency. That, too, was scrapped after a couple of months because the workers, left to themselves, built more amps in less time than they could under the so-called modern time/motion system! Maybe some of the big companies in the U.S. could learn a lesson from this. Treat people like people; you'll do much better than treating people like machines.

Anyway, back out on the production line, the cabinets and speakers went their way and the chassis went a separate way finally meeting at final assembly. The cabinets were made at Woodworking then went on to Covering and Trim. Did you ever try to cover a cabinet yourself? It only takes minutes! The covering was precut and ran through a machine that put hot melted glue on the back. The person who applied the covering only had until the glue cooled and set to get it all on and straight, which was matter of minutes! The grille cloth, handles, logos, trim, covers, speaker connectors, feet, and so on, were then added. Also a matter of minutes! That just always amazed me. When I bought my 2x12 Portaflex with the BT15 head, I watched it get covered, grilled, all hardware, and latches done in less than 10 minutes. The chassis were built mostly by highly skilled women, then sent to electronic test. If they passed all tests, including an electrical safety test, they were sent to final assembly. After that, they were sent to final test for a real life workout with a real guitar.

SOME INTERESTING AMP MODELS:
GEMINI AMPS

The Gemini series of Ampegs was named after the space flights of the 60s. Many Ampeg amp names came from the space program. These were among Ampeg's best jazz amps. The Gemini I had a 12" speaker and about 22 watts of power. It had two channels, reverb, and tremolo. The Gemini II had 30 watts of power and a 15" speaker but otherwise was similar to the Gemini I. The Gemini VI was a single channel version of the Gemini II but retained 4 input jacks so, at a quick glance, you might think it a dual channel amp; it wasn't. The Gemini II was the most popular.

The original handwired, blue check covered amps used 7591 tubes and a CTS or Jensen speaker as stock with an option for a JBL. The reverb on these amps used a two spring "C" reverb tank as opposed to the two spring "F" unit used on Fenders. I've read that people have said Ampeg used the "C" tank because it didn't require a driver transformer and therefore was cheaper to build. I was there and Ampeg used the "C" reverb tank because its higher input impedance allowed it to be coupled to/from the driver tube with a capacitor. Ampeg's capacitor driven reverb has less distortion than the Fender reverb circuit because the signal doesn't have to go through a puny low grade driver transformer. It's interesting to note my Kendrick reverb unit, which is transformer-driven like an original Fender outboard reverb unit, has a much larger and higher quality driver transformer than Fender used. In any case, the two circuits give a different reverb sound. Ampeg's better suited for jazz and Fender's better suited for surf music. In between those two uses, the choice would be personal tastes, but Ampeg wasn't thinking money when they chose the "C" tank.

THE VT40

This one is easy. The early VT40 is the one used by the Rolling Stones. It has four 10" speakers and 50 watts of power. Later VT40s had an awful built-in distortion circuit and are best avoided.

REVERBEROCKETS

This is a whole series of amps, the early ones being totally different from the later ones. I like the earlier models that used 6SL7 and 6SN7 tubes in the preamp stages. Most used 7591 output tubes but one version used 6V6 output tubes and is a real nice sounding amp under overdrive. The amps are laid out like a Fender Tweed in that the chassis is on the top back of the cabinet. Later amps are through the top front like a Fender Blackface amp. Most Reverberockets sound pretty good but the later ones have more of the Ampeg typical sound and most of the early ones have their own atypical sound which you'd never peg as Ampeg if all you ever heard was the mid-60's and later amps.

THE V4 AND VT22

These are the 100 watt Ampegs—Ampeg's answer to the Fender and Marshall high power amps of the day. The VT22 and V4 have the same chassis; the V4 being a head and the VT22 being a 2x12 combo. The V4B was the nonreverb V4 which was usually sold as a bass amp. They all had the same tone circuits. Loud and clean, they made good bass amps but were hard and brittle for guitar work. My personal V4 gave powerful crunch power chords but did nothing else well for my guitar except make it too loud. However, the V4 guitar cabinet, a rugged 4x12, straight, front-loaded monster, is a great sounding box once the fiberglass is removed. I've rarely heard old Celestions sound as good in an old Marshall cabinet as they do in a V4 cabinet. If only Ampeg had made a V4 slant cabinet too!

THE JTM45 MARSHALL OR HOW THE FENDER TWEED BASSMAN SOUND EVOLVED INTO THE MARSHALL SOUND

It seems most guitarists know that the first Marshall amplifiers, the famous JTM45 model, evolved from the 5F6A Fender Bassman. However, why these two amplifiers sound so different from each other remains a mystery to the vast majority of people I talk to. In taking the time to explain why the JTM45 has its own unique sound, I hope to inform the reader as to how little details and even blind luck can make major changes in the way an amplifier performs.

In 1962 when Jim Marshall decided to manufacture and market his own amplifiers, he chose to copy the Fender 5F6A Bassman. In 1962 the 5F6A Bassman was not being produced by Fender anymore so the circuit Fender abandoned was to become the first Marshall. The first JTM45 amps were built by Ken Brand and Dudley Craven at Jim Marshall's Music Shop in Hanwell, England.

I've worked on many original JTM45 amps and Bluesbreaker Combos and can state with 100% certainty that these amps are not modified hot rods, but in fact contain dead stock Fender 5F6A circuitry right down to the use of 5881 output tubes and the GZ34/5AR4 rectifier tube. Even the hand-wired layout is pure Fender. Yet, even though these amps were undeniably Fender clones circuitwise, they didn't sound like Fenders at all. Why does the JTM45 have its own voice? Some people will say it's due, in part, to the capacitors and resistors used in these amps. While it is true different brands and types of resistors and capacitors have an effect on sound, there aren't any major sound differences between the brands and types in the 5F6A and the JTM 45. If you cross switched these parts in these two amps, the Fender would still sound like a Fender, and the Marshall would still sound like a Marshall. There would be a minor tone shift but not British to American or the other way around.

While on the subject of capacitors, it should be explained that the Fender 5F6A schematic shows the tone shaping caps as one 250 pf (mmf) or .000250, and two .02 mfd caps. While I have seen a very few 5F6As with two .02 caps, most have one 250 pf, one .02, and one .1 cap. Using two .02 caps gives more detail and it's not known to me if the use of the two .02 caps in all JTM45s was blind luck by copying from the schematic or if Ken and/or Dudley had their hands on the rare Bassman with two .02 caps and noticed they worked better than the .02 and .1 setup. By the way, I never put down accident or blind luck—some of my best circuit designs came about that way.

Well, the question remains, why does the JTM45 sound Marshall? Number one is the use of different transformers. Transformers vary in sound as much as guitar pickups do. The 5F6A used Triad brand transformers and the output transformer in the Fender is wound to match a 2 ohm load. The JTM45 used RS Hygrade transformers supplied by Radio Spares in England. The output transformer was wound to match a 16 ohm load. You may read in some places that the '45 used RS "Deluxe" transformers. Like I've said, I've worked on many and have only seen Hygrade. If anyone out there has a '45 with a "Deluxe," please let me know.

The first JTM45s had an aluminum chassis. This changes the sound in two ways. It's nonmagnetic and doesn't affect the transformer's magnetic field like a steel chassis does. Since all signal returns through ground, which is the chassis, and aluminum is a better conductor than steel, you get a faster response and better high end. Later JTM45s had a steel chassis and still sounded real good, but the best 45s I've ever heard have an aluminum chassis. Also, all the Bluesbreaker Combos used an aluminum chassis.

Another major sound difference between the Fender and Marshall 45 is cabinet design and speakers. The Fender used American-made

Jensens while the Marshall used British-built Celestions. These speakers have a very different sound from each other. Of course, a 4x12 closed back (or the 2x12 Bluesbreaker Combo) would sound different from a 4x10 combo. Marshall did make the rare 4x10 Bluesbreaker Combo but it was Celestion equipped also.

Time for more blind luck. The Fender 5F6A electronics plays backwards. That is, the electrical signal comes out 180 degrees out of phase or upside down from the input signal. The JTM45, being an electronic clone, also "suffers" the "problem." In the case of the Bassman, the Jensen speakers also play backwards. That is if you observe the marked polarity, the speaker cone moves backwards— just the reverse of standard speaker polarity. The net result is that the amp reverses the signal and sends it to a speaker that reverses the signal again, righting it, putting the output back in phase with the input. However, in Marshall's case, the amp reverses the signal but Celestion speakers are marked in standard correct polarity and do not invert the backwards signal like a Jensen. Yes, that's right, a 4 input Marshall plays 180 degrees out of phase or backwards! This gives the Marshall a different set of harmonics and feedback characteristics when pushed than you would get if you reversed the speaker polarity to play in phase. Another happy accident!

Of course, there's the mojo of where an amplifier is built. Kendrick makes amps in Texas that take on a Texas accent. I make Trainwrecks among the toxic dumps of New Jersey and they're all little mutants. Marshalls are made in England so they can't help but sound British.

Tube Amp Talk

VINTAGE 100 WATT MARSHALLS

By the mid-60s, with the emergence of rock 'n' roll super groups playing large venues, the need for high power guitar amplification became apparent. The race for guitar amps with 100 watts of true R.M.S. power was on. Vox developed the AC100 to back up the Beatles. At the same time, Pete Townshend approached Jim Marshall about building a 100 watt amp for his band, "The Who."

The Marshall crew immediately set about building some prototypes of a 100 watt amp. Not having a 100 watt rated output transformer at hand, they built the prototypes using two 50 watt output transformers on each amp. These first amps are very rare and of course super collectible, if you can find one.

The next step in the evolution of these amps was in locating true 100 watt power transformers. These transformers were designed for P.A. use and were very large and robust. The first ones ran on 220 volts only and were fitted to an aluminum chassis with plastic faceplates. Later on, multivoltage large power transformers were fitted to these amps for international use. At this point, three important terms used to describe Marshalls came into use. First, the plastic faceplate amps became known as "Plexi Marshalls." Second, the aluminum chassis Marshalls became known as "Aluminum Chassis Marshalls." Third, the large transformer 100 watt Marshalls became known as "Large Transformer Marshalls," or as "Large Transformer Plexis."

In a short time, as the sales of these amps increased, Marshall had their 100 watt transformers custom made to different specifications. Although these transformers were physically smaller than the originals, they were in no way inferior to the larger units. They did produce a different sound. The early transformers were noticeable cleaner and louder than the smaller units. The smaller units tend to be warmer sounding and have more crunch.

The size of the smaller transformers remained the same throughout the vintage years., but voltages and impedances varied during different time periods. I'm sure that was done both to fine tune the amp for the sounds of the time and to adjust for the use of different output tubes.

At the same time the electronics were being tweaked, Marshall was tweaking their speaker cabinets. This was done in several ways. The size and shape of the cabinets evolved. The position of the speakers on the baffles was changed. Various types and wattage rating speakers were used, most being the Celestion brand.

Different types of grille cloth were used. Grille cloth does have an effect on sound even though it's always claimed to be acoustically transparent and of course Marshall introduced us to the "slant" cabinet.

PLEXI 100S

The Plexi 100 Marshall is the most collectible of all Marshall hundreds. They came in Super Lead, Super Bass, Super Tremolo, P.A. and Power Builder versions. One little known fact is that, until the late 60s, the Super Lead, Super Tremolo, and Super Bass all shared exactly the same circuits electronically. It wasn't until the tail end of Plexi 100 production that changes were made to the circuit to give the Super Lead its own voice. At that point, plate voltage was stepped up, power filtering increased, and output transformer primary impedance dropped, making the last of the Plexis identical to the '69 to '73 metal face amps except for the plastic panels.

O.K. Here's how it goes. The large transformer Plexis had the highest plate voltage and the highest output transformer primary impedance. The mid period Plexis had lower voltage than the large transformer amps and a medium primary impedance on the output transformer. Also, some amps in this period had one large transformer and one small one. On the small transformer amps the output transformer was moved about and at one point even rotated 90 degrees, which affects the flow of electrons in the output tubes with its magnetic field.

The last of the Plexis saw an increase in voltage, but not as high as the original large transformer amps. The primary impedance on the output transformer was made very low. A low primary impedance causes two things to happen. It makes the tubes work harder and for the same plate voltage you get more power. It also increases odd order harmonics, particularly the third harmonic, which adds grit and grind to the sound at the expense of the ability to play more complex chords under distortion without everything turning into a distorted jumble. That's why the early amps do the Hendrix stuff so well and the later amps do the two note heavy crunch sound better.

METAL FACE 100S

From '69 to '73, the metal face amps were consistent, except in late '73 Marshall switched to printed circuits combined with flying leads to the tubes and controls. In '74 the plate voltage was lowered at the same time amps bound for the U.S. were fitted with 6550 output tubes. EL34s were used for most of the rest of the world. Shortly after this, the first two input factory master volume models were built.

TIPS AND TRICKS

Plexi 100s do not have individual input grid resistors on their power tubes. This causes instability when all controls are turned to 10. The cure is to solder a 5.6K resistor to pin 5 of each output socket and connect the feed from the phase inverter through these resistors. This also warms up the sound.

On some early metal face 100s the power supply ground is connected to the front of the amp. This causes excessive hum. The cure, move the ground to the power filter ground like the later amps are wired. If you're not sure of where the filter ground is, just hook up the ground to the back side of the chassis.

Metal face amps have their preamp grounds connected to the controls on the front panel. Some amps then run a ground wire from the controls to a terminal on the chassis. Some amps didn't run a ground from the controls to the chassis. On these amps, as the controls and chassis corrode, the ground connection becomes weak and all sorts of problems pop up. Simply run a wire from the controls to a good clean solder connection on the chassis behind the controls, or add your own ground terminal and run a wire from it to the back of the controls.

LEAD DRESS

To reduce hum and buzz, press all preamp tube wires to the chassis except lift all green control grid wires up into the air. With the amp open, upside down and all controls facing away from you, dress all preamp wires up and to the right, except for the purple wire going to the presence control. This wire should run down and to the left.

There are many more tricks and master volumes circuits by Trainwreck which appear in *A Desktop Reference of Hip Vintage Guitar Amps* by Gerald Weber. If you want to go beyond what I've written here, I'd suggest buying a copy of Gerald's book. To keep myself out of trouble, I should mention Marshall does make vintage reissues. I like the originals but you might give the reissues a try. After all, tone is in the ears of the beholder.

GIBSON AMPLIFIERS

I've received many requests to talk about Gibson amplifiers. It's a subject I've been reluctant to approach because while I've worked on many old Gibson amps, I'm really not an expert on when which model was built, or on Gibson amplifier cosmetics. Instead of talking about those items, I'll just concentrate on model numbers and their electronic features.

Before I get into individual models I want to point out that, in general, I am not a fan of old Gibson amps. While they did make a few good sounding models, they made many more that fell short of the mark, in my opinion. However, it's rare to find a Gibson amp that can't be modified to sound great. The problem is these are vintage amps and changing the electronics will ruin the vintage value. If you pick up an old Gibson that's not in demand today for $250 and modify it, you may be sorry later. Two years from now that model may be on a hit recording and worth $2,000. If you mod yours, it will still only be worth $250, more or less. It should be noted that certain Gibson models make superb harmonica amps dead stock. So for now I'm going to just talk about factory stock models and later on I may, or may not, talk about how to "ruin" your vintage Gibson amp by making it sound great.

THE GIBSON STEREO-REVERB-TREMOLO GA-79 RVT

The GA-79 RVT is king of the hill to most Gibson amplifier fans. The "79" is a true stereo amplifier with independent 15 watt output stages supplying their own 10" speaker. A slide switch allows for stereo operation or combines the signal and feeds it to both output stages for 30 watts mono. You can also put the amp in the stereo mode and only use one channel for 15 watts of power on a single speaker. The 79 uses 9 tubes as follows:
V1: 6EU7

V2: 6EU7

V3: 6EU7

V4: 7199

V5: 6BQ5/EL84

V6: 6BQ5/EL84

V7: 12AU7/ECC82

V8: 6BQ5/EL84

V9: 6BQ5/EL84

The power supply, which is mounted on a separate chassis, uses silicon diodes instead of a tube rectifier. This use of solidstate devices for the power supply is common to many Gibson amplifiers of this period. The 79 has two input jacks per channel plus one stereo input jack. Each channel has a volume control and a bass and treble control. It also has controls for the tremolo circuit which only works on channel one. Channel one also has an extension speaker jack while channel two doesn't. The use of an extra speaker placed several feet from the amplifier greatly increases the stereo effect. The amp has a footswitch for both the reverb and tremolo. The 79's two 10" speakers are mounted to a V shaped front baffle to increase the stereo effect when played on the internal speakers alone.

The real magic of this amplifier is the use of EL84/6BQ5 output tubes run in cathode bias and without feedback, just like an old Vox. True, they are not run in class "A" operation, and the amp doesn't have a tube rectifier, but it's hard not to get a good sound out of EL84s. The tone shaping of the 79 is pure Gibson and the output tubes add life to that. The 7199 tube used for the reverb circuit is in high demand for old tube Hi-Fi gear, so may be a bit hard to find and/or expensive if you need a replacement.

THE GIBSON HAWK AMPLIFIER

So you can't find a GA-79 RVT for $250? Well, if you still have a desire for the Gibson sound with 6BQ5 tubes you might want to try a Hawk. This is a single channel Gibson with reverb and tremolo and two 6BQ5/EL84 tubes biased in class "A." It has one 10" speaker. It does use a feedback loop which must be connected as it's part of the tone shaping of the amp. For tubes, this amp uses three 12AX7s and two 6BQ5s. The Hawk uses a sealed tone network after the volume control. This part is a "Sprague 102C84"—inside are two resistors and two capacitors. If you don't like what it does, remove lead No. 2 from ground. Then remove leads No. 1 and No. 3 and put a wire between the points where 1 and 3 were connected. Or, if you wish, you can insert your own idea of what a tone filter should do in its place.

THE GIBSON FALCON AMPLIFIER

This amp is a bigger bird; it's the Hawk with two channels and a 12" speaker. It also sprouts bass and treble controls on each channel, instead of the simple tone control of the Hawk. The Falcon has a fourth 12AX7 for the extra channel. As a result of pesticide poisoning, the Falcon has a mutation. It takes place in the form of a transistor used as a tremolo driver. The Falcon also uses the sealed tone network, only using tone that is common to both channels. The feedback loop is not part of the tone shaping on this bird. The one meg channel mixer resistors are a bit much, and reducing them perks up the tone and increases the gain. Oh, drat! I really didn't want to give out even minor tips on "ruining" your vintage Gibson amp.

THE GIBSON GA-5T

This is a little "student" amp that's really not worth talking about. If you have one, ship it to me and I'll give you your fifty bucks back! It uses two 6EU7s, two 6AQ5s, and one 6X4. The GA-5T has just a volume control, and a "frequency" control for its tremolo. 6AQ5s are a 7 pin miniature output tube, sort of a baby 6BQ5. They don't put out much power, but crunch just fine. The GA-5T doesn't have a bird name, but one thing it isn't is a turkey.

The Gibson GA-5T uses two 6AQ5 tubes in push-pull output. Gibson also made a GA-5 which uses a single 6V6 output tube in single ended operation. The GA-5 and the GA-5T are completely different amplifiers with not one thing in common, yet they both have GA-5 as part or all of the model designation. Worse yet is the practice of using the "Les Paul" name as part of the designation of many completely different models. When somebody tells me they have a Les Paul amp it is totally meaningless except to identify the amp as a Gibson. One must keep these identity quirks in mind when dealing with Gibson amplifiers.

THE GIBSON GA-5

This is the Gibson most like a Fender Tweed Champ. It uses one 12AX7, one 6V6, and one 5Y3. The GA-5 has two inputs, using one as a normal gain and the other as a low gain by use of a 47K shunt resistor. There isn't the usual grid return resistor on the input stage. Your guitar's pickups and controls act as this resistor. This means the shorting connection on the high gain jack must close when nothing is plugged in or the tube can "run away." Also, when using a mike for harmonica you should add a grid return resistor to the circuit. 5.6 meg ohms works fine. The amp has just one control besides the on-off switch. This is volume. The amp uses an 8" speaker but unlike the Fender Champ, this one is 8 ohms.

THE LES PAUL AMP

Well, here we go! This Les Paul amp is the same as the GA-5 we just talked about, except it uses a metal 6SJ7 tube in place of the 12AX7 of the GA-5. I want to get the electronic facts straight because the 6SJ7 tube was used very commonly in many brands of early amplifiers. The 6SJ7 is a sharp cutoff pentode. That means as the grid voltage rises, it quickly reaches cutoff. Since the 6SJ7 and 6SK7 "pin out" the same you could try a 6SK7 in a guitar amplifier. Warning: Do not plug a 6SQ7 into a 6SJ7 socket. It is a completely different tube and will not work and may damage your amp. If you do plug a 6SK7 into a 6SJ7 socket, you do it at your own risk. Depending on the circuit design, the 6SK7 can draw too much current and burn a resistor or two!

THE GIBSON EH-150

Let's get really rare. This is a fun amp for me because no other amp in Gibson's line ever used such a collection of rare and oddball tubes. This is very early Gibson at its finest. It is a truly classic example of many elements of early tube audio design.

TUBE ONE: The 6F5 is a triode with a metal terminal on the top of the tube. That terminal is the control grid. The EH-150 has one microphone input. This input feeds the grid of the 6F5 which is used only for the microphone channel. This channel contains only a volume control.

TUBE TWO: The 6N7 is a dual triode. In the EH-150 it has three uses. First, one triode is used as an additional gain stage for the microphone channel. Second, the other triode in this tube is used as the first gain stage for the instrument channel. The instrument channel has two input jacks connected directly to a 500K pot. This pot has two functions. One is to serve as the volume control for the instrument channel. The other is to serve as the grid return resistor, also for this channel. The 500K pot on the microphone channel also serves this dual function. The third action of the 6N7 is as a mixer tube, blending the signal from both channels together. The rest of the amplifier processes this combined signal. The output of the 6N7 is fed to the next stage by a two capacitor setup. A .0007 mfd cap is always in the circuit. If more bass is needed a .013 mfd cap can be added in with a switch. This is the only tone control the EH-150 offers.

TUBE THREE: The 6C5 tube again has three uses in this amp. It's another gain stage giving the microphone channel three and the instrument channel two in total. It is very rare to find an amplifier made as far back as the Gibson EH-150 with three gain stages. Most only had one! The 6C5 is also an impedance matcher to a transformer. It also serves as a transformer driver. No, this isn't the output transformer, but the phase inverter transformer for the push-pull output stage! The EH-150 used a transformer instead of a tube as the phase inverter. A transformer phase inverter has many advantages over a tube. It doesn't go weak or wear out. If it's wound right, it's in perfect balance and won't drift out. You can't clip it, and if it's designed right it won't saturate. Simply put, it won't overdrive so there is, in practical terms, no phase inverter distortion. Of course, these things could be looked at as disadvantages in a modern amp. A transformer as a phase inverter does have some disadvantages in other areas. I won't go into them here, as this isn't exactly an engineering paper. Let's just say the practice of transformer phase inverters was common in old radios and phonographs, but very rare in guitar amps.

TUBES FOUR AND FIVE: The 6N6 is a direct coupled power triode. The EH-150 used two in push-pull, Class "A" output. It is important to note the 6N6 is a true triode, not a pentode forced into triode duty. There is a very major difference as any audiophile will tell you. Each 6N6 contains two triodes. One is a driver and the other a power triode. They are directly coupled inside the tube itself. Direct coupling eliminates the coupling capacitor, and the problems capacitor coupling produces. A direct coupled, Class "A," push-pull triode output stage is state-of-the-art tube audio today. Who was the guy who designed this amp some 50 years ago!?!

The EH-150 has a single electrodynamic loud speaker. The output transformer has two output impedances. This is also very rare for an old amplifier. When you plug a speaker into the extension speaker jack, the EH-150 automatically puts the extra speaker in series with the built in speaker and changes impedance to match.

TUBE SIX: The 5Z3 is the rectifier tube. The 5Z3 is a four pin tube that uses a four pin socket. Again, it's a rather rare tube. The 5Z3 provides all the D.C. current needed to operate the tubes and the field coil on the speaker. The field coil is used as an A.C. ripple choke to the drivers in the 6N6 tubes. An additional choke provides current to the first three tubes and the output triodes are fed directly from the heater of the 5Z3. The EH-150 doesn't use feedback anywhere in its entire design.

I received a call from a guy with two EH-150 Gibsons who raved on about their creamy texture and tone. Then an EH-150 walked in my door, but this one has a factory bass and treble control instead of the "bass" switch like all the EH-150s I've seen before.

THE GA-40 LES PAUL AMP

The early version of this amp is a real classic. It's great for guitar and is also a top notch harmonica amp. The early amps used a single 12" Jensen P12P speaker. Later models added an 8" speaker in addition to the 12". Both channels on this amp feature pentode tube input à la Vox AC-15, and 6V6 output à la Fender Tweed Deluxe. The tubes used in the GA-40 are:

V1 and V2: 5879,

V3: 6SQ7,

V4: 12AX7,

V5 and V6: 6V6,

V7: 5Y3.

The 5879s are the input pentodes. The 6SQ7 is for the tremolo. The 12AX7 is the phase inverter. The 6V6s are the power tubes, and the 5Y3 the rectifier tube.

Channel one is straight through flat response with no tone shaping. Channel two has a fixed tone network to shape the tone to a typical Gibson voice. This channel also has tremolo with frequency and depth controls as well as an on/off footswitch. Each channel has its own volume control and there's a master tone control. The power supply has a choke and all B+ current including the output tube current flow through this three Henry choke.

Some simple easy to reverse mods for the GA-40 are...

ONE: To increase treble on the first channel, a bright cap can be installed on the volume control.

TWO: To increase gain on channel one, the 470K mixer resistor from the volume control can be reduced in value. However, the more you reduce this resistor's value, the more gain channel one will have, but channel two will have less and less. I think most people will like this amp stock, but I always seem to get requests for mod tricks.

THE GA-15

The GA-15 is sort of like an early tweed Fender Deluxe. However, it has an interesting choice of tubes. The T.V. Deluxe uses a 6SC7 in grid leak bias for the first stage of both channels. The Gibson substitutes a 12AX7 used in the grid leak bias mode. It is very rare to find a 12AX7 used in grid leak bias in a guitar amp. The next tube used is a 6SL7 as a phase inverter. The Fender uses a 6SC7 for this purpose. The 6SL7 and 6SC7 are both eight pin octal dual triodes of the same gain structure. However, they pin out differently. Also, in the case of the 6SC7, both cathodes are tied together in the tube which limits the circuit designs to ones sharing a common cathode. The GA-15 uses two 6V6 tubes in push pull and a 5Y3 rectifier. It has one tone and two volume controls. The tone control affects both channels. The 200 ohm cathode resistor for the 6V6 tubes is unbypassed. A 25 mfd/25 volt electrolytic across this resistor will give more punch. Don't forget to put minus to ground and the plus end toward the capacitor. If you need more highs, the tone setup from a 5F3 Fender Deluxe will work just fine.

THE GIBSON SUPER MEDALIST

The Gibson Super Medalist has a lot of features in common with a Blackface Fender Pro Reverb. That is two channels, each with two inputs. The amp has reverb and tremolo. It also uses two output tubes and two 12" speakers. That's as far as it goes. The Super Medalist is electronically its own animal. It uses four 6EU7 tubes, two 12AU7 tubes, one 12AX7, and two 7591 output tubes. The rectifier circuit is solidstate. I should point out that because Ampeg and Gibson, plus many Hi-Fi companies, used 7591 output tubes, they have the reputation of being very clean sounding tubes. Indeed they can be, but in the right output circuit, with the right preamp circuit they will out rock any 6L6 tube ever made! They can be made to sound a lot like a Mullard EL-34. I never change 7591 tubes to anything else unless the bottom line is to go ultra-cheap. The Super Medalist uses two chassis; one sits on top and the other sits on the bottom connected by wires.

O.K... in my opinion this is one amp Gibson got really wrong. All the right parts are there, but it features the classic wrong tone shaping and gain structure that keeps most Gibson amps in the pawnshop prize category. This amp is fairly complex so I won't go into extensive mod options here. I will say that this amp can be turned into a killer amp by someone who really knows what he's/she's doing. The same can be said for the Gibson Duo-Medalist, too. The Duo is a lower power, single speaker Super Medalist. I did work for a guy for years who blew trumpet through a stock Super Medalist and it sounded quite nice. I don't know how many people are hunting for a trumpet amp!

VINTAGE CELESTION SPEAKERS

Too often overlooked in the quest for the ultimate guitar sound is the selection of the best speakers for the task. Your choice of which speaker to use plays a crucial role in the character of your final tone. Some speakers can give you crisp, ringing, bell-like clean tones. Others have a rich warm distortion tone with singing sustaining harmonics. A few rare speakers do it all.

Over the years one of the most popular guitar speakers continues to be the British-built Celestion. In the U.S.A. we were first introduced to Celestion speakers by the Beatles and their Vox amplifiers. Later on, Eric Clapton and Jimi Hendrix, playing Celestion equipped Marshalls, nailed down Celestion's reputation as being the definitive British speaker. In this section, I'll cover some of the most notable Celestion speakers starting with the earlier models and perhaps some current speakers later on.

All Celestion speakers have a model number on them that starts with the letter "T." In some cases I'll use the "T" number, and in cases where it could be confusing, I won't. For example, a speaker with a 16 ohm voice coil would have a different "T" number than exactly the same speaker with an 8 ohm voice coil. A "T" number change was often used only to indicate a change of paint color on the frame. With thousands of "T" numbers in Celestion's records, it's not practical to go into all of them here.

The first Celestion I'll talk about is commonly known as the "Vox Bulldog." It's the speaker used in AC15 and AC30 amps made by Jennings Musical Industries (JMI). This speaker was made for Vox by Celestion, although the label on the back says Vox. This speaker was used first in the AC15 with an unpainted frame in 1957. In 1960 the speaker was given the model number T530, stamped on the front paper gasket. Its frame was painted "Vox Blue" and the cone was up-graded to H1777 status. Although these speakers, to this day, are called "15 watt" Celestions, my best information indicates that they became 25 watt speakers in 1960. In any case, a good running AC30 can put out 50 watts under distortion and a pair of these speakers held up better in that amp than the electronics did!

I should also point out that the Bulldog label didn't show up until the mid-60s, mostly on Thomas Organ Vox speakers that were not built by Celestion. The gold frame, black magnet cover speakers had these labels. Some said "Made in England," some had the Utah speaker code, but the only gold frame Celestion "Bulldog" was made for Selma amps.

Later on in the early 60s, Vox had Celestion change the color of the "Bulldogs" from "Vox Blue" to "Poly Grey." The speaker became the T1058 and is commonly known as the "Silver" Bulldog. The early "Silver" and blue speakers were the same. Later on, changes in the "Silver" version made it sound a bit harsh, and so the blue T530 became more sought after as it took no special knowledge to know you were getting the better version. The Celestion T530, a.k.a. "Blue Vox Bulldog," has become the Holy Grail of vintage speakers and a clean original example can sell for several hundred dollars. Also, one should keep in mind that Marshall used these speakers in some early equip-ment. I've seen them in early Marshall and Park cabinets for example.

Now that we've covered some history, let's get into some details of the speaker itself. First and foremost, these speakers used Alnico magnets. Alnico magnets give a speaker a unique voice that cannot be had with the modern and much cheaper ceramic magnets. Ceramic magnets do make a fine sounding speaker, and can give tones that Alnico can't, but if you want the Alnico voice, you just plain have to use Alnico. The Bulldog voice coils were wound on a white paper voice coil form. This paper gave a different tone than the brown paper voice coil forms used in early ceramic model Celestions. Also, the cone edge treatment was much lighter than the edge damping treatment used on ceramic magnet speakers.

The T530 and T1088 are 8 ohm speakers, although 16 ohm ver-sions do exist. The front gaskets on the Bulldogs were paper as opposed to the cork gaskets on the ceramic Celestions.

As for tonal characteristics; the word is clean. The Bulldogs have a smooth, clear, bell like voice with more chime than any other speaker I know of. It has a very strong, thick midrange, almost hinting of a Wah pedal in the middle of its range. It has a nice bottom end, although not quite as deep or solid as some current high power speakers that have much less tone all around. Compared to the Alnico speakers made by Jensen, it has a far better bottom end so I hope that places it for you. Under distortion the speaker has a smooth crunch. It is not the heavy metal crunch of modern speakers. With a P90 pickup they'll growl like a jungle cat. They don't fuzz and fizz—thanks to the cone design and Alnico magnet. If you like Brian May's tone you'll love these speakers for your solos.

Be careful. There's a 100 watt "Vox" speaker being made by Fane of England. They do not have the original tone. Not a bad speaker for 100 watt rating.

I'll try to highlight some of the more popular Celestion speakers in the "M" and "H" series. Unlike the "Vox" speakers, all these speakers have a ceramic magnet in their design. Keep in mind that Celestion made many more models of guitar speakers than I can cover here. Also, Celestion speakers were always subject to fine tuning and upgrading by the engineers, so the same model speaker can sound different from different time periods. Celestion used cones made by several manufacturers, wound voice coils different ways and on different materials, varied paper formulas, cone weights and thicknesses, damping treatments, cone treatments, spider stiffness, magnet size, and other factors. So here are two rules when hunting down vintage Celestions:

RULE ONE: The speaker should be all original in good working order. No recones! You may ask, why no recones? I have found that it's rare to find a Celestion that has been reconed with the original part. Even if it is a Celestion part, it's a good bet it's a recone kit from a later model speaker.

RULE TWO: Listen to the speakers you're about to purchase. If you like their sound, then they're good speakers. I have my favorites, but you may not agree with my choices. It really is a matter of personal taste, but the speakers I'll talk about seem to please people the most.

THE G12 M 20 WATT

This speaker has the smallest magnet size of the "M" series. A small magnet yields less bass response, volume per watt input, and power handling ability compared to larger magnet speakers. However, the "Greenback" (named for the color of the plastic magnet cover) 75 Hz cone resonance version of this speaker is a warm, rich sounding speaker even though it's not much on bottom end. The earlier "Pulsonic" cone versions of this speaker are what to look for. You can tell these cones by the part number on the back of the paper cone, stamped in white block numbers that end in "003" (brighter) or "014" (deeper sounding). While no one knows for sure how Eric Clapton's Marshall was set up for the "Beano" album, I've worked on lots of those old Bluesbreaker Marshalls, and KT66 tubes with the 20 watt speakers have given me that tone. Also, most of the grey stripe early Marshall cabinets had these speakers, although I've seen Alnico Celestions and a few other types in these cabinets as well. If a 4x12 cabinet is too loud for your use, give these a try. Four of the 20 watters are about as loud as two 30 watt Celestions. Later versions used a "1777" cone by a different manufacturer (Kurt Muller) and sound nice but a bit harsher. Again it's a matter of taste.

THE G12 M 25 WATT

Much more common than the 20 watt speaker, the 25 has a bigger magnet, more bottom, a very thick midrange, but less note separation than its little brother. Again, the earlier cones were the "003" and the "014." Later the "1777" appeared, and even later the green back was changed to black plastic. A black back 25 was built with a 55 Hz cone, part number "444." Those are very dark sounding. The 75 Hz speakers with Pulsonic cones make excellent lead speakers because of their thick tone and excellent harmonics. The "1777" cone speakers have a more aggressive chord sound, but lose out to the earlier version on smoothness and their clean sound. The "1777" version is the one you'll find in mid-70s Marshall cabinets. I had a 1967 Marshall cabinet with 25 watters, cone part number 2102003. They were my personal favorites of all the 25s I've ever heard. The 25s also came with other color backs, such as grey (very bluesy, soft, creamy texture), white and so on. Again, if you run into these give them a try.

THE G12 H 25 WATT

These speakers have the standard 25 watt cone but with a larger magnet. The 25 H has more high end and note separation than the

25 M and is worth a listen. As for my personal taste I'd skip these and go for the 30 H but, again, it's all subjective.

THE G12 H 30 WATT

This speaker has the largest magnet of all the Celestion 12s of the time. Right off, I can tell you it's my personal favorite ceramic magnet Celestion when equipped with the 102/014 Pulsonic cone. Loud, dynamic, big bottom, crisp highs, great note separation and a sparkling clean sound. Less mids than the 25 M but, aside from the Alnico 15 (which has the strongest Celestion mids), it's a fairly thick sounding speaker. The 003 cone gives more highs (overkill in my opinion) but give them a try too.

Sometimes these speakers came with no label on the green back. Also, many times they came with no green backs at all! When you watched Hendrix play through cabinets with the 100 watt logo in the corner, you were hearing these speakers. Later on, Celestion switched to black backs with a "1777" cone and a fiberglass instead of standard paper voice coil form. Very high brightness speakers! Too much for a Marshall, but they sound killer in a blackface Fender Deluxe! A 55 HZ 444 cone version exists also. The white back 30 H has a killer power chord sound and is very loud.

CURRENT SPEAKERS

G12M VINTAGE 25 WATT GREENBACK

I own original late 60s G12M Celestions and to be honest, these reissues of the G12Ms sound different. Less bottom and mids and more high end. I like the originals better myself, but *Guitar Player* magazine poll winner Henry Kaiser has his Trainwreck Rocket parked in Jerry Garcia's studio on top of four reissue G12M speakers in a Vox 4x12 cabinet. Really hot Bay Area players, such as Henry and Steve from the band Zero, tell me this rig has absolutely killer tone. In the *Guitar Player* combo amp shoot out, they reported the Matchless 30 combo sounded great on its own speakers and did not improve by plugging it into an old Marshall cabinet. The Matchless combo had the reissue G25M as one of its speakers. I guess I'm not a good judge of this particular model. Give 'em a test drive and decide for yourself.

THE G12 VINTAGE 30

A very popular modern Celestion. Although called a 30, its high temperature voice coil can handle 70 watts. Unlike the original G1230H which had a 30 watt paper voice coil, this speaker can be used almost anywhere—a pair in a Fender Twin, for example. Personally, I like this speaker better than the 25 watt reissue. In fact, my brother, Scott, uses these as his main speakers. This is an even, warm speaker and is an excellent choice in a medium power loud-speaker. The original G1230H is the Hendrix/Clapton speaker and the G12 Vintage 30 is the closest current Celestion in that tone vein. Anyway, the current Vintage 30 is really worth checking out.

THE G12M70

This is the standard speaker for Heavy Metal. Pretty, clean tones? Forget it! This is the speaker for raw, crunch, balls out, Metal shredding. O.K., I like vintage stuff, but a hot humbucker through a Marshall JCM 800 100 watt head on a stack of these gives a power chord fix that's hard to match.

THE G1280

A harder edged Vintage 30 is my impression. I owned some original vented magnet versions. These are now the Boogie 90s and the current Celestion 80s are non-vented. Eric Johnson uses 80 watt Celestions, so obviously they have tone.

THE G12100H

This Celestion has the big bottom. It's the Billy Gibbons speaker. Recycler is a 12 watt solid state Marshall on just one G12100H. I've heard that Billy will hook up a JTM 45 Marshall or his 'wreck to a single G12100H in open back, so this speaker can really hang tough. The classic Celestion sound in a high power rating. If you like bottom, this Celestion is best.

Well, Celestion has many more models in 12s and other sizes but I've tried to cover the most common and the ones I know best.

Tube Amp Talk

HIWATT AMPLIFIERS

The original series of Hiwatt Amplifiers were designed and manufactured by Dave Reeves in his factory near London, England. Hiwatt Amplifiers were designed to be a very upscale product. Indeed, no other British amplifier of that time period could match Hiwatt for quality of materials and workmanship. Even today, or should I say, today more than ever, when a technician gets his first look at the inside of a Hiwatt, stunned amazement (at the level of craftsmanship) is the common result. The original DR (for Dave Reeves) Series of Hiwatts came in 50, 100, 200, and 400 watt versions. They also came in versions designed for P.A. use.

The first of the 50 watt units were equipped with a tube rectifier. Later 50s and all 100, 200, and 400 watt models used silicon diodes in their power supply section. There was a 50 watt combo Hiwatt called the "Bulldog 50," which was equipped with a 50 watt Fane "Bulldog" speaker. At the same time Vox was using a "Bulldog" speaker not made by Fane. I guess the British do love their Bulldogs, but then so do I, both the speakers and the canine!

Hiwatts have their own sound. This is one camp that's not trying to be a clone of something else. Clean and very powerful with tons of headroom. If you want loud and clean, nothing comes close to a Hiwatt. I would take two or three Fender twins to keep up with just one DR103 100 watt head on its Fane-equipped 4x12 cabinet!

All Hiwatts were equipped with a master volume control. This is not to be confused with the modern master volume control used to generate distortion. On the DR Hiwatts you would adjust the individual channel volumes when using more than one instrument in the same head. Once you got your two instruments in balance with each other you could raise or lower the total volume with the master without affecting that balance. Turning the preamp volume to 10 and the master down yields no usable distortion in an old Hiwatt.

Because of their high voltage and high current transformers, Hiwatts need ultimate respect. If you touch the wrong point in a Hiwatt and you're lucky enough to live, at minimum your fillings will have melted from your teeth, your zipper will be welded shut, and you'll wish you'd been wearing a diaper.

The 50 watt Hiwatt is the smallest of the group. It uses four ECC83 tubes in the preamp and two EL34 output tubes. Like all Hiwatts, it uses Partridge transformers, which are bulletproof. It's rare to see a burnt Partridge transformer. The 50 shares the same preamp circuit with the 100, 200 and 400 watt models. They have four inputs, a high and low gain input for the normal channel and a high and low gain input for the brilliant channel. Each channel has its own volume control. Both channels share a common set of treble, middle, bass, presence controls and a common master volume.

The 100 watt model uses four EL34 tubes for increased power output. Early 100s use four ECC83 preamp tubes and later 100s use a five tube preamp. The extra tube is used as a driver tube in the phase inverter section to increase power and headroom. All 200 and 400 watt amps use five preamp tubes except the P.A. models.

The early 200 watt heads use six EL34 output tubes. Later on four KT88 tubes were used, and due to the higher power output of KT88 tubes, the power stayed the same. The six EL34 tube amps are warmer and the four KT88 models have more bottom end and are also well suited for use with a bass.

The 400 watt heads use six KT88 output tubes and again work well for both guitar and bass.

If Hiwatts have a weak point, it's the use of low wattage screen resistors on all models. I like to replace these resistors with 5 watt wirewounds to increase reliability. Also, the 100 watt heads contain a 10 watt 470 ohm resistor feeding the screens. I like to substitute a 25 watt rating in its place. Biasing a Hiwatt can be tricky on some models and I won't cover it in this article.

The one layout flaw in all Hiwatts is this: In order to maintain an ultra neat look inside the amp, all preamp grounds are bootstrapped under the preamp terminal board. Then they run to the front of the amp through a single ground wire. This causes ground loops and hence buzz at high volume and treble settings. I like to separate these grounds and run them individually to the front of the amp. This not only quiets the amp, but improves its tonality as well.

Hiwatt 4x12 speaker cabinets contain Fane 50 watt speakers with cast frames. The cabinets are rear ported and are super heavy duty and just plain heavy. The early cabinets use the blue/gray Marshall-style paper weave grille cloth. The cloth is rotated 90 degrees so the grain direction gives it a slightly different look. The side handles on the original cabinets are merely cuts into the thick sides of wood. Hey, I did an article on Hiwatts and didn't mention Pete Townshend once. I must be slipping!

TREMOLO — WIGGLE TO WIGGLE

During the 50s and early 60s, tremolo and the vibrato became parts of the circuit in many guitar amps. Now in the 90s these effects are popular once again. When it comes to guitars and amplifiers, the terms "tremolo" and "vibrato" are often interchanged. First, let's get the definitions correct. Tremolo is an alternating change of volume. Vibrato is an alternating change of pitch. Some examples of incorrect, but common usage of these terms would be the "Tremolo" system on a Stratocaster. The system on a Strat produces "vibrato" as it changes the pitch of notes. Another example of incorrect usage of these terms would be a 60's Fender Super Reverb or Twin. These amps contain a system that changes "volume" in an alternating way. Yet these amps have that feature labeled "vibrato." Of course an alternating change in volume is "tremolo."

At this time I'm only going to talk about tremolo systems and leave this more complex vibrato systems for some other time. I should also tell you that these electronic tremolo and vibrato systems may be built into an amplifier, or made as an outboard unit. This is true both in the past and with current designs. Often a guitarist would/will combine the sounds of the vibrato system on his guitar with tremolo. An excellent example would be Duane Eddy. Along with Duane, I grew up listening to the Everly Brothers, Roy Orbison, Chet Atkins, and many other who use tremolo to great advantage in their music. Tremolo has three main parameters. One of these is how much of a change in volume the tremolo produces. Terms used for this include "depth," "intensity," "shake," and "strength." Most amps have a control for this function, but cheaper amps often didn't. Some amps, like a Vox AC-30, have a preset inside the amp which you can adjust with a screwdriver.

The next parameter is the rate at which the volume changes. Tremolo can be very slow or fast. "Speed" and "rate" are the common terms for amps that have a control for this function. Again most amps do, but some very inexpensive amps only had tremolo "on" and "off," with no controls for adjustment of the effect.

The last parameter is "waveform." This is a very important parameter, and in my mind affects the tremolo to the point where a bad waveform (also known as "waveshape") can render a tremolo circuit useless. The waveform is the shape of the volume envelope. If you turned the volume change on-off, on-off, as with a switch, it would not sound very good. But if you change the volume in a gradual manner the tremolo will sound better. An example is a blackface Twin (bad waveform) vs. a blackface Vibrochamp (better waveform). If you get the waveform just right, along with several other factors, the tremolo will have a magical, shimmering quality and be very musical. I've had guitarists tell me that tremolo was a useless, goofy effect, only to play through a Trem-Rocket and say, "Sign me up!"

So what's involved in a tremolo circuit? First, there's the oscillator circuit. The oscillator controls both the speed and shape of the waveform. The shape of the waveform can be changed after it leaves the oscillator, sometimes in a negative way and sometimes in a positive way. More on that in a bit. The other part of the tremolo circuit takes the oscillator waveform and modulates your signal volume to produce the tremolo effect. The two common methods used to accomplish this task are as follows: One is to modulate the bias voltage of a tube or transistor with the oscillator waveform voltage. Changing the bias affects the gain and thus the volume the tube or transistor gives to your guitar signal.

The other common system is called "photoelectric." A light bulb or LED is lit to varying brightness levels by the oscillator waveform. The bulb or LED is placed next to a photo-resistor, which is a special kind of resistor that changes its value depending on how much light shines on it — sort of like a solar cell, except it doesn't produce electricity, it just changes resistance value. This photo-resistor can then be used to modulate the signal by passing or bypassing a signal, depending on circuit design, through its varying resistance, which is light intensity controlled. The system used in tube amps is often one using a neon light. Neon bulbs can change intensity faster than incandescent bulbs like the type that light your home. This is because neon is a gas that changes in response to voltage, and doesn't have a hot glowing filament like an incandescent bulb. In an incandescent bulb, voltage is applied and it takes a small amount of time for the filament to heat up and glow. If you remove the voltage the filament takes some time to cool down and stop glowing. This gives it a slow response. The famous Univibe pedal uses the slow response time of incandescent bulbs to good effect.

On the other hand, neon bulbs are faster to respond to voltage changes but have some negatives to deal with. One is the neon bulb doesn't produce any light at all until a certain "turn on" voltage is reached. At that point a good amount of the light it can produce is already present. From that point the change of voltage vs. change of light output is also not linear. I said earlier a waveform can be changed after the oscillator, and the neon bulb is the bad actor in this regard. True, you can try to modify the oscillator waveform to try to overcome this, but the result is often the dread large blackface amp tremolo. Ampeg had better results with this system. Personally, I don't like it at all. If an incandescent bulb is used, a smoother waveform can be had, but as you increase the speed of the waveform, the filament has no time for the heat/cool cycle and its light output becomes more constant, reducing the tremolo effect. I favor non-photoelectric systems myself.

Where the tremolo is introduced into the circuit can also affect the results. Of course outboard units add the tremolo before any of the amplifier's circuits. I've heard many good sounding outboard units. However, I feel that onboard tremolo can be made to sound the best of all. That's to not say onboard is not without its problems, which I won't go into here. I will say an onboard, non-photoelectric, tremolo circuit, with the right waveform, injected at the right point in the amplifier's electronics, gives the best tremolo I've heard to date. Of course for an amp without onboard tremolo, an outboard unit is the way to go. You can use the outboard unit to add tremolo to any amp you own, a big plus. There you have it. If you've heard any amazing tremolo systems, drop me a line and let me know.

Tube Amp Talk

PURCHASING A VINTAGE AMPLIFIER

I receive many calls and letters from musicians with questions about purchasing vintage amplifiers. This is a complex subject, but I hope I can shed some light on it. My definition of a "vintage" amplifier is any amplifier that is no longer in production in its original form. For example, a Fender Model 5FGA 4x10 Bassman amplifier made in 1958 is a vintage amplifier. A Fender 4x10 Tweed Bassman made last year is a "reissue." A Kendrick 2410 4x10 Tweed is a "reproduction" amp. There are also "collectible" amps. These may be vintage or current production amps that have properties that make them desirable to people who collect equipment for their own reasons. For example, a Fender Model 26 Deluxe is both vintage and very collectible because it was the first of the Fender Deluxes and came standard in a hardwood cabinet. A JTM 45 aluminum chassis Marshall is both vintage and collectible. Vintage for obvious reasons, and collectible because it was the first in the long line of Marshall amplifiers. It's a part of amplifier history. Examples could be given for Vox and other brands but you get the idea.

Modern collectible amps such as Dumble, Trainwreck, and others, become collectible because they have a unique sound, are very scarce, both of the above, have a unique design, or are made collectible because of their use by a famous artist or artists. An interesting wrinkle is that modern collectible guitars and amplifiers seem to be increasing in value at a higher rate than the majority of vintage gear. It's really too soon to tell if this is a trend or a government plot to balance our trade deficit. As far as I know, none of the makers of modern collectible equipment had any idea that this phenomenon would happen. Anyway, the selection of the right piece of vintage gear is still the benchmark for investing your money if profit is your motive. Most of us purchase vintage gear to enjoy rather than as an investment. The rest of this article is geared towards those who buy vintage gear for the joy of owning and using it.

CHOOSING AN AMPLIFIER

When looking for a vintage amp there are three main ways people select the piece they want. By brand name alone, such as Fender, Marshall, Vox or another. By brand and model, and sometimes by year also. An example would be a 1964 Vox AC-30 top boost. Or they might not have any particular amp in mind but are looking for something that catches their ear and eye. For example, my friend purchased a Kalamazoo Model One. It's a cross between a Tweed Fender Champ and a Vox AC-4. Sounds great, costs little, and how many people would start out thinking, "I think I want a K-200 Model 1?" So before you start the quest for the amp of your dreams ask yourself, "Do I know just the amp I want or am I on a hunt for a certain sound, and trying to find it in whatever comes along?" If you know exactly what you want, that narrows the search. If you're on a hunt, that may take a bit more time and work.

WHERE TO BUY A VINTAGE AMP

When on a general hunt for your yet unknown amp, guitar shows are an excellent source to scope out the different less-known brands of guitars and amps. You can purchase amps at flea markets, through local ads, musicians, publications and so on. When buying an amp over the phone or by mail, find out if you can return the amp if it doesn't meet your expectations. Speaking of expectations, guitar amps are usually sold by some of the following factors.

DESIRABILITY: For example, a Fender Tweed 3x10 Bandmaster is more desirable than a Fender Blackface Bandmaster and therefore, the price of the Tweed Bandmaster is higher than the later version.

CONDITION: The better the amp's physical condition, of course, the more it's worth. The better the condition of the covering, for example, the higher the value of the amp. There is a catch...beat-up factory original tweed, again as an example, is worth more than being recovered in fresh, perfect tweed.

That brings us to...

ORIGINALITY: The closer to having all the same parts it left the factory with, the more it is usually worth. Reconed speakers are worth less than ones with original cones. If the speakers have been replaced with other than what came with the amp, usually the amp's value will go down. The reason I say usually is because on rare occasions this doesn't hold true. If you have an 1980 Vox AC-30 reissue that has had the factory stock speakers replaced with Blue Vox Alnico speakers, the amp would increase in value. In fact, those speakers might be worth as much as the whole amp! Replaced or rewound transformers are a big negative factor on resale value also. Also negative are add-ons such as master volumes, effects loops, channel switching and other non-factory modifications. Here's a grey area item. Amps contain power filters. These have a limited life-span, and then they fail. Do changed filters hurt the value more than filters that don't work? In other words, an amp that plays with changed filters versus an amp that doesn't play but has original failed filters. As a general rule, the more original an amp is, the greater its value.

RARITY: Some amps have a value not related to anything but the fact that they are very rare and hard to find pieces. Sometimes these rare amps are also great sounding and that of course is a bonus.

CELEBRITY VALUE: How much would you pay to own a Marshall that belonged to Jimi Hendrix? How much is the Marshall Blues Breaker Combo used by Eric Clapton worth? You get the picture, I'm sure.

MAKING THE PURCHASE

When purchasing an amp by phone, there's a numbers game that is often used. Amps are rated on a scale of 1 to 10. Ten is "Dead Mint," or an amp in the same condition as the day it left the factory. One is either an amp or a leftover jelly donut, but is not in good enough condition to tell. This scale is linear in condition terms but non-linear in terms of determining selling price. For example, to go from a Marshall Plexi 50 that's a 5 to one that's a 7.5 might cost you 50% more. To go from one that's a 5 to one that's a perfect 10 might cost you three times more. There's another catch. The seller often sees his amp as a higher number on this scale than the purchaser would assign it.

Another problem is that this scale is usually based on condition and not on sound. If you purchase an amp rated a perfect 10 but it sounds awful, the dealer can only be held to account if the amp is not in the condition he promised. Asking a dealer to judge sound for you is not a good idea. What sounds good to the dealer may not match what you think an amp should sound like. Everyone has different tastes, right? Also, many times a bad sounding amp only needs a new tube, cap, or resistor to bring it to life. Even a simple bias adjustment can make a world of difference. Unless the dealer sells you the amp "fully serviced," you're the one who will be responsible for its setup to proper specs. Any "not working" or "as is" deals are also at your own risk.

Also, keep in mind that it's much harder to return an amp purchased overseas, so be careful that you and the seller have all the facts and details correct before you use that credit card. Going to amp shows at least lets you see the amp before you buy it. There's generally no return privilege, so check the amp out carefully. Most shows won't let people crank an amp out and that is one downside to such a purchase. Most local private sales will be at a place where you can crank out the amp, so that's a plus. Phone or mail private sales usually go very well but be aware there are a few sharks in that ocean.

REISSUE AND REPRODUCTION AMPS

Many favorite vintage amps are now available as reissues or reproductions. The reissues are of course made by the same brand manufacturer as the original. A reissue Fender, Marshall, or Vox amp has the companies service and reputation behind it. Reproduction amps are made by independent companies and range from poor to excellent. Their resale value tends to drop, a point to be considered. Most people think the originals sound better. These days that's not always true. The advantage of a reissue or reproduction amp is that it's brand new and shouldn't require the maintenance of an original. It should hold up better on the road and is replaceable if stolen. The original may indeed sound better in many cases and hold, if not increase, in value. So there you have it. Buying a vintage amp should be fun. Make your purchase in a educated manner and you're more likely to be satisfied with your purchase.

VERY BASIC ELECTRONICS

This course in electronics is for the many of you out there who have no electronics background. The idea is to give you a basic understanding of electronics as used in guitars and amplifiers without the need for heavy math or physics. I will try to keep this course very simple and will use many examples and analogies and skip the quantum physics and any math much harder than two times two.

Technoheads can stop reading at this point. I don't need any letters making advanced complex statements about how something I've said is not exactly 100% just so. I'm sure you would be correct but as I've stated this is a very basic electronics course. You will not be able to build a nuclear reactor after taking this course. Come to think of the typical nuclear reactor, maybe you will!

WHAT IS ELECTRICITY?

We all know everything is made from atoms. Atoms have many parts to them. The ones we care about are electrons which carry a "negative" charge and protons which carry a "positive" charge. This paper is made up of atoms which contain both electrons and protons. The number of electrons and protons are equal and in perfect balance so there is no "potential" for electricity to flow in this paper. However, if one atom contains a surplus of electrons and another atom contains a surplus of protons these particles tend to want to equalize. Since electrons are attracted to protons (or to put it another way, opposite charges attract), the electrons will flow towards the protons if an easy path is provided. We'll get to this easy path in a moment but first let's see how we can create an electrical charge.

One example is the battery that powers your effects. Note that one terminal is marked + (plus) to indicate a positive charge caused by a surplus of protons. The other terminal is marked - (minus) to indicate a negative charge caused by a surplus of electrons. This is caused by a chemical reaction inside the battery. Another way to cause this effect to happen is with a magnetic field. That's what happens at the power plant that supplies your home with electricity. A magnetic generator strips electrons from the atoms in a coil of wire and puts the surplus electrons on a terminal. The wire is left with a surplus of protons and is connected to another terminal. When you turn on a switch in your home you are providing an easy path for the electrons to reunite with the protons and electrical current flows. There are other ways to free up electrons, such as friction which causes lightning, but the battery and generator are the ones important to us now.

CONDUCTORS

Since we now know electrons will flow toward protons if an easy path is provided the question becomes what makes an easy path. It turns out that many materials, particularly metals, have an atomic structure that will let electrons jump from atom to atom of the material with little "resistance" on their way to meet up with protons. Copper is one of these materials and is used for wire because it offers low resistance or opposition to the electrons flowing through it. Some materials have less resistance to electrical flow than others. These make better "conductors." Silver is a better conductor than copper for example. The reason copper is used more often for wire than other materials is because it's a very good conductor and is relatively cheap. Silver wire is a better conductor but too costly for most users.

INSULATORS

An "insulator" is a material that offers a high "resistance" or opposition to the flow of electrons. The atomic structure of an insulator is such that an electron does not have enough energy to make the jump from atom to atom of the insulating material and therefore electric current cannot flow through it. Some examples of materials that make good insulators are glass, plastic, dry air, and rubber.

An example of the use of an insulator would be the power cord going to your guitar amp. When you plug your amp into a wall outlet, two or three copper wire conductors are completing a path for electricity to flow into your amp so that it can do its job. If these wires were left bare, chances are one would touch another completing the circuit and allowing the electrons to flow up one wire to the point where the wires touch. From there the electrons would flow down the opposite polarity wire right back into the wall outlet without ever reaching the amp. This is called a "short circuit." Just as it sounds, a short circuit is when a circuit is completed short of where the electrons are needed in the normal circuit. If you were to touch the bare wires, you would become the short circuit and receive an electrical shock or worse.

To avoid this, the conductor is surrounded by an insulator. This keeps the electrons flowing to where they're needed and keeps you from becoming a french fry. Plastic and rubber are common insulators used on modern wire power cords. Other insulators may be used on wire used for different jobs. For example, the pickups in your guitar contain wire insulated by a form of varnish or varnish-like synthetics. Cotton insulation was used in some amps as wire insulation. Cotton insulation is not very good, but it doesn't buzz and rattle, so it was used in lots of guitar amps to reduce the tendency for resonant buzzing caused by the chassis being vibrated by the speakers.

VOLTAGE

Electricity has many parameters used to describe how it behaves. One of these parameters is called "voltage" which is measured in a unit called the "volt." Voltage is electrical pressure. A good analogy would be the speed of an automobile. If you drive your car at 9 mph you can think of that as low voltage. However, if you drive your car at 120 mph that would be thought of a high voltage. Hence, the 120 volts at your wall outlet has a higher voltage (electrical pressure) than the 9 volt battery that powers your Fuzztone.

CURRENT

"Current," which is measured in a unit called the "ampere," is the rate of electron flow. If a certain number of electrons flow past a point in one second, one ampere of current flow is produced. Using our automotive analogy, a Honda Civic would be low current and an 18 wheel truck would be high current.

WATTAGE

"Wattage" is a measure of electrical power. The unit of measurement is called the "watt." To find out how much wattage is being produced you simply multiply the voltage times the current. Using our automotive analogy our Honda going 9 mph would be low wattage. The Honda going 120 mph (with a tailwind!) would be a higher wattage. Our 18 wheel truck doing 120 mph would be higher wattage still. Increase either the voltage or the current and wattage will increase.

POLARITY

"Polarity" is best understood by looking at a battery. One terminal is marked + (plus); this indicates a positive polarity. The other terminal is marked - (minus); this indicates a negative polarity.

DIRECT CURRENT VS ALTERNATING CURRENT

To use the battery example again we know the battery has a positive terminal which is always positive and a negative terminal which is always negative. We know that current flows in one direction from the negative terminal to the positive terminal. This is known as "direct current," or D.C. for short. If we had an electrical source whose terminal could change polarity; that is, the negative terminal could become positive while the positive terminal became negative and the terminals could repeat this pattern over and over again— you would have an electrical current that keeps changing the direction of its flow back and forth. This is called alternating current or A.C. for short. This change of polarity occurring from the start, reversing and then going back again to the start one time is called a "cycle." The electric outlet in your home does this 60 times per second and therefore is called "60 cycle alternating current."

FREQUENCY

When you strike the "A" string on your guitar, it vibrates at a rate of 110 times per second. The old term for frequency was "cycles per second" or "CPS" for short. Today the term "hertz" or "Hz" for short is more often used. What happens is the string moves one direction and then moves the opposite direction 110 times per second in the case of the "A" string. As it vibrates back and forth it alternately puts pressure on, then removes pressure from the molecules of air surrounding the string. These waves of increased and decreased air pressure radiate outward from the string and eventually reach your eardrum. Your eardrum then vibrates at 110Hz and sends a electrical signal to your brain and that's how you hear sound.

We learned that changing the magnetic field around a coil of wire can generate electricity. This is how the magnetic guitar pickup works. The pickup consists of a coil of wire, a magnetic field supplied by a magnet, and pole pieces that bring the magnetic field to the strings. If you strike the "A" string it vibrates 110 times per second. The string contains iron which influences the magnetic field around the pole piece and hence the coil the pole piece runs through. By changing the magnetic field by the amount of times per second the string vibrates, an alternating current matching the frequency of the string is produced. Now we have an alternating current that matches the string which we can amplify to vibrate a speaker at the same frequency. The speaker "shakes" the air at the same frequency as the string, only at a much higher volume.

Now let's say we want an "A" note but at a higher frequency. If we press the "G" string at the second fret, we get an "A" note that vibrates at 220Hz or one octave higher. If you press the high "E" string at the fifth fret, you get an "A" note that vibrates 440 times per second or "A440," the standard guitar tuning reference note.

RESISTORS

A resistor is a component that has as its main characteristic a resistance to the flow of electricity. The unit of measurement of resistance is called the ohm. Resistors used in guitar amps can range from a fraction of an ohm to several millions of ohms. Kilo is the unit to describe thousands of ohms. For example 2.2 kilohms (K for short) is 2200 ohms. Mega is the unit to describe ohms on the million scale. For example 2.2 megohms (M for short) is 2,200,000 ohms.

Besides the ohms rating, resistors are also rated in watts. As resistors generate heat in operation we have to be sure the power rating measured in watts is high enough so the resistor doesn't overheat and burn out. Typical resistors used in guitar amplifiers range in wattage from ⅛ watt to several tens of watts.

Another type of resistor is called a potentiometer or pot for short.

They are used as the volume and tone controls in guitars and amps. A regular resistor has two wires on it so you must use all of its resistance. The potentiometer has a third connection that connects to a "wiper." This element slides up and down the resistance element and allows you any resistance from zero to the maximum valve of the pot by turning a shaft or pushing a slider.

Resistance is used to control voltage and/or current. A resistor for our purposes is blind to A.C. vs D.C. and is also blind to frequency. That is to say those factors do not affect the way the resistor behaves in doing its job.

CAPACITORS

A capacitor has as its main characteristic capacitance. This is measured in a unit called the farad. Since this is a very large unit, most capacitors are measured in units such as microfarad or micromicrofarad. Microfarad is mf for short and of course mmf would be micromicrofarad. These days mmf has been replaced with picofarad or pf for short. Some techs use the slang term "puff" for picofarad. A capacitor is formed from two conductors separated by an insulator or "dielectric." Picture two sheets of foil separated by a plastic film rolled up into a cylindrical shape with one wire connected to each layer of foil. That's one way to make a capacitor. In fact, if you've ever looked inside your vintage amp you'll see exactly that device sealed in plastic for protection.

Besides its capacitance value, a capacitor has a voltage rating, typically from a few volts to several hundred volts. If you exceed the voltage rating the insulator will break down and the capacitor will be ruined. Capacitors do several things. For one, they can store an electrical charge. Another thing they do is block D.C. electricity. Don't forget the two conductors are separated by an insulator. A capacitor will let A.C. pass through it because the changing field on one foil will change the electron field on the other foil.

Capacitors are also sensitive to frequency. A large value cap will let all frequencies pass through it but as the value of a capacitor gets smaller it blocks lower frequencies as if they were D.C. and will only let the higher frequencies through. This can be very handy if we want to control certain frequencies in an amplifier. You would use a larger value cap to send low frequencies to the bass control and small value cap would allow your treble control to affect only the higher frequencies.

Another type of capacitor is the electrolytic capacitor. These are the caps commonly called "filter caps" or "bypass caps" in a guitar amp. Unlike resistors and other caps, the electrolytic usually has a polarity and must be installed in the correct direction as far as positive and negative voltages are concerned. There are electrolytics made without a polarity. (A common use for those would be in the crossover network in your stereo speakers.)

Even though we are very early on in this course, let's put some of what we learned so far to use.

The potentiometers used in guitars are almost always the ½ watt type. The two things we really are concerned about is value and taper. The taper is how much change occurs with a certain amount of rotation of the control shaft. Most guitars use audio tapers but a few use linear tapers. The value is how many ohms the pot is rated. The value of the volume pot will affect the volume and tone of your pickup. For example most Fender Jazzmasters use a 1 meg (1000K) volume pot. These guitars usually sound bright and thin. If you use a 250K pot (the Standard Stratocaster value) in place of the 1 meg, your guitar will sound thicker and have more beef. If the 250K takes away too much high end, try a 500K.

Another example is modern Les Paul guitars. Vintage Gibsons always used 500K pots for both tone and volume controls. In recent years, Gibson made many of these guitars with lower value controls for both tone and volume. If your newer Les Paul sounds muddy, check the value of the pots. If they're not 500K, try changing them to 500K to bring the guitar back to life. If you use super high output pickups, 1 meg controls will usually give more output and high end. As for tone pots the rules are simple: lower value pots spread the workings of the control more evenly from one end of rotation to the other. Higher value controls give you more high end but don't do much until most of the way down.

Capacitors are used along with your tone controls to cut high frequencies on your instrument. We learned that small value caps only affect the higher frequencies, and larger caps go lower into the frequency range. Typical values of tone caps are .02 mfd, .022 mfd, .047 mfd, .05 mfd, and .1 mfd.

If you don't like the way your guitar sounds as you turn the tone control down, you can change the value of the tone cap and the control will react in a very different way. Your ear is the best judge of tone cap values but .01 to .1 are pretty much the useful limits. Another use of small value cap is on your volume pot. If you lose too much high end as you turn the volume down on your guitar, a small cap from the "hot" to the center terminal on your volume pot will keep the highs from fading. The values I've found that work well are from 50 to 220 pf for guitar, up to 500 pf for bass. If you use too small a value, the guitar will still darken as you turn down the volume. If the value is too large the guitar will get brighter as you turn down the volume. When the high end is the same at any setting you've got the right value. Rick tips: Many Rickenbacker guitars and basses use a series capacitor on the bridge position pickup. If you have one of these, a jumper across the cap will give more bottom and output. A Rick bass with that cap jumped out through an Ampeg SVT will downright thunder!

VOLTAGE DIVIDERS

The type of voltage divider we will need to know about for guitars and guitar amplifiers is the resistor type. An example of this is the volume control in your guitar or amp. One end of a carbon strip is connected to a voltage source, such as a pickup, and the other end is connected to ground which is zero voltage. As you move down the carbon strip from the voltage input towards ground, the voltage become lower as you go towards ground.

As we know from earlier lessons, the volume control has a moveable element that contacts the carbon strip and can be moved anywhere between the input of this strip and ground via a shaft and knob. Let's say your pickup has an output of two volts. This two volt signal is applied to the input of the control's carbon strip and ground. If you turn the volume knob so that the sliding element of the control is exactly half way down the resistance path of the carbon strip, one volt will appear between the input and the slider, and one volt will appear between the slider and ground. What you've done is divide the voltage in half on two equal one volt segments. If you take the signal from the slider and ground and feed it to your amp, the amp will get one volt of signal. By dividing the voltage in this way you can control the volume.

In a guitar amp there are many variable voltage dividers, such as your volume and tone controls. In other places in the circuit, voltage dividers are used which don't need to be varied. In this case fixed value resistors are used.

For an example, let's use the case where a voltage from one stage of an amplifier circuit is too high for the next stage to operate properly. Let's say we have 10 volts and the next stage can only handle 3 volts. We now have to reduce the ten volts by 7 volts, or 70%. Without getting into Ohms law or any serious math I'll tell you the ratio of value between two fixed resistors in series is also the ratio of voltage division. O.K. then, let's put a 7 ohm resistor in series with a 3 ohm resistor. If we input the 10 volts on the free end of the 7 ohm resistor and connect the free end of the 3 ohm resistor to ground, the voltage from the input (free end) of the 7 ohm resistor to the junction of the 7 and 3 ohm resistor will be 7 volts; in addition, from the junction of the 7 and 3 ohm resistor to ground, 3 volts will appear. So we've divided the voltage by a 7 to 3 ratio to obtain the voltage we need.

OSCILLATORS

An oscillator circuit generates a continuous signal. When you press the keys of an electronic keyboard you are turning on an oscillator. Each key may have a separate oscillator for each note or a single oscillator whose tuning or frequency is changed by which key you press. In a guitar amp, an oscillator is a part of any vibrato or tremolo circuit. The control labeled "speed" controls the frequency of the oscillator. That frequency is used to cause the volume (tremolo) or pitch (vibrato) to change in sync with the frequency of the oscillator. Oscillators are very common in everyday objects. Your electronic watch uses one, your radio, TV, and CD player all use oscillators. So does your microwave oven.

REVERB CIRCUITS

There are many ways to generate reverb. Some are 100% mechanical, such as singing in the shower. Some are 100% electronic, such as a digital reverb. The reverb in your guitar amp is typically comprised of both electronic and mechanical elements.

Let's take the reverb circuit found in a vintage Fender amp and use it as an example. A portion of the signal from the preamp stage is divided off using both resistors and capacitors. This signal is applied to a tube called the "reverb driver tube." This tube turns the voltage signal it receives into a power signal. This power signal is fed into an impedance matching transformer. The output of this transformer, often called the "reverb driver transformer," is then sent to an electromechanical device known as "the reverb tank."

This is how a reverb tank works. Inside the tank are suspended some springs (usually two or three). On each end of each spring there is a magnet. On the input end of the tank there are metal laminations, much like those of a transformer, that surround the magnets on the springs (without touching). At the end of the laminations is a coil of wire. If we apply the signal voltage from the driver transformer to this coil the magnet field generated in the coil will travel down the laminations and cause the magnets on the springs to vibrate at the same frequency as the drive signal. The magnets cause the springs to vibrate but, as it takes time for the vibration to travel from one coil of the spring to the next, a mechanical delay is created from the time the input ends of the springs are vibrated until the vibrations reach the other end of the springs. On the output end of the springs another set of laminations with a coil of wire is placed around the magnets. Now when the vibration introduced at the input end of the springs finally reaches the output end of the springs, the vibrating magnets on the output end of the springs induce a signal voltage in the output coil. This signal is the same frequency of the input signal, only delayed by the time it took the signal to travel through the coils of the springs.

A couple of more factors are also involved here. First, the vibrations in the springs "bounce" back and forth from end to end like a pendulum creating additional delays. Second, the springs are not identical in length so the delay time of each spring is different. Many delays, each with a different delay time, are created and that's what reverb really is.

Now that we have the reverb signal at the output end of the reverb tank, we can add a stage to amplify the reverb signal and then mix it back in with the original signal, giving us both the dry original signal, plus the signal from the reverb circuit to feed into the power amp section. The reverb control, which is a voltage divider, lets us control the ratio of reverb to original signal.

THE FUSE

The fuse is an electrical safety device. A fuse contains a strip of metal that melts at a low temperature. If a fuse is placed in series with a circuit and the circuit shorts out or overloads, the fuse strip will melt and disconnect or "break" the power to the circuit before any major damage can occur. The fuse has a value which is carefully chosen so it will not "blow out" during normal operation but will blow quickly when something goes wrong. When I used to do amp repair, the number one cause of major damage or total death beyond repair to an amplifier was by a fuse blowing and some goof installing a larger fuse to "get through the rehearsal or gig." Trouble is, the amp will blow up and not finish the rehearsal or gig, and now, instead of requiring a $1 part, needs hundreds in repairs or is damaged beyond repair.

I once got a Vox AC30 from the Record Plant that Aerosmith set on fire! That's a real fire with flames out the vents! Of course it had a too large fuse. I know Aerosmith is a hot band, but that was really livin' on the edge!

SOLDER AND SOLDERING

Soldering is the way electrical connections are completed inside guitars and guitar amps. Solder, which is made from lead and tin and sometimes may also contain silver, is melted on the electrical connections heated with a soldering iron or gun. The solder forms a molecular bond with the conductors being soldered, making a positive connection and adding strength to the connection. Commonly the solder contains a nonacid flux which helps clean the conductors to assure a good connection. Often the conductors to be soldered, the leads of a resistor or capacitor for example, are first cleaned with an abrasive to remove any oxidation that would hinder a good connection. When soldering, eye protection should be used. Also, soldering fumes are harmful and you should avoid inhaling them.

SOLDER

If you're going to do any soldering, a good grade of electronic solder should be used. For guitars and amps, a good solder with a flux core designed for electronic work is best. The alloy of lead and tin commonly used for this type work is Sn60, also known as 60/40 solder.

Solder is sold by the ounce or pound and comes in different gauges. For small connections, or where little space exists between different points in a circuit (the printed circuit board in your effects pedal for example), a fine gauge solder should be used as not to bridge connections that should be separate. Larger connections, such as a solder pool on a control or chassis ground, are best made with a thicker gauge of solder.

SOLDERING GUN AND IRONS

Soldering guns heat up in seconds and are O.K. for the oneoff quick soldering job, but I don't recommend them for serious work. They're not made for continuous work and can overheat. They're not temperature controlled. If used near magnets, such as the magnets used in guitar pickups, they can demagnetize the magnets and ruin your pickups. They're also heavy and bulky.

The soldering iron is a much better choice. Do not buy the $3.99 type if you're serious about soldering. A temperature controlled soldering iron is the pro choice. At Trainwreck we use the Weller EC2002 with an assortment of tips for most work. There are several brands on the market and most work quite well. The iron I just mentioned allows you to set the exact temperature you want and has an LED readout to let you monitor the tip temperature, which is electronically controlled. When working on solidstate devices, your iron should be "ESD Safe" (a red circle with a red line crossing a lightning bolt also means ESD Safe). ESD Safe means a damaging jolt of static electricity cannot be conducted to your work through the iron.

Why would you want to control the temperature? For many reasons. Most first time solderers fail because the iron they use is too hot. If the iron is too hot the connection will oxidize from the excess heat before the solder can bond. Also excess heat "boils" and oxidizes the solder itself. Excess heat can melt the insulation on your wire and damage your components. Excess heat will cause the traces on a printed circuit board to peel off, ruining it.

With an adjustable iron you can use lower heat on a P.C. board (I use about 650 degrees), medium heat for most connections (700 degrees or so), and high heat for grounds (750 degrees to 800 degrees).

Also, a high wattage, uncontrolled iron is good for those big chassis grounds like in the blackface Fender amp.

SPEAKERS

Everyone knows the speaker turns the electrical power from your amplifier into sound. It really is a simple device in terms of how it turns power into sound. It's also a complex device when viewed from the point of producing the tone that you want to achieve.

To understand the speaker, we have to understand magnets and magnetic polarity. Magnets, in very simple terms, attract iron and certain other metals and a magnet's field can influence the flow of electrons. Let's take a pole piece from a Strat pickup for an example. The magnet used as a Strat pole piece is a long cylinder shaped design. Each end of this cylinder has a polarity. One end is called north or "N" polarity, the other end is called south or "S" polarity. Like magnetic polarities repel each other and opposite polarities attract each other. If you bring two north polarities close to each other they will push apart. The same holds true for two south polarities. If you take the south polarity end of a magnet and place it near the north polarity end of another magnet, they will attract each other and try to mate. They will lock together with the invisible force we call magnetism.

So how do we use this knowledge to make sound? We start with a magnet which we place on a frame to which all the other parts of our speaker are mounted. Then we wind a coil of wire on a hollow cylinder. We call this cylinder of wire the "voice coil." If we pass an electric current through this coil it generates its own magnetic field. The polarity of this field depends on which polarity the incoming current happens to have at the moment. If we put a cylindrical gap into the magnet structure—let's call it the "voice coil gap" into which we can suspend our cylinder of wire or "voice coil"—we've created the heart of a speaker. The fixed magnet has a constant polarity. Our voice coil will change polarity with changes in the polarity applied to it. In addition, the strength of the voice coil's magnetic field depends on how much power is applied from your amplifier. The voice coil is suspended in such a way that it is free to move back and forth lengthwise in the voice coil gap. If we apply the alternating polarity signal from the amplifier to the voice coil, it will vibrate in step with the signal supplied.

The last step is to connect the voice coil to a paper cone that is also free to vibrate back and forth. As the voice coil vibrates back and forth, so does the cone which vibrates the air, which vibrates your eardrum, and thus you hear the sound.

LIGHT DEPENDENT RESISTORS

The light dependent resistor, also known as LDR or vactrol, is a device in common use in today's guitar amplifiers as a switching device. In days past, many channel switching amps used an electrically controlled electromechanical device, called a relay, for channel switching and other remote switching functions. Today, the LDR has taken over for the most part. The LDR is simply a device with a light source inside, usually a light emitting diode or LED for short, and a photo-sensitive resistor. That is a resistor whose value changes when light strikes it. These are sealed together in a lightproof case. When no light hits the photoresistor, its value is very high, almost an open circuit. When the LED is lit, the photoresistor value becomes very low, almost a completely closed circuit. When you step on the footswitch of your Swiss Army knife type amp, you send a signal to a logic chip. This logic chip then turns on and off the various LDRs in the circuit needed to perform the function you choose, channel switching for example.

HEAT PROBLEMS USING VACUUM TUBES

In normal operation vacuum tubes generate heat. Output tubes in particular tend to generate considerable amounts of heat because all of the output power passes through them. Transformers also generate heat in a tube amp, but the amount is small in a well designed amp compared to the tubes. Heat is the enemy of most of the parts used in a guitar amp. For example, the typical electrolytic capacitor used in a guitar amp is rated for a maximum of 85 degrees Centigrade or 185 degrees Fahrenheit. Some 500 volt electrolytic capacitors are only rated at 65 degrees Centigrade or 149 degrees Fahrenheit. Tubes can generate much higher temperatures than that, so a well designed guitar amp must take that into account.

Since heat rises, if you place the tubes upright above the chassis and provide for ventilation, as in a Marshall head for example, things are just fine. If you hang the tubes under the chassis without a way for the rising heated air to escape, Fender style for example, the chassis gets cooked by the heat of the tubes. In the real world there are trade offs in the construction of guitar amps and most combo style amps trade off handling heat in the best way possible for ease of construction and to make a more compact package. In low wattage amps this is less of a problem but in compact high wattage amps it can be a real source of trouble. Some companies will use a fan to cool the amp. Of course then you have to deal with the noise of the fan itself. Also, many people feel that fan cooling the tubes causes them to run too cool with a loss of tone.

In a class "A" amp, heat is even more of a problem. Class "A" amps normally run much hotter than class "AB" designs. Note the single row of vents on a 1960 (first year) Vox AC30. By the time a couple of years and a high number of flameouts occurred, the AC30 sported two rows of vents. Better still, but these amps still ran very hot. The current Korg/Vox AC30 now sports vents on top that are more serious. One company builds class "A" amps today with the tubes hanging down, Fender style. They can be seen going up in smoke at your favorite club or on the Nashville Network. So here's a Trainwreck amp-buying tip. Play the amp you're interested in buying for at least one-half hour. If the control panels or metal switches and such feel uncomfortably hot to the touch, chances are the amp will suffer heatstroke on stage. It doesn't matter how good an amp sounds if you can't keep it running.

INDUCTORS

Inductors are coils of wire which have the property known as inductance. The unit of measurement for inductance is the henry (h for short) smaller unit of inductance are the millihenry (mh) and the microhenry μ. Inductors can be air core or have core of magnetic material such as iron or ferrite.

TRANSFORMERS

One type of inductor is known as the "transformer." We've all heard the term power or output transformer used when talking about guitar amps. The transformer uses two or more coils of wire wound around a iron core. When you feed alternating current into one coil, the coil generates an alternating magnetic field in the core. This alternating magnetic field induces a voltage in the other coils subjected to this field. Depending on the turns ratio of these coils you can "transform" the voltage into a higher or lower voltage.

In a guitar amp, several different voltages are required for operation. For example, when you plug a Fender Super Reverb into the wall, the "primary winding" is connected to 120 volts A.C. The tubes need 6.3 volts to light up so there's a 6.3 volt secondary winding. The GZ34 rectifier tube needs 5 volts to light up so there's also a 5 volt secondary winding to take care of that need. The output tube plates run at about 460 volts D.C. so there's a higher voltage secondary winding which produces several hundred volts A.C. which the GZ34 tube changes from A.C. to D.C. There's also a tap on the high voltage winding at a point that allows a diode to provide the 50 plus volts needed to bias the output tubes. So to make it simple, the power transformer converts the wall outlet voltage to all the various voltages needed to run the various demands inside your amp.

The "output transformer" does two jobs. Since all the coils are insulated from each other, the output transformer blocks the D.C. voltage on the plates of the output tubes from reaching the speaker where it would do harm. Also the transformer matches the higher impedance of the output tubes to the lower impedance of the speaker. Just as the transformer can step up or step down voltage, it can also step up or step down impedance. We won't cover impedance here, but think of it as something you'd want to match in a transformer output circuit to develop maximum power. The coil that connects to the plates of the output tubes is the primary coil and the coil that connects to the speaker is the secondary coil of an output transformer. The coils are also known as "windings."

THE CHOKE

On many amps, there's a single coil of wire wound on an iron core called a "choke." A choke doesn't like its magnetic field to change, so it lets direct current of a constant voltage and current flow through it but opposes a changing magnetic field such as produced by an alternating current. This is useful in smoothing out fluctuating D.C. and also for blocking A.C. flow where using an insulator would also block the D.C. component you'd want to flow.

Inductors such as chokes or various other types of coils are frequency sensitive. The larger the value of an inductor or the higher the frequency of signal, the more an inductor opposes the A.C. electrical flow. The inductor in a Vox Wah Pedal is a 500 millihenry ferrite core coil. It is used as a "tuned" circuit along with a capacitor and resistor. You control the resistance with your foot when you rock the treadle back and forth. This changes the tuning and hence the frequency that is cut and boosted. That's how the mid-range control on the Ampeg SVT works also. The graphic EQ on your vintage Boogie works the same way. Mesa used several pre-tuned circuits using individual resistors, capacitors, and inductors for each frequency; then by using a slider pot, you could add or subtract each frequency you wanted to tailor your sound. Inductors are also used in the crossover network in stereo speakers. A series inductor to your woofer will keep the high frequencies from reaching it and a small capacitor in series to your tweeter will block low frequencies from reaching it. This way the woofer and tweeter receive only the frequencies they can reproduce and handle without damage.

RECTIFIERS

A rectifier changes A.C. current to D.C. current. The rectifier can be a tube, silicon diode or, in some old amps, made of selenium.

HOW A TUBE WORKS

I'm going to explain how a tube works using a "triode" or three element tube as an example. First let's apply 6.3 volts to the heater (filament) of a triode to generate heat. This heat will cause the first element called the cathode to glow. The cathode is coated with a chemical that gives off electrons when hot. We know from lesson one that electrons are attracted to a positive charge. Let's put a metal "wall" around the cathode and call it the "plate" (this will be element two). If we make the plate positive by putting a positive charge on it from an outside source, electrons will start to flow to this element. If we provide a return path which will be outside of the tube, such as a small resistor from the plate through the power supply and back to the cathode, the electrons will have a return path. This setup provides a continuous path for electron flow. Also a voltage will appear across the "plate resistor." If the electron flow increases, the voltage across the plate resistor increases. If the electron flow decreases, the voltage across the plate resistor decreases. Now here's where it gets interesting. Let's take a fine wire and wind it in a wide open spiral and place it between the cathode and the plate. Let's call this third element the "control grid." If we put a positive charge on the control grid it will attract lots of electrons from the cathode. By the time the electrons reach the grid they will be attracted even more by the higher positive voltage of the plate. Because the grid has so little surface area, most electrons will fly right by the grid without striking it, almost like it wasn't there. The plate (being of solid construction) will "catch" all these electrons. The large increase in current flow will cause a large increase in voltage across the external plate resistor. Next, let's make the control grid more negative. Like charges repel each other. Now we have the negative cathode trying to send electrons to the plate but they are repelled back to the cathode area by the negative grid. Very little current flows and the voltage across the plate resistor decreases.

Here's how the tube amplifies. Since the grid is wound close to the cathode very small changes of voltage at the control grid cause very large changes of voltage at the plate. For example, a tube with an "amplification factor" of 40 means that a 1 volt change of voltage at the grid will cause a 40 volt change at the plate. The British call a tube a valve. Let's run this through again, only in English! If you turn on the faucet of your kitchen sink full blast, and then try to stop the flow of water by placing your finger over the end of the faucet, you'll find it impossible to stop the flow. However, if you turn the handle of the faucet it requires little effort to stop the flow. Think of the handle as the "control grid" even though it's really a "valve." A negative voltage called "bias" is applied to the control grid of a tube with no signal being fed in. This is to set the idle point of the tube to what might be called center of tube plate current curve. What all that really means is the tube is set so it reacts the same amount whether the voltage of a signal of the same strength is negative or positive. In a pentode, used commonly as an output tube, there are two more grids. A screen grid connected to a positive charge placed after the control grid helps to pull more electrons to the plate. The suppressor grid placed after the screen grid connects to ground which is negative. By the time electrons from the cathode reach the screen grid they are strong enough to knock unwanted electrons from the screen grid should some strike it. The negative suppressor grid will repel these "slower" secondary emission electrons back to the screen grid without slowing down the higher energy electrons flowing to the plate. "Beam power tubes" such as a 6L6 use beam forming elements to "focus" the electron flow for greater efficiency.

THERMISTORS

A thermistor is a device whose resistance changes with temperature. A thermistor can have a negative temperature coefficient (NTC) or a positive temperature coefficient (PTC). Negative temperature coefficient thermistors decrease in resistance as temperature increases. While thermistors are not a part of modern guitar amp design, many amps of the 50s and 60s used themistors or devices with a similar function as part of the circuit. An example of a thermistor equipped amp would be many early Ampegs including the early Ampeg B-15 bass amps.

This is what they were used for and how they work. When you first turn on your amp, there's a large surge of current as the cold heaters of the tubes have a low resistance until they heat up. It was thought that if you could limit the current to the heaters and let them warm up a bit before hitting them with full voltage, the life of the tubes would be increased. A thermistor was placed in series with the transformer primary winding. When you first turned on the amplifier the NTC thermistor was cold, had a high resistance and limited the voltage and current to the amp's circuits. As current flowed through the thermistor, it heated up and its resistance dropped, allowing full voltage and current through so the amp operated normally.

BRIMISTORS

The brimistor is a thermal device consisting of a fixed value resistor(s) and a thermal relay. They were used in a whole bunch of J.M.I. Vox amplifiers. While the thermistor used in amps like the Ampeg was used to prevent heater shock, the brimistor was used in Vox equipment to prevent shock in the high voltage circuits. That is to protect the power filters, tube plates, and output transformer. Don't forget Vox amps didn't use a standby switch and amps like the AC-50 and AC-100 most commonly used silicone diodes, which provided full high voltage the instant the amp was switched on.

The way the brimistor worked was it was placed in the high voltage circuit between the diodes and the rest of the circuit. When cold, the resistor was in series with the circuit keeping the voltage down. As things heated up, the relay contacts would close, allowing full voltage to flow. The current flowing through the relay warmed it, keeping its contacts closed. I've always replaced brimistors with straight wiring. They tended to be finicky, very hard to get, and the amps seem to do fine without them. Of course, you could always add a standby switch. If you want brimistor protection but can't find one, a timed relay/resistor circuit will work just fine.

AMPLIFIER POWER RATINGS

The power output of your audio/guitar amplifier is measured in a unit called the "watt." This should make it easy to compare the power produced by the many brands and types of guitar amplifiers on the market. This is not the case, however. In the stereo audio market, the manufacturers all comply to a universal standard. That standard is, "watts R.M.S. before clipping occurs." However, there are many ways to measure A.C. wattage (an audio/guitar amp puts out A.C. wattage). For example, a 100 watt R.M.S. amp could be called a 141 peak watts amp, or even a 282 watts peak-to-peak amp. The power rating could be of the averaged or instantaneous type and so on. Worse yet is the fact that there's nothing to stop an amplifier manufacturer from picking any number they choose out of thin air and using that for their wattage rating! Many of them do just that! After all, 60 watts sounds better than 35 or 40 to most people when shopping for a new amp. It would be great if all companies would use true R.M.S. power before clipping, but bigger numbers mean bigger sales, and hence, the numbers game.

Also, tubes have a designed maximum power rating. For example, a pair of 6L6GC tubes is designed for a maximum of 55 watts in a push-pull circuit. When someone claims to give you 70 or 75 watts from a pair of these tubes, beware. The power rating is not likely to be true (by the R.M.S. standard). Also, if you could force these tubes to produce 75 watts R.M.S. the cathode coating would self destruct in a very short time and the tube would be ruined. Also, no matter how hard you run any tube, there's a point where it just cannot produce any more power. The cathode can supply only so many electrons. The plate can only handle so many watts without burning up.

Example time: Korg/Vox, Matchless, and Trainwreck all build amps using four EL-84 tubes. These tubes are designed for a maximum of 8 watts per tube or a total of 32 watts. None of the above companies claim more than 33 watts for their amps, an honest R.M.S. rating. Other companies claim 50 watts R.M.S. from four EL-84 tubes ...in their dreams! You can increase the power by increasing the number of output tubes. Going from two to four 6L6GC power tubes will let you go from 55 watts R.M.S. to 110 watts R.M.S. Going from four EL-84 tubes to twelve EL-84 tubes would let you go from 32 watts to 96 watts R.M.S. In reality 100 true watts R.M.S., or even a few more than that, would not be out of line for a twelve EL-84 output stage.

Twelve EL-84 tubes and we can have a 100 watt AC-30 right? Wrong! When you start to connect many output tubes together, bad stuff starts to happen. One bad thing is the "Miller effect." That occurs when all the capacitance on the input of the tubes is amplified by the tubes and transferred to the tubes' output. This messes with a lot of stuff, but let's just say the sound suffers. Six seems to be about the most tubes you can connect together in a guitar amp and still get a decent sound. It also depends on the tubes used. I once tried a six EL-84 tube design myself for a 45 watt output. I lost some of the EL-84's magic chime and jangle. The problem—EL-84s amplify more than most output tubes, amplifying the problems of multi-tube output stages.

One last thing to consider is projection. That is, you can take two different amps with the same R.M.S. and average power ratings and one amp will cut through and the other won't. This is caused by many factors such as frequency response, slew rate, its ability to transfer power to a speaker as the speaker impedance changes with frequency and so forth. Wouldn't it be nice if all guitar amp companies used the true "R.M.S. wattage before clipping occurs" method of rating power. I think so too!

BIASING THE OUTPUT STAGE OF A FIXED BIAS AMPLIFIER

It seems that with the introduction of Gerald Weber's book, *A Desktop Reference of Hip Vintage Guitar Amps*, I stirred up an old controversy about the "correct" way to bias an amplifier. In 1989, I wrote a 29 page article for the Angela Instruments catalog called,

"The Trainwreck Pages." These pages have been reproduced, exactly as they originally appeared in that 1989 catalog, in Gerald's book. One of the subjects covered in 1989 was biasing a fixed-bias amplifier. Before I go on, I should tell you that this might get a bit technical and that the following info is my personal opinion. I don't always agree with my friends in the amplifier kingdom, but I do respect them all, and their opinions. Guitars and amps should be for making music and having fun. Let's not make a war out of any of this stuff, O.K.?

This is how the controversy re-started... On page 177 of Gerald's book (page 25 of the Trainwreck Pages) under the title "How To Adjust the Bias On A Fixed-Bias Amp," I wrote the following: "There is an incorrect, yet commonly recommended method for biasing amps being circulated by many sources, including the manufacturers of tubes and amplifiers. This method instructs the repairman to connect the amp to a "dummy load," attach it to an oscilloscope and put a signal into the amp with a signal generator. Then the instructions are to bring the amp to full power, raise the bias voltage until a crossover notch appears, and lower the bias voltage back until the crossover notch just disappears."

In my opinion this method was wrong in 1989 and is just as wrong today. Electronics is a science using physics, math, and measurements among other things. Here are my objections to the signal generator/dummy load/oscilloscope/crossover notch method of adjusting bias.

NUMBER ONE: With this "crossover notch" method you are not taking any measurements. That's right, you do not get even one measurement of the many parameters involved in the operation of the output and driver stages...not one!

To use an analogy...when you put air in your car's tires, you use a pressure gauge to take an exact measurement of the air pressure inside the tire. Could you imagine trying to guess the pressure just by looking at the shape of the tire? Of course not. Well, that's what the crossover notch people do when they look at the shape of a waveform on an oscilloscope. They don't measure anything but "it looks about right."

NUMBER TWO: The dummy load is usually a resistor. A speaker is an inductor. The crossover notch will appear at a different bias voltage for a resistor than it will for a speaker load. Even beyond that, the crossover notch will move around on a speaker load depending on frequency. Sounds like the crossover notch method ignores the laws of science, doesn't it? You bet!

NUMBER THREE: A good engineer takes many factors into account when designing a tube power output stage. He must consider power output, harmonic distortion, intermodulation distortion, linearity, frequency response, stability, reliability, drive requirements, slew rate and a host of other parameters. Biasing by crossover notch ignores all of these factors.

THE ELECTRONIC FACTS OF LIFE

FACT: As you increase the bias voltage, the plate resistance of the tube increases, requiring corresponding changes in the output transformer. That is, if you want maximum performance.

FACT: As you increase or decrease the bias voltage from what the engineer has selected, based on the science of using the plate curves of the tube in question, you will move the curve out of its linear region. This changes all the other parameters including the proper specs for the output transformer, the drive requirements, the types and amounts of distortion and on and on.

FACT: The crossover notch method will always bias an amp to the cold side—too much bias voltage. Aside from the extra distortion this creates, which some heavy metal players like because of its hard aggressive (real harsh) sound, most guitarists will not be happy with the results. Even for guitarists who like an over-biased amp, this is still better done with measurement than an oscilloscope.

FACT: All the high end tube stereo amp manufacturers measure the bias by the "current" method. Sometimes they install a small resistor between the cathodes and ground. The voltage developed across this resistor depends on the current flowing through it. This means that although you will measure a voltage, what you are really setting is the current. An example of this would be the Dynaco Stereo 70 or an Ampeg SVT.

SETTING THE BIAS BY EAR

This involves plugging in a guitar and playing while someone else who is qualified adjusts the bias to the sound you want. (When you're holding the guitar you're grounded and should never reach into an amp unless you want to die.) Then you must take measurements to be sure you don't exceed the design limits of your amp or tubes.

Do not adjust a stereo amp by ear, as it will not meet its specs and you may put it in grave danger. Well, I've tried to keep this simple and I hope I've shed some light on the subject. By the way, I do know some amp techs who use the crossover notch method and when they get the notch out, tweak the bias control a bit more and get the amp into a good range. This is a skill they develop with years of working on amps and a novice couldn't get it right in a million years. A novice can get perfect bias every time from the first time using measurement. As I've said the choice is yours. If you've had your amp adjusted by the crossover notch method and like it, fine. If you think you could use more beef, try running 30 or 40 mA per tube through those 5881s...they can take it and you might be in for a tonal surprise.

I guess I should mention one more thing about the oscilloscope and signal generator. They are useful when setting up an output stage to check for balance. That is to make sure both sides of a push-pull output stage reach clipping at exactly the same time. The oscilloscope will also show up anomalies in the output waveform. I'm not anti-scope...I own two...one for each eye! I just don't bias amplifiers with them.

AMPLIFIER MODIFICATION

Do not attempt an amp mod unless you have the skills needed for such a project. Messing about inside a guitar amp without the proper skills or knowledge can leave you or your amplifier permanently dead. Use good judgement. If you want to learn how to mod amps, find a person who's qualified to teach you the right way to do it. To mod amps you must be good with hand and power tools, know how to solder, know how to use basic test equipment, and have a basic understanding of electronics.

SHOULD YOU MODIFY YOUR AMP?

Almost every amp ever made can be modified to better suit your needs. However, before jumping into modification of your favorite amp you should ask yourself some basic questions. For example, how will modification affect the value of your amplifier? If you own a vintage Fender or Marshall or Vox, altering your amp from its stock form will reduce its value, usually by a large amount. Sometimes, in rare cases, it won't. Doing a proper top boost mod to a non-top boost Vox won't always result in a loss of value. This is due to the demand for top boost AC 30s. However, a factory top boost is worth more than a recent top boost add on. A poorly done top boost add on will surely reduce the value of your Vox. And if your non-top boost Vox is in mint condition, I wouldn't touch it.

On the other hand, recent, super-mass-produced amps, or older amps from smaller companies that never reached vintage collectible status generally won't have their resale value harmed by a good mod. Even companies that have made amps with collectible value such as Fender also made models that do not have collectible value at this time. You never really know, who would have thought a 70s Strat would be collectible! Who knows what your silverface Fender amp will be worth in the future?

The other point of view is that you own a vintage amp, and you want to mod it to do what you want, and don't care about its resale value. That point is valid too.

REALISTIC EXPECTATIONS

If you decide that you are going to modify your amp you should have a realistic view of what it can be made to do.

A classic example is that many people think a Fender Bassman head can be made to sound just like a Marshall. While a Bassman can be modded to a more Marshall-like sound, owing to the different designs of the transformers they will never sound the same. If you own a 50 watt Marshall head and want to convert it to a jazz amp, that wouldn't be realistic. If you own an Ampeg B15N and want to turn it into a heavy metal rig, that's not going to happen either. With modification you can expect a basic improvement in tone and/or performance. Also, you may bend the amp a bit more towards the style of music you play. Of course, some mods have other goals in mind. Some of these might include adding an effects loop, or reverb, or upgrading the transformers or speakers. You might be interested in cosmetic mods...a tweed Ampeg Jet anyone?

START SMALL

If this is going to be your first try at amp modification I suggest that you start small. Starting small can be looked at two ways. You can start with a small, simple amp like a Champ, or you can do a small, simple mod on a big amp.

THINK FIRST

It doesn't matter if you buy a mod kit, or do a mod out of a magazine article. You should study the instructions or article until you are sure you understand everything. If you buy a kit, make sure all parts are accounted for, and if the kit states that additional parts are required, make sure you purchase them before you start the project. If you're doing the mod from an article, make sure you have all the required parts on hand before starting. There's nothing worse than getting halfway through a mod and then figuring out you can't find a needed part anywhere, or that the part you need to finish the job and make your amp play again is on a six month backorder. Also, make sure you have all necessary tools and hardware on hand before you start. I know this all seems like a lot to keep in mind, but once you get into it, you should find amp mods both easy and rewarding. "Now where is that .002 mf disc cap I just had my hands on?...Drat!."

SPEAKER STUFF

People keep asking me, "What's the best speaker for my amp?" That's always a hard question to answer because speakers are so much a matter of personal taste. As most of you know, old Celestions are my favorite speakers.

When picking out a speaker, keep the power rating in mind. Many current vintage type speakers use a paper voice coil which makes for great tone but have a power rating of only 15 to 30 watts. When replacing speakers, keep in mind speaker polarity. Most speakers use what is called "normal" polarity. That is, when you apply a positive voltage to the red or plus (+) terminal, the speaker cone will move forward or away from its magnet. Some speakers such as old Jensens and JBLs have what is known as "reverse polarity." That is when a positive voltage on the red terminal causes the cone to pull in towards the magnet. If you mix normal and reverse polarity speakers in the same amp or cabinet and hook up the hot wire to all the red terminals you'll have some speakers pushing and some pulling. This is called "wired out of phase." The sonic result is low volume and a lack of bottom end. You can use both types together, but you have to hook the hot wire to the negative or ground terminal of either the normal or reverse group. Then you can make the entire cabinet normal or reverse by reversing the wires on the cabinet jack, or the plug that goes to the back of your amp.

To test polarity, put a 9 volt battery to the plug going to the speakers, with the battery's positive terminal to the positive speaker lead. If the speakers move forward, they're wired normal polarity. If the speakers move in, they're wired reverse polarity.

Again, you can reverse the wires at the plug or jack to change the polarity of the whole speaker group. If any one speaker moves the "wrong" way, simply reverse the wires on that speaker. When replacing the speakers in a vintage amp that used Alnico Jensens, keep in mind they were reverse polarity and most current replacements are normal polarity. To keep the amp sounding vintage, reverse the wire going to the replacements.

For example in a 4x10 tweed Bassman, the white, or hot wire went to the red terminal on the Jensen and the black to the negative, or ground terminal. On a normal polarity speaker you must hook the black wire to the red terminal and the white to negative. Catch 22...some amps run a ground wire to the speaker frames and to the negative terminal of the speaker. Don't reverse wires on these amps or you'll short circuit the output. If you use an extension cabinet, you may want to check to make sure it has the same polarity as your amp's main speakers. If your amp has its original speakers wired properly, don't change the amp's speaker polarity, change the extension speaker polarity to match your amp.

CLASS "A" AMPLIFIERS

Class "A" amps have the best tone going for those who put tone first. Yes, Class "AB" amps have more crunch, and because the best sounding Class "A" amps use EL84 tubes, in groups of two or four tubes, their power is limited compared to the other types of output tubes in common use in high power "AB" amps. I've been making Class

"A" EL84 amps for almost twelve years. Since I keep exact records on every amp I build, including "current owner," I looked back to see how many were ever resold—five Liverpools and no Rockets. That would say to me that guitarists are very happy with their Class "A" EL84 amplifiers. Besides Vox and "Wrecks," there are some more great sounding Class"A" EL84 amps now being made. There's also a great new Class "A" 5881 amp made especially for blowing harp from Kendrick Amplifiers, and of course Dave Funk makes high power Class "A" amps for those who demand high power along with fine tone.

John McIntyre also makes Tube Coolers for those of you that have amps that run too hot. If you turn a Fender Twin or similar amp into a Class "A" amp, Tube Coolers will help you deal with the extra heat.

Now for the guy who said, "more Ampeg," but didn't leave his name on my machine...try the Ampeg R12R Reverborocket with 6V6 output tubes. Ampeg didn't use 6V6 tubes very much at all, and the old R12R used both 6V6 tubes and 6SL7 and 6SN7 preamp tubes for a very unique voice. Do you play bass? Look for the very early Ampeg New York Bass Amps with stock British Mullard EL37 (yes, that's 37) tubes. They're tall skinny Coke bottle shaped output tubes.

TUBE PLATE VOLTAGE

Some time back I was asked by *Guitar Player* magazine how tube plate voltage affects the sound of the output stage of a guitar amplifier. They also asked several other well-known amp experts this question and printed our replies in an article about the subject. As I always say, I respect these guys and their work, so the following is my opinion on the subject, and I can't help it if I'm right! Try this stuff for yourself and you can be the judge.

The voltage applied to your output tubes can affect many things. Voltage can affect power, headroom, tone, dynamics, feel, tightness or looseness, wet or dry, distortion, intermodulation and a host of other parameters. It can also greatly affect tube life. Different tube types and different brands among the same type will react to a particular voltage in different ways. Whether the tube is used in a Class "A" or Class "AB" circuit has a direct affect on plate voltage parameters.

O.K. let's take a typical Class "AB" amp such as a Fender Twin or Marshall 100 watt. The first Blackface Twin, model AA763, ran about 460 volts on the plates. The dread CBS Fender Twin model AA769 only had 405 volts on the plates. The 135 watt Silverface Twin had 500 plate volts. They all used four 6L6GC output tubes. We all know that the AA763 and AB763 Twins were the ones with tone. The AA769 had many changes, but even converted to AA 763 specs, it lacks punch, dynamics, and headroom. 405 volts just doesn't cut it for the design of a Twin. The 135 watt Twin has mondo volume and headroom, but a colder, harsher sound. Too much voltage and the Twin becomes stiff and unyielding.

So is 460 volts the best voltage for 6L6GC tubes? Here's where it gets tricky. 460 volts works great in the Twin, but 405 volts doesn't. Now let's use the Fender Tweed Pro model 5E5 A as an example. This amp only uses 385 volts on its 6L6 tubes, yet it sounds great. What is going on? Well, the answer is that the circuit and transformers in the 5E5A are designed in such a way that when all parameters are considered, 385 volts on the plates of the 6L6 tubes put them at their optimum operating point for this circuit.

Let's move on to a '67 Marshall Plexi 100 vs. a '69 Marshall 100 watt head. The Plexi runs about 460 to 480 volts on four EL34 tubes. The '69 Marshall 100 runs about 505 to 520 volts on four EL34 tubes. The Plexi has a warmer, smoother tone and the '69 has more volume, headroom, crunch, and a tighter bottom end. It's also a harsher sounding amp. I'm using my terms here, some may say harder sounding instead of harsher.

Also interesting to note is that Siemens' brand EL34s will run forever in the Plexi, but die the first time the '69 is cranked to ten. If you use Mullard or Brimar EL34s like Marshall did, the tubes will live in both amps. The "one brand can take more voltage than another" problem is very much alive today. A Blackface Fender Deluxe Reverb runs its 6V6 tubes over the maximum design voltage. American brands such as RCA can take this with no problem. The 6V6 made in Russia won't hold up in the Fender Deluxe Reverb or a Trainwreck Express because the circuits push them too hard. They work well in lower power amps though. I'm told China now makes a 6V6 that will take anything thrown at it. I haven't listened to them yet so I can't report on their tone.

Amps that use EL84 tubes normally run at lower plate voltages than other amps. This is because the EL84 has the lowest design maximum plate voltage rating of any tube used in the output of current amp designs. I really love the tone of the Sovtek Standard grade EL84. The only EL84 with better tone is the very rare old German Telefunken EL84 tube. Many companies run EL84s at around 400 volts. That's too much voltage and destroys their tone, in my book. I guess some companies value maximum power and tightness over maximum tone, but it is a matter of taste. Mike, of Dr. Z Amps, dropped the plate voltage of his prototype EL84 amp from 400 volts to vintage Vox/Trainwreck levels. He then brought his amp down to the local music store and smoked every EL84 amp on the floor.

Gerald at Kendrick Amps did some tricks with voltage to get his harp amp to rip it up. The point is, one should keep an open mind on this subject. While we're on the subject of tube voltage, the voltage on the plates of your preamp tubes also affects their performance. However, it does it in different ways. Higher preamp tube voltage, with all else being equal, translates to more gain. Also higher voltage usually gives you more headroom. Preamp designs being so variable makes it hard to give general rules about how they will react to voltage. For example, in most preamp designs more voltage equals less compression. But in some multi-stage master volume preamps, more voltage gives more compression.

STABILITY

One characteristic an amplifier should possess is stability. This means the amp should be stable during operation in several ways. One way would be that all voltages remain within design limits during use. This would also apply to parameters such as currents, component values and such, staying within their limits. Also, factors such as frequency response and power output must stay within their design limits. Imagine playing through an amp that starts with 100 watts of power, but by the second set, after it gets hot, drops to only 50 watts. Or picture playing through an amp that, as the night wears on, gets darker or brighter sounding or playing through a bass amp whose bottom end fades during a gig. We've all played through an amp in need of repair with one or more of these problems, but there have been amps manufactured with this kind of stability flaw built in at the factory. An example would be when the first transistor guitar amps hit the market, most suffered a problem called "thermal runaway."

What would happen is, if the amp was played hard, the output transistors would heat up. As the output transistors got hotter the heat would allow them to draw more current. More current would make the transistors run hotter still. Of course hotter still means even more current would flow causing yet more heat. This cycle would go on until the transistors reached the heat level that would cause them to fail. The problem was lack of "thermal stability." Tube amps can also suffer thermal stability problems. Even today there are tube amps made with thermal stability problems. Some of these designs suffer with power and tone change problems. I know of more than one design that will run fine for months or even years and then suffer catastrophic failure.

There's another kind of stability problem that can affect an amp that's a bit harder to explain. It's a kind of unwanted electronic feedback. We all know if you place a mike in front of a speaker you'll get acoustic feedback. In a guitar amp, if a signal from a later stage gets sent back to an earlier stage, an electronic feedback may result. The frequency of this feedback may be in the range of human hearing, or above or below it. It can be caused by bad design or layout, stray capacitance or inductance, or a bad component.

Too much gain can affect stability, so can plugging in some effects pedals. Your amp may be fine until you plug in an overdrive box which then will whistle and squeal even with your guitar turned down. A common variation of this electronic feedback is the dreaded "parasitic oscillation." In the case of parasitic oscillation, the amp is fine until a note is played. The note increases the strength of the electronic fields inside the amplifier to the point where electronic feedback can occur. This type of instability can have many sonic results. One is the amp can break into oscillation and continue to produce its "squeal" even after you stop playing. Another affect of a parasitic can be the sound of an insect riding up on top of your note. A third common effect is the note is played with the amp up, but there's hardly any volume from your speakers. If you look at your output tubes, they're glowing like mad but still almost no sound. What is happening in this case is the amp is oscillating at a frequency out of the range of human hearing. The amp is producing full power, you just can't hear it 'cause it's all going to the parasitic. The note you do hear is the bleed from the guitar signal which is much weaker than the oscillation frequency signal.

It takes a really good tech to track down an instability problem and is not a do-it-yourself job. Of course, if the amp design was wrong from the start, modification, if possible with the design in question, is the only cure.

BATTERIES

Many musicians I've talked to use battery-powered effects. What most don't know is that the brand and type of battery you use in an effect can have a mild to severe affect on the way the unit sounds or performs. The only batteries available in the early days of battery powered effects were of carbonzinc construction. Today the most popular batteries are the alkaline type. Also new to the market is the very long life lithium battery. There are also two types of rechargeable batteries on the market today, nickelcadmium and metalhydride. Of course many current effects have an input for a plug-in battery eliminator.

There is really no way to second guess which battery will work best with which effect. I'd advise buying several brands in both the carbon/zinc type and the alkaline type. Keep in mind some companies make carbon/zinc batteries in both regular or "classic" and heavy duty versions, so do try both types.

Take one effect at a time and run all the batteries through it to find the sound you like the best. If you have two of the same effect, for example two TS808s, you might find one likes one brand or type and the other box likes something else. Also, be aware that in the 60s most of your favorite artists used carbon/zinc; the Everready transistor model (heavy duty carbon/zinc) being the most popular. Run this type through a Fuzz Face or Vox Wah and you'll see what I mean.

I don't like the sound of lithium batteries myself. I find them very harsh sounding. However, if you're into punk, grunge, or hardcore metal you might want to give one a try. Don't use rechargeables—yeah, I know, save the planet and all that, but nicads don't fade away; instead they die in the middle of a song. A battery eliminator is the most earth friendly, but may or may not sound as good as the best battery. Yes, different brands of battery eliminators sound different. Warning: do not try to recharge a carbon/zinc or alkaline battery. They can explode in a battery charger.

TUBE SHIELDS

Tube shields are those little metal cylinders you find over the small tubes in your amplifier. They're used to reduce the chance of picking up hum and noise. Sometimes they're also used to hold the tubes securely into the tube sockets. Tube shields have an adverse affect on the tone of most amps. Here's the reason why. The tube has an element called a "cathode" which gives off electrons. Electrons are attracted to a positive charge and repelled by a negative charge. The cathode is in the center of the tube element called the "plate." The plate has a positive charge which it uses to attract electrons. OK, so now we have a cathode flowing electrons to the positive plate and all is well. Then we install a tube shield that connects to the chassis which is negative in relationship to the plate. Confused? So are the electrons that see this negative charge past the positive plate. Also the plate and the tube shield form a small value capacitor.

To test this for yourself start with the tube shields off. That way you won't be dealing with hot tube shields later in this test. Warm up the amp and play, listening to the tone with a critical ear. Next put the tube shields back on. Be careful, the tube shields will be cool but the tubes will be hot. Do not change any settings on the amp or guitar and listen to the tone again. I rest my case. If your amp makes too much noise or does any nasty tricks with the tube shields off, do put them back on.

GUITAR CABLES

There's been a lot of voodoo lately about guitar cables. All kinds of claims are made by many companies about the wire that links your guitar to your amplifier. The cable you use can have a major affect on your tone. Besides the tone factor, some cables are prone to problems such as hum and noise, microphonics, inability to reject radio signals, inability to handle static electricity, and outright failure.

I've tested many cables that cost from five bucks to several hundred dollars per cable. I've tested cables made from linear crystal wire, oxygen free copper, and even a cable with pure silver conductors. The pure silver cable never made production but it was a great sounding wire. Its big problem was with microphonics. As this cable was pulled along the floor, it was almost like dragging a microphone on the deck. Some of the most expensive cables turned out to be dead and dark sounding with major signal loss. The linear crystal wires I've tried have all been very fast in response but had a very brittle glass-like high end.

After all my testing I found I liked George L's cable the best. I use it as my reference cable here at Trainwreck. Of course, like speaker tone, the sound you get through your guitar cable is personal taste. If you'd like to experiment with cables you might pick up a George L's for a baseline sound and compare other cables to it. Always compare cables of the same length as that also affects the sound. If you're going to spend over $30 for any cable I'd ask if it can be returned for a refund. You wouldn't want to get stuck with a $250 turkey. To be fair, a cable I wouldn't own might work for you. For example, if you own a dark sounding guitar and a dark sounding amp, a linear crystal cable might bring them to life. If you have a very bright Tele and a Twin with JBLs, the dark sounding cable might work for you. My advice: Don't believe the cable ads, use your ears.

SPEAKER WIRE

The wire used to connect your amp to its speakers can also have an affect on your sound. The same holds true for the wire used inside your speaker cabinets. Just like guitar cables there is a trend toward voodoo products in the speaker wire market. Many of these voodoo products do some good in stereo systems, but keep in mind your guitar, amp, and guitar speakers are not made for exact reproduction of a pre-recorded signal. Instead they are used to generate great guitar tones and many hi-fi rules do not apply when trying to reach that goal.

The current thinking is that the shortest length of the heaviest wire is the way to go. My personal testing has shown that this isn't always the case. Short lengths of thick wire will give less volume loss, more high end, and a tighter more controlled sound. However, all guitar amps generate some trash in their sound when pushed to distortion. Longer and thinner wire can sometimes help to filter out this "static" and make the amp sound smoother and less harsh. Also the longer or thinner wire can reduce damping of the speaker cone motion. This can provide for more "ring." For example, I use 20 feet of 18 gauge vacuum cleaner power cord to my favorite JMI AC30 speaker cabinet. I get the most chime and jangle with that wire. If I use 3 feet of 16 gauge lamp wire I get a tighter sound with more high end but less chime and warmth. Lamp wire is cheap and works well but there are also many pre-made wires that also work well. Let your ears be the judge.

As for inside the speaker cabinets, I've found that the stock wire works fine in most cases. Again, you can wire your cabinets with thicker or thinner wire and judge the results for yourself.

ELECTRONICS 101

THE TRIPLE SOUND TAPPED HUMBUCKER MOD

If you play a humbucker that has a tap between the coils, this will be real easy. If your pickup does not have a tap you can find the wires that are soldered together, taped, and tucked in, and add a tap wire to that point. The main parts needed for each pickup you wish to get triple sounds from are a .02 mfd capacitor, and a single pole, double throw, center off switch.

The three sounds we will be going for are full humbucker, single coil, and full humbucker bottom with single coil high end response. This gives the pickup an almost acoustic quality when used with a P.A.F. type winding. Higher output pickups will also work well with this wiring trick. Be aware you'll have to install extra switches on your guitar so don't do this on a 1958 Paul! See diagram one.

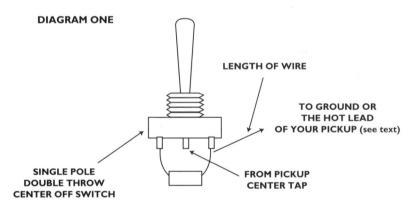

DIAGRAM ONE

LENGTH OF WIRE

TO GROUND OR THE HOT LEAD OF YOUR PICKUP (see text)

FROM PICKUP CENTER TAP

SINGLE POLE DOUBLE THROW CENTER OFF SWITCH

INSTRUCTIONS

Install the switch on your guitar. Be sure it will not interfere with other controls or playing the instrument. Solder the pickup center tap wire to the center terminal of the switch. Solder one end of the .02 cap (any voltage rating will do) to either of the end terminals of the switch. Solder the other end of the .02 cap, along with a length of wire to the free terminal on the other end of the switch. Here's where it gets interesting. You can solder the length of wire from the end terminal to either ground (back of pots) or to the terminal that the hot wire from the pickup is soldered to. One way will give you the single coil towards the bridge, and the other way will give you the single coil towards the neck. Which is which depends on pickup brands and polarity but here's a simple test. Before you solder the length of wire to either point take an alligator lead from the center tap and connect it to ground. Tap both coils' pole pieces with a screwdriver (lightly) with the guitar plugged into an amp. The loud coil is the one that's on for single coil. If the coil is the one you want, solder your length of wire to ground. If you'd rather have the other coil, solder your length of wire to the pickup's hot lead terminal point.

HOW IT WORKS

With the switch in the center (off) position, the center tap is not connected (open circuit) and you get the full humbucker. In one end position the center tap "short circuits" one coil leaving only one coil working. Don't worry about the coil that's not working being short circuited by the switch. A pickup does not generate enough voltage or current to harm itself when shorted out in this manner. In the other end position, the capacitor "shorts" out only the highs from one coil and lets the low frequencies continue to flow. You get the bottom end component from both coils, but the high end from only one. If you want to reverse which mode occurs at each end of the switch you can, of course, turn the switch 180 degrees or move the added length of wire from one end terminal to the other. You could also add another single pole double throw switch, to switch the length of wire between ground and hot. This would let you choose which single coil you want to use with the three-way switch for a total of five sounds! If you did this with a two humbucker guitar and added a phase switch, you would exceed a hundred combinations and that's not including volume control blends! What's that? You haven't had enough? O.K., Tele time!

Obtain a .002 mfd (2000 pf) cap and a single pole single throw switch.

INSTRUCTIONS

Disconnect the hot lead from your bridge pickup at the selector switch. Solder one end of the .002 cap to the switch terminal that you just removed the lead from. Solder the bridge pickup hot lead to the other end of the .002 cap. From one lead of the .002 cap, solder a length of wire to one terminal of the single pole, single throw switch. From the other lead of the .002 cap, solder a length of wire to the empty terminal on the switch. Now you can switch the .002 in and out of the circuit.

HOW IT WORKS

On the bridge pickup alone, switching in the cap cuts the bottom from the pickup. I don't like this sound myself, but some people find it useful on bass heavy amps. The sound I like is both pickups on, switched in. This gives a sound like a Strat on the neck and middle pickup position. It's caused by the thinning out of the bridge pickup and the phase angle shift caused by the capacitor. By the way, you can do the same thing by installing a Strat five-way switch in your Tele. Then you only need the .002 cap and no other switch! I'd tell you how it's done but your guitar tech should enjoy working on the "five-way Tele switch puzzle!" Hey, don't call me... call the Fender Custom Shop!

AMPLIFIER POWER SUPPLY SECTIONS

Here's a subject I get questions on all the time, so I'll cover it in a bit more detail and hope it will answer your particular problem.

One thing I hear now and again is, "I have an amp and when I first hit a note or a chord, the volume fades down for a second and then returns to normal. Do I have a bad power transformer?"

The chance of a power transformer causing this problem is about one in a thousand! It can be caused by the dreaded parasitic, or quite a number of other problems. Before replacing an expensive power transformer, have a qualified amp tech check the transformer in action using an oscilloscope to check the amount of transformer sag as the problem occurs. A good rule to follow with both power and output transformers is if there are no obvious signs of failure such as shorted or open winds, or smoke or a burnt smell, jump in a known good transformer. If that solves the problem then go ahead and change the original.

Transformers are replaced on amps at a rate of about twice what is really needed, from what I've seen over the years. If your tech doesn't have enough substitute transformers to work on most amps, find yourself a tech who does.

Here's another question I get, "Ken, you said I can add a tube rectifier to any Vox AC-30 amp that didn't come with one. My Vox doesn't have the needed center tap on the high voltage winding of the power transformer. How can I use a rectifier tube?"

Some of the Vox AC-30 reissues and many other amplifiers don't have a center tapped high voltage winding. Instead of the standard full-wave rectifier you see in most amps, they use a full-wave bridge rectifier. This doesn't require a center tap transformer, but does require at least four diodes. Of course a full-wave rectifier tube such as a 5U4, 5Y3, 5AR4, or GZ34, only contains two diodes, not enough for a full-wave bridge. So how can we use a tube rectifier in a full-wave bridge circuit? It's really very simple. Install the tube rectifier. Hook up the 5 volt transformer for the tube filament. Then read this warning: tube amps contain dangerous voltages even when unplugged. Only qualified personal should ever attempt these modifications. Now hook up the two high voltage leads, one each to plates of the rectifier tube. The filament or cathode of the rectifier tube goes on to the B+ of the power filters. To complete the bridge obtain two IN4007 silicon diodes. Connect the banded end of one diode to one plate and the other end of the diode to ground. Take the second diode and do the same thing for the other plate. You now have four diodes in a full-wave bridge rectifier. Since all forward current, the current that the amp draws in use, goes through the rectifier tube, you have rectifier tube performance.

I know what you're thinking! "But Ken, my amp has a voltage doubler power supply. Now what?" This too can be handled. Here's the easy way: Let the silicon diodes in the circuit do all the work and once you have your rectified and filtered D.C. you can feed it through a rectifier tube with its plates in parallel to obtain the tube rectifier sag. Just don't forget to use about a 40 mfd power filter with a high enough voltage rating on the cathode of the tube rectifier to handle signal bypass. If you do this to a high power amp (100 watts or more) you should use two rectifier tubes in parallel to handle the power. Notice I haven't given exact step-by-step details here. To a qualified person everything they need to complete the task is in this text. If you don't understand it, bring this article to a qualified tech to have the work done. A wrong move here is very dangerous to you and your amplifier. Enough said!

THE CHOKE

The choke is the third "little transformer" on some amps. A choke is not really a transformer but a coil of wire used to smooth D.C. and decouple various stages in the power supply. They don't often go bad but, if one should fail, the single most important factor in substituting a different choke, other than a stock replacement, is the current rating. The Henry rating is not critical. For example, the Vox AC-30 uses a 15 to 20 Henry choke. An 8 or 10 Henry choke would work just fine. A Fender amp might use a 4 or 6 Henry choke, but a 20 Henry choke would also work fine. A choke also has a D.C. resistance rating. This is usually not important unless the resistance rating is much higher than the original. In that case you would lose some voltage to the tubes that come after the choke and output tube screen voltage. Of course, an exact replacement is always nice but we don't always have that option on a vintage amp.

POWER FILTERS

Here's a common question I get on power filters. "Ken, I've just come across brand new, never-used power filters for Fender... Vox... Marshall from the 60s, should I buy them?"

My response, "Absolutely not!" Power filters deteriorate with age whether they are used or not. Always use fresh power filters when working on your amp. Also be aware that different brands of electrolytic capacitors (power filters) can make your amp sound different. Also, changing the mfd value can alter the sound of your amp. Sometimes a higher or lower mfd value will be of benefit. For example, an old under-filtered amp (many old low wattage models) will hum and give false harmonics. A little more filtering can solve this problem. Let's take a 1967 Plexi Marshall vs. a late 1968 Plexi. The late 1968 has a lot more filtering, particularly on the pre-amp. Lowering the filter values to 1967 values will warm up the sound but you will have a bit more hum and subharmonics at high volume. In any case, always be sure to use a high enough voltage rating. A burst filter will spray a metal-eating chemical you don't want inside your amp.

Tube Amp Talk

AMPLIFIED ELECTRONICS

SINGLE ENDED VERSUS PUSH-PULL OUTPUT STAGE

In the last few years, more new brands of guitar amplifiers have hit the market than in the past twenty before that. I'm always glad to see these new amps appear. The more tube amps there are, the more demand there is for tubes and related parts, keeping tube amps alive and well. Also, with all the "cottage industry" builders out there, many more ideas are tried and amplifier design advances. I personally do not like the term "retro" when applied to guitar amps. I've been building my amps since 1983 and have never changed my designs. All of a sudden, Trainwrecks and many of the new amps are put into this "retro" category. The fact is that many of these amps are far more advanced than the amps of the 40s, 50s and 60s. While many of these amps are less complex than others as far as the amount of features packed into one amp, they are by no means all "simple." There are some so-called "retro" amps that use printed circuit construction, master volumes, channel switching, and effects loops. Pardon me, am I missing something? Isn't that just what we've had from the mid-70s until now?

Anyway, the subject is push-pull vs. single ended amps. Many current new amps use what is called "single ended output" for their power section. The claim is push-pull cancels even order, particularly second order, harmonics. This is somewhat true as far as the output stage is concerned. Single ended amps do not cancel this even order harmonic. Since even order harmonics are thought of as "musical harmonics," a single ended amp should sound more musical. Things are never that easy! First thing we'll need is some background info.

SINGLE ENDED OUTPUT

In a single ended amp traditionally one output tube was used to provide the power to drive the speaker. A Fender Champ amp is a classic example. Originally the main reason for this design's use was because it is cheaper to build. Push-pull requires at least two output tubes and a more complex output transformer. For an inexpensive "student" amp, using a single output tube was the way most companies went. This applied to TV, radio, "Hi-Fi," and economy stereo gear as well. When using only one output tube, it must amplify both the negative and positive portions of the alternating current audio signal. Because of this requirement all single ended audio amplifiers run in class "A." Single ended amps have some technical problems such as unbalanced D.C. in the output transformer and low power output, but that's not important to know for our tonal comparison.

PUSH-PULL OUTPUT

A push-pull output stage typically uses two, four, or six output tubes. In this design the signal is split into negative and positive signals. Using the two tube design as an example, one tube amplifies one half of the total signal and the other tube amplifies the other half of the signal. Again, I don't want to get technical here, so I'll just tell you the result is you get more power, and the output stage cancels out some of its second harmonic distortion. After all, in Hi-Fi, isn't the goal the least amount of distortion?

Well, many audiophiles today like the sound of single ended amplifiers over push-pull. To get more power some amps run two or more tubes in parallel. Still, two of the same tubes running in single ended parallel will not make as much power as the same tubes push-pull. Also, it should be pointed out that audiophiles do not drive their amps to the point of clipping (output stage overdrive). Other audiophiles like the power ("beef") and control over the bottom end that push-pull amps display. The single ended guys talk of airy openness and space between notes.

But let's get back to guitar amps. Some makers of single ended amps state that push-pull amps cancel second harmonics and therefore are less musical sounding. There's only one truth in this statement, and here's what it is — push-pull output cancels some of the second harmonic created in the output stage. It's very important to note that I said the output stage cancels some of the second harmonic and not the preamp stages. Any second harmonics produced in the preamp stages will pass through a push-pull output stage unharmed! The fact is, in some push-pull amps over 95% of second harmonic content is produced in the preamp and not the output stage at all!

Now, let me say I have great respect for the guys who build single ended amps. Many of them build really great sounding amps. However, a small minority come up a bit short on practical guitar amp knowledge when they claim you can't get second harmonics from a push-pull guitar amp. Better than 98% of all your great guitar music from the Beatles to Jimi, to Clapton, to pure jazz, to fusion, to blues were done on push-pull amps. Think about it guys! Now, what about single ended guitar amps? I think they sound great. Who wouldn't love a tweed Champ or one of the better new single ended amps? They have a unique sound. When a push-pull amp clips its output stage, both sides of the waveform clip in an even manner. In a single ended amp that doesn't happen. The waveform shape clips in a different way on top than it does on the bottom. While the second harmonic doesn't cancel, and you can add that to the preamp second harmonic, the output also becomes rich in odd order harmonics. The crunch and grind of a single ended design is a result of those harmonics.

Look at it this way, the second harmonic is an octave of the original note. If you play the "A" note on the 3rd string, second fret you have A-220. The second harmonic would be A-440. A-440 is the high "E" string at the 5th fret. Play those two notes together clean and it's like the A-220 is the root note and the A-440 is the "second harmonic." Sounds fuller, but no crunch, no grind and no excitement. Those odd order harmonics are the spice of your tone. A single ended guitar amp's real source of its different tone is the different balance of odd order harmonics it produces. Both single ended and push-pull amps can sound great. It's up to you to pick the spice you like.

AMPLIFIER TROUBLESHOOTING

First up is a recent phone call from a gentleman with three Ampeg SVT heads. One of these heads was a 1969 SVT converted from 6146 tubes to GE 6550 A tubes. The SVT amps with 6146 tubes used a 10 ohm, 5 watt plate resistor on each output tube. The 6550 amps used a 5 ohm, 5 watt resistor instead of the 10 ohm as in the 1969 head. The main purpose of this resistor is to act as a fuse for the tube to which it's connected. The question was, "Can I use a 5 ohm resistor in my converted head in place of the 10 ohm? Also, would it be O.K. to use a 10 watt rating?"

SVT amps go through lots of those resistors. When I was servicing them, I'd buy the 5 ohm 5 watt resistors by the hundreds. I'd use the 5 ohm resistor in both the 6146 and 6550 amps with no problems. As for the 10 watt rating — don't forget it's used as a fuse. Doubling the wattage rating is using too large a value "fuse" for the job. Stick with the 5 watt rating.

But there's even more to the story. The resistors used in the Ampeg were 5 watt resistors commonly called "sand block resistors." They are rectangular in shape, three sides are the same and one side looks like it's filled with concrete. While that resistor is a true 5 watt normal power handling unit, when overloaded it is very quick to fail. When used as a fuse that's just what we want. However, these 5 ohm 5 watt resistors also come in silicone coated and vitreous enamel (glass) styles. These are usually round in shape. While still rated at only 5 watts, these take a lot more overload before they will burn open. Not a good "fuse"! All three types come in 5 watt 5 ohm ratings, but experience, with both the three types of resistors and SVT amps, teaches that the sand block is the type to use.

Of course, there just isn't any electronics book or electronics course you could take that would teach you that the sand block type power resistor is the one to use in an Ampeg SVT. There are thousands of things about amps like this example you learn from experience, but this stuff is mostly in people's heads and not on paper. While on the SVT, replacing the bias pots with quality 10 turn types will make biasing the amp much easier and keep the bias much more stable.

DO YOU LIVE ON BAKER STREET? WELL, DO YOU?

Let's play detective with a British amp I fixed for a friend a few months back. The amp is a Sound City type LB-50 Plus. It's a 50 watt head using two EL-34 output tubes. My friend Mike, who's a good tech himself, bought this pristine head at a giveaway price because nobody could get it to run. Here's what the complaint was — turn on the mains (power) switch and everything was fine. Tubes all lit up, B+ voltage was fine, bias voltage was right on, and everything looked good. As soon as the standby switch was moved to the play position the fuse would blow with a bright flash. The first thing one would suspect is a shorted output tube. Substitute EL-34s — fuse pops again! The first power filter after the standby switch has a 15K resistor feeding it. It can't be a shorted filter because that would cause the 15K to burn and it's fine. Next suspect, an output transformer primary to ground short. Check the output transformer and it's fine! Time to check everything inside by eye. The two .047 coupling caps from the phase splitter to the output tube grids have been resoldered. Obviously a previous tech had suspected a leaky cap was feeding positive voltage to the output tube control grids, unsoldered the caps to remove that possibility and found the amp still blew fuses.

Spotted a very burnt 4.7K resistor in the bias supply. The Sound City has an independent bias wind on the power transformer. The diode and bias filter were fine as we had the required minus 36 volts bias in standby. Replaced the 4.7K resistor and it went right up in a cloud of smoke. Very strange — the 4.7K resistor has two 10K pots between it and the bias supply and the 4.7K completes the path from the bias pots to ground. Yet despite this arrangement the bias pots were unharmed while the 4.7K resistor almost burst into flames. It would take more than 36 volts to fry the 4.7K resistor in this manner. More like several hundred to get this response. Yet why wasn't the bias pot harmed? Pull out the output tubes, new 4.7K resistor in place, voltmeter across the 4.7K, put the standby switch to play and the resistor blows, but not before the meter captures the peak voltage which happens to be the exact B+ voltage of the screen supply!

O.K. now we know a few facts. The screen doesn't get voltage until the standby switch is in the play position. Also, as soon as the screen voltage is on, it appears in the bias section only across the 4.7K resistor. The Sound City uses terminal boards. The screen supply and the bias supply share a small terminal board inside the amp. A bit of testing with an ohmmeter reveals that a wire running through an insulation tube that connects one of the 10K pots to the 4.7K resistor runs past a screen supply terminal and is physically touching this terminal. Pushing the bias wire away from the screen supply terminal revealed what the meter found. The insulating sleeve on the bias wire had carbonized and put the entire screen supply across the 4.7K bias resistor. This caused the resistor to burn. It also put positive 410 volts on the control grids of the output tubes. This caused them to draw huge amounts of current, thus blowing the fuse. Case solved! A new 4.7K, a new wire, and the amp was fine!

Well, that's how it's done. Fuses and resistors are cheap. I could have used mains current limiting, but the very dramatic way the 4.7K resistor burned was an important clue I might not have seen with current limiting.

I should point out that it takes more time to read this column than it took in real time to fix this "un-fixable" amp. In any case, you now have an example of how I troubleshoot a one-of-a-kind problem. Pass me my fiddle and my coke — leave my cane alone!!

AMPLIFIER TROUBLESHOOTING

I'll try to cover some basic techniques. Unless you are 100% qualified to do this work yourself, we strongly urge you to contact an expert repair center and arrange to have them service your amplifier.

TROUBLESHOOTING BASICS

I should tell you that this chapter applies to most amps using 100% tube circuits only. Warning, solidstate and hybrid types (part solidstate and part tube) should not be serviced with the system that follows.

To use this system you will need at least a volt-ohm meter (VOM) or a digital multi-meter (DMM) of decent quality. A signal generator and an oscilloscope are a major plus. Also, you will need hand tools, spare tubes and parts, soldering equipment and so on.

GO WITH THE FLOW

Question number one is, when you plug in your amps and turn on the power switch, does the amp draw power? That is, do the tubes light up? Does the pilot light come on? Can you hear the power transformer hum? In other words, is there some activity or is the amp stone cold dead? Let's cover stone cold dead first. Item one is check for a blown fuse. Be aware some amps have internal backup, or other internal fuses which could be blown also. Never change a fuse with the amp plugged in. Let's say all fuses are O.K. To be sure a fuse is O.K., even if it doesn't look open, check it for continuity with your meter.

If the fuses are O.K. and the amp has no power, place your ohmmeter across the plug of your amplifier — main power prongs, not the ground prong. With the power switch on, check for continuity. This checks the plug, power chord, fuse holder, fuse, voltage selector (where one is provided), power (mains) switch, and power transformer primary all with one simple test. If you have an old Vox or amplifier with similar slow start up system it also checks these parts as well. If you have continuity, the problem is going to almost always be from the secondary side of the power transformer and beyond. If you don't get continuity, check each item in the primary side power chain one at a time. From one prong of the plug check for continuity to the inside of the amp. Do the same for each prong including the ground, if so equipped. Don't forget to wiggle the power cord at the strain relief. Often a wire will break at this point. Check the switch for continuity. Then check the power transformer primary winding. Once you find the part with no continuity, repair or replace it and chances are the amp is fixed. On amps with multi-voltage power selectors, don't overlook the possibility the amp is simply set for the wrong voltage. Let's say the circuit to the power transformer primary is complete and the primary also is O.K. but the amp doesn't light up. Now we check the voltages on the power transformer secondary at the transformer lead or terminals with the amp plugged in and turned on. If you find any of the voltages missing, disconnect that set of leads to isolate them from a possible short down the line. Check for voltage again. No voltage on a secondary winding with an energized primary indicates a bad transformer. If all voltages are present and the tubes do not light up, a broken wire or bad connection running to the tube sockets would be the likely fault. Check for these as well as cold solder joints. Light the tubes and the amp should work. Also while inside, check for all proper D.C. voltages at the main points in the power supply system. Tubes that are lit but have no D.C. voltages will not work.

THE FUSE BLOWS

Maybe instead of not turning on at all, your amplifier blows its fuse every time you turn it on. Step one, remove all the tubes, put in a new fuse of correct value and turn on the amp. If the fuse doesn't blow under these conditions the next step is to plug back in the rectifier tube if your amp has one. If the fuse now blows check the rectifier tube, power filters, output transformer and other items in the B+ path for shorts. Don't forget the choke if your amp has one. Find the short — cure the amp. Let's say plugging in the rectifier tube doesn't blow the fuse. Install the power tubes one at a time. The one that blows the fuse is bad. Replace the entire set with matched tubes and re-bias the amp. Preamp tubes don't blow fuses as a rule.

Let's now say your amp uses silicon diodes, not a tube rectifier. First pull out the tubes and try a new fuse. If the fuse doesn't blow, put the power tubes in one by one as before. By the way, I assume you've already checked for proper bias voltage before reinstalling your tubes. No bias will blow the fuse with the power tubes installed. Some amps, such as Marshall, have a high voltage (HT) fuse in the output circuit which will blow in case of shorted tubes or no bias voltage. On cathode-bias amps check the bypass cap for shorts. If the amp blows its fuse with no power tubes installed, check the diodes and everything described in the tube rectifier section of this article check list. Find the short — cure the amp. Let's say the fuse blows and you find no shorted items. Disconnect all transformer secondary leads. If the fuse still blows, a power transformer failure is most likely.

EVERYTHING LIGHTS, THE FUSE DOESN'T BLOW, BUT THE AMP DOESN'T WORK

First check the speaker, and check for continuity between the speaker and output transformer. Repair/replace any bad items. It's important to do this step first because putting a signal into a tube amp without a speaker or dummy load connected can cause damage to the output stage. The easiest step at this point is to substitute tubes since most tube amp failures are just bad tubes. If this doesn't work, check for voltages. Eyeball the insides for burnt parts, broken wires or leads, and anything that doesn't look normal. To check for missing voltages start at the secondary of the power transformer, on to the diodes or rectifier tube, choke, resistors between filter stages, standby switch, and so on down the line until you find where the voltage is interrupted. Repair/replace any part, wire, or connection that doesn't go with the flow.

WORST CASE

Let's move on to the worst case in troubleshooting an amp. Everything lights, all working voltages are correct, but the amp still doesn't work. We've checked the speaker and its continuity to the amp, we put in a known working signal be it from a signal generator or guitar to no avail. Tried new tubes and so on. There are now two ways to troubleshoot this amp. From front (preamp) to back (output stage) or from back to front. With a signal generator and an oscilloscope, most people use the front to back approach. Let's get into this method first.

Plug a dummy load into the speaker jack. Plug a signal generator into the input jack, turn on the amplifier. Be sure to turn up the volume and tone controls a bit. With the oscilloscope probe, check for signal at the first gain stage (input stage). If you get a signal fine, move on to the very next stage and look for a signal. If you do not get a signal, check voltages yet again. If all is O.K., check for bad caps, resistors, solder connections, substitute the tube and so on. Look for shorted, opened, and burnt or out of value parts. Do not go on until you get this stage working. The idea is to go stage by stage from input to output, one at a time, stopping to fix any stage that doesn't do its job. By isolating each stage one at a time, you will have a simple circuit with only a few components to check.

Now let's say you get to a stage and the voltages needed aren't there or are way off. By "way off," I use the figure of plus or minus more than 25% of what's called for in the schematic. However, that's not a hard and fast rule. A schematic may call for 1.4 volts on the cathode of a preamp tube. You might measure 2.6 volts, yet the circuit may be fine anyway. Different brands of preamp tubes will self-bias at quite different voltages and still work fine. In any case, if you have no B+ or plate voltage, start at the beginning of the power supply and work forward towards the front of the amp to find the interruption. On a preamp tube, measure both sides of the plate resistor for voltage. There should be a drop in voltage on the plate side. No drop in voltage may indicate an open cathode circuit. Too much drop in voltage or lack of a small positive voltage on the cathode may be caused by a shorted cathode cap. An overly high cathode voltage along with a overly large drop across the plate resistor may indicate a positive voltage being applied to the control grid. This may be caused by a leaky or shorted coupling cap from the previous stage.

Back to front troubleshooting works the same way except you hook the oscilloscope to the output and inject the signal in at the output stage and work towards the input to find the non-functional stage.

USING A METER TO TROUBLESHOOT

This is a good system to use when you're on the road and don't have a scope and signal generator handy. The speaker becomes the "scope;" the "signal generator" is an alligator clip to inject a hum "signal" and the meter checks voltages and continuity. This system must be used back to front. Clip the alligator lead to the control grid of the tube section in question, starting with the output stage. Let the other end of the alligator clip hang free. Be careful not to let the loose end touch anything. In fact, taping it up is a good idea. The lead will pick hum out of the air. When you reach the stage that does not hum with this lead on the control grid, you've found the problem. One warning, some amps such as Marshalls, Fender Bassman and Twin Tweed use a cathode follower stage that has B+ on its control grid. Clip your lead to the grid of the stage prior to the cathode follower for testing.

REVERB AND TREMOLO

Many problems in the reverb circuit are caused by open or shorted leads and reverb tanks. Bad footswitch leads, plugs, and the switch itself cause many failures of both reverb and tremolo. Be aware that some amps need their footswitch for these circuits to work. For example, on a Fender Blackface amp the footswitch completes the circuit for the tremolo function. No footswitch plugged in, or a open circuit in the footswitch unit, and you'll have no tremolo. On the other hand the reverb will work without the footswitch. In the case of the amp's reverb, the footswitch "shorts out" the reverb signal. A short in the footswitch will kill the reverb function. Putting it simply, a short in the reverb footswitch will keep the reverb off, and a short in the tremolo footswitch will keep the tremolo on! The footswitch works the same in every brand of vintage amp. That is it either completes a circuit or "shorts out" a circuit. Testing the electronics of the reverb and tremolo circuits is done the same way as any other circuit.

THE CARE AND FEEDING OF YOUR VOX AC30

The first AC30 amp introduced in 1960 had a black control panel with gold lettering. The original black panel AC30, along with its little brother the AC15, used a 6267 tube, also known as a EF86. This tube was used as a preamplifier for one channel in these amps. This tube is a pentode or five element design and was dropped from all other AC30 amps in favor of triode or three element design preamp tubes. The problem for owners of an AC15, black panel AC30, the smaller AC4 and AC10 amps, is that decent 6267 tubes have not been made for years. Current production 6267s are low gain, microphonic and have no tone. Unless you can find working originals, your pentode channel just won't perform well. This is also a problem for Matchless amplifiers which use this tube. If you have a working original 6267 in your amp, don't change it. All one can do is search out this tube among old and vintage tube dealers.

Before we delve into the other tubes in an AC30, let's do some basic tune up work.

All guitar amps contain electrolytic capacitors commonly called power filters. They are also used in other parts of your amplifier for other jobs. Electrolytic capacitors have a limited life and should be checked every couple of years once an amp reaches old age. If one bad electrolytic is found in an old amp, change them all. It's like riding on bald tires—if one blows out the others will soon follow. The main power filtering on an AC30 ranges from a low of 16 mfd on early amps to 160 mfd on some reissues. The first power filter stage is the most important. You can go up in value at this stage if hum and subharmonics are a problem but don't exceed 80 mfd on rectifier tube models. Keep in mind that as you increase the value here the amps tone will change. An over filtered amp sounds awful!

In all other power filter stages keep the stock value. If you can't find 16 and 32 mfd caps you can use 20 and 40 mfd with no tonal change. I like to use Mallory TC series electrolytics myself. Be aware that different brands can alter the tone of an amp to a noticeable extent. An AC30 contains several more electrolytics in functions, such as cathode bypass. Change all these with the same value. A 22 mfd will replace a 25 and a 220 mfd will replace a 250.

Now it is time to check the entire amp for burnt out resistors, wires, loose or broken parts and such. Clean and retension the tube sockets, inspect the transformers for signs of burning or overheating. Early amps use transformers potted with beeswax. Big blobs of beeswax under a transformer is a bad sign. Some amps used a device called a Brimistor. It's a controlled warm up device. If it's bad, simply bypass it. Vox used many brands of transformers in the AC30. Albion and Woden are my favorites; the Parmeko is my least favorite. If a transformer is bad, have it rewound. Since, on the AC30, the transformers hold the vertical and main chassis together they're hard to substitute. Also, the original brands varied in size and a Parmeko and Albion output transformer won't fit in each other's place, for example even tracking down an original, which is almost impossible, may require sheet metal work to install. Rewinding is the hot tip.

Now back to tubes. The early Vox AC30 amps used a 5AR4 (GZ34) rectifier tube. The Korg Reissue AC30, Trainwreck Rocket and Matchless 30 also use this tube. If you have a solidstate rectifier reissue and you want to make it sound more like an original, you can install a 5 volt rectifier transformer, 8 pin octal tube socket and a 5AR4 tube in any Vox. Keep in mind that if you have 160 mfd as your main power filtering, cut it down to 80 or less or you'll pop a lot of 5AR4s. Don't use a 5U4 in an old Vox. The 5AR4 has a controlled warm up which lets the other tubes fully warm up before high voltage is applied. The 5U4 warms up faster than the other tubes and "shocks" them every time the amp is turned on. This is not a problem on other brands of amps, such as Fender, which employ a standby switch which lets the main tubes and the rectifier warm up independently. The output tubes in an AC30 are EL84s, also known as 6BQ5s. The best current EL84 for Vox use is the Russian regular grade (nonmilitary) version. They should be matched sets. These also work well in Trainwreck amps but do not use them in a Matchless amp. Matchless amps like GE 6BQ5 tubes best, so consult Matchless on retubes for their equipment. All other tubes in your AC30 are ECC83s, also known as 12AX7s or 7025s, except for one ECC82, a.k.a. 12AU7, in trem channel. Pick the brand whose tone suits you.

Here's the trick to reduce hum and buzz on early AC30s without a hum cancel control. Check along the preamp filament line until you find a ground terminal on the chassis with one side of the filament line grounded to it. Disconnect the filament line from this terminal but leave the line connected to the tube, of course. Obtain two 100 ohm, ½ watt resistors; solder one between pin 9 and the ground terminal. Now take the other and solder it between pins 4 and 5 (they're connected together at the socket) and the ground terminal. Instant huge noise reduction!

Any non-top boost AC30 can be converted to top boost in a couple of hours. The schematic for this model is in *A Desktop Reference of Hip Vintage Guitar Amps*, under the Vox listings. Also, a top boost can be made non-top boost with equal ease.

Finally we come to speakers. Vox used two 8 ohm speakers wired in series to make 16 ohms. If your amp has original T530 or T1088 Alnico Celestions, great. If it doesn't have these speakers, obtain the new Celestion G12 Alnico Blues (model no. T4427) and install them wired in series. These speakers and the rectifier tube trick will make the sound of a reissue Vox AC30 come alive. Also, check that all nuts and screws that hold the amp together are in place and tight. Never overtighten the speaker—you can warp the frame and cause voice coil rub. If your amp is missing back panels, make replacements; they affect the tone in a significant way. Never try to make your AC30 Combo closedback, it will overheat and die. If you have an AC30 extension cabinet (the one that's the same size as the combo but with no chassis) it can be made open back for more chime and jangle. Take the original back off and store it. Cut one panel to go across the bottom of the back. I make mine 7¾" high and I don't use a panel up top.

Well, there you have it. With a bit of loving care, your Vox should reward you with some of the best guitar tones that can be had.

VINTAGE VOX TUNE-UP GUIDE

If there's one brand of amplifier that has truly developed a mystique about its servicing and maintenance, the vintage Vox is that animal. Amp techs who can service a Marshall or Fender amp blindfolded, shudder when they see a J.M.I. Vox walk through their door. This reputation is based on two realities. First is the fact that Vox amps of the 50s and 60s were more prone to dying in a cloud of smoke than other brands of the same day. The other reason for their hard to service reputation stems from the fact that, outside of England, there just aren't a lot of people who have worked on enough Vox amps to gain the extra experience these amps require for their maintenance.

I'm going to cover several of the more popular vintage Vox models in this chapter in the hope that it will keep more of these great amps alive and running as they should.

GENERAL INFORMATION

Like any amplifier, vintage Vox amps are subject to the failure of parts due to normal wear and use, and old age. In some models of Vox amplifiers, parts may fail sooner because of design flaws, particularly those relating to excessive heat.

A good starting point is to change all electrolytic capacitors. Clean and tighten all tube sockets. Replace any that are broken or burnt. Clean all controls. Replace any that are worn out. Replace any broken switches. Replace any obviously defective or broken parts. Check all tubes for proper operation. In other words, go through the general checklist you would use on any other brand of amp you would normally service.

THE AC-4

The AC-4 is the "baby" of the Vox line up. Think of it as a Tweed

Champ with a vibrato circuit and an EL-84 in place of the Fender's 6V6 output tube. It's a single ended design, and uses a 3.2 ohm 8" speaker.

This amp uses four tubes. These are as listed with the European tube numbers, followed by the American numbers.

1: EZ80/6V4
2: EF86/6267
3: ECC83/12AX7 or 7025
4: EL84/6BQ5

An interesting feature common to the AC-4 — and on one channel of the following two channel Vox models such as the AC-10, AC-15, and the first year (1960) AC-30/4 — is the use of an EF86/6267 pentode first gain stage. Far more common is the use of a triode first gain stage in most guitar amps.

The AC-4, being such a simple amp, does not suffer the breakdown problems of the larger Vox amps. Fresh electrolytics, good tubes, and the amp is happy. This amp has a 115/230 volt selector. Besides the one amp main fuse, there is a two amp backup fuse located in the main's voltage selector plug. I've seen AC-4s have their power transformers replaced because a tech didn't know about the second fuse, and thought he had a dead transformer. This is a classic example of why it takes a Vox experienced tech to handle the work on your prized Vox amplifier. It really is a shame that there isn't a school for guitar amplifier servicing. Amp techs are forced to learn about vintage amps either by on the job experience, reading the limited number of books on the subject, or knowing a person who has the needed knowledge, and is willing to share it.

THE AC-10

The AC-10 is the next step up from an AC-4. The AC-10 uses six tubes, has a push-pull output stage, and moves up to a 10" speaker. It also has both a normal and vibrato channel. The tubes used are as follows:

1: EF86/6267
2: EZ81/6CA4
3: ECF82/6U8
4: ECC83/12AX7 or 7025
5 and 6: EL84/6BQ5

The AC-10 has an input voltage selector for 115, 160, 205, 225, and 245 volts. Always be sure it is set for the correct voltage for your country. Weak points on the AC-10 include a 200 ohm 5 watt resistor that connects to pin 3 of the EZ81 tube. All the B+ voltage of the amplifier flows through this resistor which can open up, shutting down the amp. Also, there's a 100 ohm resistor on each of the power tubes, connected to the plate (pin 7). These should be checked also. The ECF82 tube controls the vibrato circuit and the normal channel. If this tube dies, the normal channel won't work and you'll have no vibrato. The AC-10 contains a speaker lead junction block with an 8 and 16 ohm speaker tap. This is a common feature on many other Vox models of the time. If a set-screw vibrates loose, or you over-tighten the set-screw and cut a wire, the amp won't work. Always check the speaker junction block when working on Vox amps. The speakers used in this model were not heavy duty, and I've seen many with open voice coils from being pushed hard into the overdrive range.

THE AC-15

This is the little Vox that started it all! An eight tube screamer with a 12" speaker. This model has both vibrato and tremolo. The tubes used are as follows:

1: ECC83/12AX7 or 7025
2: EF86/6267
3: EZ81/6CA4
4: ECC82/12AU7
5: ECC83/12AX7 or 7025
6: ECC83/12AX7 or 7025
7 and 8: EL84/6BQ5

The AC-15 likes to have a well matched pair of output tubes at all times. Under distortion the AC-15 exceeds the design limits for maximum power of its power tubes. Rule #1 for an AC-15 is a well matched pair of EL-84s that don't draw excessive current at idle. This amp contains an internal pre-set trim pot for its vibrato/tremolo strength. If the vib/trem doesn't work, first check the ECC83 tube (V6) which is the oscillator for both. Next check the footswitch, and then check the pre-set. If the pre-set is turned down, the vib/trem, of course, won't work. Why would the pre-set be turned down? People who don't use the vib/trem circuit will sometimes turn it down so they can't accidentally engage this feature. Some rough boys will actually cut off the footswitch so it doesn't "get in the way!" On the normal channel of an AC-15 there's a "brilliance" switch. If this switch adds too much high end for your taste, an easy trick is to put a resistor across it to reduce its effect. Start with 220K and reduce from there to cut highs, or increase the value to bring more highs back. Do this with the top cut control at the full treble position. If you need more high end, or a different type of high end, you can bypass the volume control (on either channel) with a 100 pf cap from hot (input) on the pot to the wiper (center terminal). The AC-15 runs hot so always place it so that air can flow around it at all times.

THE AC-30

The Vox AC-30 isn't one amp, but in fact a series of amps with the AC-30 designation. Most people know the AC-30 as a three channel amp with a Class "A" EL-84 output stage, and two 12" speakers. The first AC-30 to see the light of day was a "T.V." front amp with two EL-34 tubes in its output. This amp came out in 1959. I've seen this version in both a single 12", and a single 15" speaker combo. In 1960 the AC-30/4 was introduced. This amp used four EL-84 tubes in class "A," and has two channels. One channel retained the EF86/6267 pentode of the AC-15 model. The control panel is black and gold. This amp was soon replaced by the AC-30/6 model. This model has three channels. The control panel is a bronze color, also referred to as "brown" or "red."

The AC-30/6 was built in bass, normal, and treble versions. Since at this point the amp had no tone controls, except for a "top cut," Vox used different capacitor values and circuit variations to "pre-set" the frequency response curves to please Vox users. Any tech could change the internal values to turn one version into another. If you own a non-top boost Vox, this is the way you should adjust your amp if its frequency curve doesn't suit you. In 1961 the top boost circuit became an option. This gave you a bass and treble control, and extra gain on the top boost channel. At first, the top boost was added to the back cover panel. Later the circuit would become standard, and was moved to the control panel. Reverb was also an option early on. The first amps to use reverb used a piezo reverb tank. There are no current replacements for this unit. If it dies, a rework to the later reverb design is the only cure.

As time went on, the brown grille cloth was changed to black, and the control panel went from bronze to grey. The AC-30 suffers from overheating problems. There just isn't enough airflow to keep it cool. (Not a problem on the current production Vox amps.) This becomes a problem if you cut up your Vintage Vox to give it some airflow; you'll ruin its vintage value. Without increased airflow, your already well cooked amp will suffer even more.

When the Beatles became popular, Vox developed higher power amplifiers for them to try to compete with thousands of screaming fans. While these fans proved they could make more noise than even the mighty AC100, the result was a line of amps that are powerful even by today's standards. These amps also sounded great as an added bonus!

The first of these amps was the AC50/2, followed later by the AC50/4. The 100 watt lineup started with "The 80100 Watt Amplifier," followed by "The 100 Watt Amplifier." Then came the AC100/2. All of the four output tube models are known now as AC100 amps. The 80 100 watt amp is a 1963 model. That means in 1963, going by true R.M.S. output, Vox was the first amplifier company of note to break the century mark with a guitar amplifier. In the Beatles first movie, "A Hard Day's Night," there's a scene where George Harrison leans against a Vox cabinet, almost falling backwards as it starts to fall over. That's an AC50/2 cabinet, early version, with two Vox "Bulldogs" and a single midax horn.

THE AC50/2

This amp used two EL34 tubes, three ECC83/12AX7/7025 tubes, one 12AU7 tube, and a GZ34/5AR4 rectifier tube. The 12AU7 was selected for the first gain stage of all AC50 and AC100 amps. Often I see a 12AX7 substituted for more gain, but the amp's true voice is the 12AU7. This amp has a normal and a brilliant channel.

The AC50/2 uses fixed- and cathode-bias in combination. One of the few amps to ever combine both types of bias and yet sound good! Because both types of biasing are used, adjusting the bias can be a bit tricky. The factory recommended that the single fixed-bias pot be "preset" to a certain voltage. Since I've seen different recommendations from the factory, and these were for 60s British EL34s, I won't give a voltage to use. Let's just say these amps like to run hot, and should be set up by a person who knows these amps inside out.

A little age dating tip...the very early AC50/2 amps had a little neon bulb inside (Hey, it's a Vox!). Besides the main power fuse, the AC50/2 used a fuse in the center tape of the high voltage winding to ground. This internal fuse is a ¾ amp, also known as 750 mA or .75 amp, fuse. Always use the correct value. In my personal Vox AC50s I used a ½ amp fuse, but it blows before you can fill the room with smoke.

THE AC50/4

Replace the 5AR4 with silicon diodes, give each individual output tube its own bias pot and you get the AC50/4 model. The big trick here is when you adjust the bias pot for one output tube, it changes the current flow in the other output tube. What this means is you have to adjust both bias pots several times to get the output both in balance and at the proper current levels. OK, I admit it, when I owned an AC50/4 I used just one of the bias pots, connected to both output tubes, and used matched tubes. Of course the AC50/4 is a very clean amp, and an unbalanced output by misadjusting the bias was favored by some as a way to get a bit of dirt. Hey! Isn't that what 50 watt Marshalls are for?

THE 80100 WATT AC100

This model is one of my all time favorites for a big clean class "A" cathode bias, with no feedback loop sound. Four EL34s, two 12AU7/ECC82s, and one ECC83/12AX7/7025 tube. A single channel AC30 top boost blown up to DR103 Hiwatt power, before there ever was a Hiwatt! The original JMI Vox sound at its loudest. No, the 80100 would not overdrive into the complex harmonic distortion of an AC30, but if you wanted the Vox sound loud and clean, or needed a Vox tube head for bass, 1963 was a very good year.

Each output tube in this amp had its own separate cathode-bias circuit. This amp also had a problem with fire, as in "Hey my amp's on fire, get me an extinguisher quickly!" If you own one of any of the AC100 Vox models, you'll have this problem. The problem is caused by making the AC100 head cabinet so compact, they actually had to drop the output tube sockets below chassis level so that the output tubes would clear (most of the time and depending on brand) the top of the cabinet when sliding the chassis in or out for service. This ultra-compact head left no room for the high heat of 100 watts of class "A" power to escape and thus the amps often reached their flash point.

If you own an AC100, and plan to use it for more than a conversation piece, I'd suggest you take it out of its original cabinet. You can put the original aside for its vintage thing, and have a custom head cabinet made about 2 inches taller and loaded with ventilation. That will reduce the flameout factor to almost zero.

THE 100 WATT AMPLIFIER AND AC100/2

These two heads are very much the same; the /2 model adding a brimistor to the high voltage supply and the 1 amp H.T. fuse common to all 100 watt Vox amps became a slow blow type. The 100 and AC100/2 used fixed-bias and increased voltage to the first gain stage. A bit louder than the 80100 amp, but a tad less sweet sounding. The tubes were the same: an ECC82/12AU7 first gain stage, a ECC83/12AX7/7025 top boost stage and another EC82/12AU7 as the phase inverter. The four EL34/6CA7 tubes were in fixed-bias with the bias voltage set at 35 volts with Zener diodes. These amps run very hot at 35 volts bias, and I often change the bias to 42 volts, especially with the EL34s being built today. Never use EL34s from China in these amps as they will melt within hours.

THE V125 LGAD AMP

The V125 is not a J.M.I. Vox but is worth talking about. The heads sound pretty good, but their speaker cabinets were awful. The tubes used are:

V1, V2: ECC83/12AX7/7025,

V3: ECC82/12AU7,

V4: ECC81/12AT7,

V5, V6, V7, V8: EL34/6CA7.

This amp also uses a 2N3819 transistor on the input of each of its two channels. The V125 has a tube driven, 5 band rotary control, and "graphic" EQ. The EQ ranges are 100Hz, 250Hz, 500Hz, 1600Hz, and 3200Hz. The V125 also has a sensitivity control for gain, a volume, and a master volume control. It doesn't use the classic top boost circuit of the AC30, AC50, or AC100 but relies on its graphic EQ for tone control. It also uses a very un-Vox-like feedback circuit using very Fender Blackface values. Still these amps don't sell for much and, while not an AC100, will do very well for many guitarists.

To finish up on Vox tube gear, I'd like to cover some later, less well-known Vox tube amps in smaller sizes that also sound great. These are from the Thomas Organ period but don't let that put you off. The amps I will cover are pure Vox and are the last from the "Golden Sound" period of manufacture.

THE PATHFINDER AMP

Think of this model as a Blackface Fender Vibro-Champ, but with a single ended class "A" EL84 output stage. This amp uses four tubes:
V1: EZ80/6V4 Rectifier
V2, V3: ECC83/12AX7/7025
V4: EL84/6BQ5
This single channel model has high and low gain input jacks, volume, bass, and treble controls, tremolo with speed and depth controls with an on/off footswitch and an 8 ohm 8" speaker. I've always liked the Fender Vibro-Champ and the Pathfinder is the Vibro-Champ with a Vox voice. The Fender has a bit more volume, and the Pathfinder has those great EL84 harmonics. If you like one, you'll like the other. The only service problem on the Pathfinder is the can-type electrolytic power filter. It's a 40/30/20 at 350 volts. Kendrick sells a 20/20/20/20 at 475 volts for a Fender Champ that works just fine for the Vox. This requires you to enlarge the hole for the original filter. Use two 20 mfd sections for the Vox 40 section and one 20 section for the Vox 30 section. Use the last 20 section for the Vox 20 section. The lack of 10 mfd in the middle section will make no noticeable difference in the sound of the Vox. The fresh Mallory filter will only make the amp sound better. Like most Vox amps, the Pathfinder does not use negative feedback which gives it a very open tone.

THE PACEMAKER

The Pacemaker model adds another EL84/6BQ5 to the Pathfinder to give it a more powerful 15 watt push-pull output. One section of a 12AX7 is used in the tremolo section of the Pathfinder. This section remains for the tremolo in the Pacemaker and the other section of the tube is removed from tremolo duty to use as a phase inverter. It should be noted that R17 in the factory schematic is shown as a 5.6K when in fact it is a 56K in the amplifier. The schematic is wrong; 56K is the correct value. The Pacemaker also upgrades to an EZ81/6CA4 rectifier.

The Pacemaker has the same power filter as the Pathfinders so the same information of power filter applies to both. The Pacemaker uses a 10" speaker and is an excellent example of a 15 watt amplifier. If you are looking at some of the current 15 watt amps using 2 EL84s you may want to try one of these first, if you can find one.

THE CAMBRIDGE REVERB

The Cambridge Reverb is a seven tube reverb model with a single 10" speaker. The tubes are:
V1: EZ81/6CA4
V4, V5: EL84/6BQ5
V2, V3, V7: ECC83/12AX7/7025
V6: ECC82/12AU7

This is a Pacemaker upgrade with reverb and a "longtail" phase inverter (two tube sections) for more drive to the output stage: it uses a transformer-driven reverb tank with a 12AU7/ECC82 as the driver tube. This is a really fine sounding amp. This model again uses that same power filter common to all amps in this series.

THE BERKELEY II SUPER REVERB

This is the same amp as the Cambridge except that it is equipped with two 10" speakers. It's my personal favorite out of the group for its increased volume and bigger sound. I'd put this one right up there with an 18 watt Marshall combo, and the rare but righteous Selma Truvoice Constellation Twenty Combo.

So there you have it. Any of the Vox models I've covered sound great, and many haven't yet risen out of reach of the average player.

VINTAGE MARSHALL TUNE-UP GUIDE

I'll try to cover the basics of maintaining or restoring a vintage Marshall amplifier to perform to original factory specifications. In some cases I'll include upgrades — both factory recommended or my personal tricks and tips. If you own a vintage Marshall, or just have an interest in these classic instruments of tone, I'd strongly suggest that you obtain a copy of *The History of Marshall* by Michael Doyle. It's not only a history of these amplifiers, but contains all kinds of highly useful technical information, schematics, and specifications on all of the various models. Keep in mind when working on vintage Marshalls, many of the original parts are very difficult or impossible to find. In many cases you will have to make do with the best possible substitute parts. Don't forget that the people at Marshall have always stood behind their products and are a valuable resource for certain parts and information.

The Marshall book I mentioned also lists good sources of parts, tubes, and information, so whatever you do please don't call me! Also, I don't sell parts.

STEP ONE

Determine if your amplifier is stock. This article is about factory spec amplifiers only. If you have a hot-rodded Marshall and you don't want to return it to stock operation, I'd suggest that you talk to your mod tech before trying anything contained herein.

I think that you will find there are very few twenty- or thirty-year-old amplifiers that contain 100% original parts. Most have had a repair, mod, circuit added (such as master volume), and that's to be expected. Also to be expected—parts do wear out. What's not always expected is that certain parts can "wear out" even if the amplifier was never used. If you have an amplifier that has 100% original parts and has never been touched, you might want to consider that in that particular case the amp may be worth more to a collector (working poorly but untouched or even not working and untouched), than the same amplifier properly repaired and set up. We've all heard of the case of a vintage Stratocaster with factory original strings selling for huge sums of money. Of course the guitar can't be played with the original strings, and a fresh set of playable strings would drop the guitar's value by thousands. This also happens with vintage guitar amps. If you do have a 100% original, untouched guitar amp the choice is yours. Collect or play, it's up to you.

ELECTROLYTIC CAPACITORS

Commonly known as filter caps, bias caps, bypass caps, power filters, and so on, these are electrochemical devices and have a limited life span. When I referred to parts that can "wear out" without being used, electrolytics are in that group. They can, "dry out," leak, short out, "open up" (that is, act as if they weren't connected), leak electrically (act like a partial short), fail under voltage, change value and have a host of various problems.

The electrolytics used in guitar amps typically have an expected life of around eight to twelve years. Many twenty- and thirty-year-old amps still contain many original electrolytics, but if these are removed from the circuit and tested for all their original design parameters, they almost always fail to meet their specs. Many people hate to change these parts unless they fail completely because the new part often changes the sound. This is true, but if a quality part is used the amp may regain its original "as new" sound. If economy parts of low quality are used, it's to be expected the sound will change for the worse. Also, old electrolytics may fail at any time, taking with them a valuable part that's no longer in production. I replace any electrolytics in the amps I own at the twenty-year-old mark. I can only tell you what I would do; you can do the same or not.

On any Marshall that I own, I change all the electrolytics on my amp that are twenty years old or older. In 1995, any 1975 or earlier model would get fresh electrolytics from me. My favorite brands are Mallory, Sprague, and Cornell Dubilier. In the late 60s, Marshall started using double section 50 mfd at 500 volt caps. The current brand used by Marshall is LCR. Personally, LCR is not my favorite, but you can no longer get the Erie or Hunts brands, which were my favorite dual 50s in older Marshalls.

Now with a fresh set of electrolytics, we can get on to other important Marshall tune-up items.

GROUND CIRCUITS

Most ground connections in older Marshall amps were mechanical, as opposed to solder connections directly to the chassis. The first Marshalls, just like my amps, used an aluminum chassis which can not be soldered. The grounds get soldered to a terminal which is connected to the chassis with a nut and bolt. This system works fine, but as the amp ages, the grounds may suffer from corrosion or the bolts and nuts may loosen from years of vibration. It would seem that solder connections to a steel chassis would be better, but in fact, I've found more soldered grounds broken by vibration in blackface Fender amps than loose grounds in old Marshalls. In any case, clean and tighten any loose ground terminals in your Marshall amp. On amps that use dual 50 mfd filters, a loose ground on a preamp ground gives you double notes.

On some early metal face 100 watt Marshalls, the ground ran from the power diodes to the front of the chassis. The amps have an A.C. buzz you can't seem to get rid of. The Trainwreck trick is to move the bridge diode ground to the main power filter ground terminal, or just to the back of your amp and the buzz will be gone. Another old Marshall ground problem is that many grounds run to the back of the controls, and the controls complete the ground path to the chassis. On some amps, a wire runs from the ground buss on the back of the controls to a terminal on the chassis. If your amp doesn't have this last setup, install it. As the controls age, they corrode and make a poor ground connection, resulting in hum, buzz, double tones, or other performance problems. A Trainwreck trick is to take a new control "star washer" and solder a wire to it. Replace the star washer on one of the volume controls with the new one and solder the free end of the wire to the ground buss. The amp will still look factory stock but have a solid ground connection.

Also, check the A.C. power ground from your plug to the chassis.

A weak point is the old round style power connector that runs to the chassis on many older Marshalls. The pins often become loose or missing altogether on both the plugs and sockets on these amps. Once the grounds are checked we move on to...

CLEANING THE CONTROLS AND TUBE SOCKETS

Buy a good brand of control and contact cleaner at your electronic supply house. Make sure every electrolytic is completely discharged. Spray a short burst in each control while you rotate its knob end to end. Spray the tube sockets and tighten any loose pin gripping socket terminals. If you spot any cracks or burning on a socket, replace it. On early Marshalls the presence control will always make a scratching noise when turning it with the amp on. This is normal. Any control or socket which does not function well after cleaning needs replacement.

SCREEN RESISTORS

All 100 watt Marshalls and later 50 watt units contained 5 watt 1000 ohm screen resistors. If your JTM-45 or 50 watt Marshall lacks these resistors, install them. This is a factory Marshall service update. Some early metal face 100 watt units used 2 watt 1000 ohm resistors on the screens. If these haven't yet burned open, replace them with 5 watt ratings.

GRID RESISTORS

On early 100 watt Marshalls, I put a 5.6K ½ watt resistor on every output tube control grid terminal and feed the signal in through these resistors. It makes the amp more stable, and "browns" up the sound. This can also be done to a 50 watt unit. Try both 1.5K and 5.6K on these units. Be sure to use the same value on each output tube.

50 WATT BIAS FIX

On some early 70's 50 watt Marshalls, the bias feed was connected to the cold side of the standby switch. When the main power is turned on, no bias voltage builds up as the amp warms up. When you put the standby switch in the play position the tubes get plate voltage for several seconds before the bias builds up. This kills tubes and fuses. If your early 70's 50 doesn't have bias voltage on pin 5 of the output tubes with the main power on and the standby switch in the standby position, move the bias feed wire to the hot A.C. terminal—that is the terminal above (or sometimes below) the terminal it was connected to, on the standby switch.

BIASING THE AMP

Here's where I always get flack! My friends at Marshall and Groove Tubes don't agree that my method is the only correct biasing method. Mitch at Korg, and Aspen at Groove Tubes are my friends, and we help each other out whenever we can. Listen to what they say—respect it—and do it my way!

A true story: Several years back, Marshall held a "Best British Blues Contest"—any one could send in a tape of their best British blues playing as long as it was done with a Marshall amplifier. First prize was a reissue Blues Breaker combo and a Les Paul Standard. The winner, based on his playing and getting the best Marshall British Blues tone, was my friend and very hot guitarist Ken Dubman. I set up the amp he won with. I installed Groove Tubes and biased it my way. Groove Tubes and the guys at Marshall didn't know I had set up the winning amp, but the guys at Marshall picked it as the best. Any questions?!! I wrote about biasing an amp in the August 1994 issue of *Vintage Guitar*. Gerald Weber's book, *A Desktop Reference of Hip Vintage Guitar Amp,* gives exact details. For now, let's just say 30 to 40 mA per tube for EL-34s and 30 to 50 mA per tube for 6550s. Treat 6L6 and 5881 tubes like EL-34s for bias specs. Always adjust your vintage Marshall by current for best tone and consistent performance.

A Trainwreck trick: To use the German Siemans brand in a Marshall with a plate voltage supply of more than 450 volts, but less than 520 volts, reduce the output tube control grid return resistors (also called bias feed resistors) from 220K to 100K each. In most old 100s this keeps the Siemans brand from meltdown.

OUTPUT AND VOLTAGE SELECTORS

Most old Marshalls used two pin plug-in selectors for both voltage and to select 4, 8, or 16 ohms output. These corrode, burn and loosen. These should be cleaned and tightened. Burnt ones should be replaced. The correct voltage selection for the USA is 120 volts, not 110 volts. Running on the 110 volt tap is like running your amp at 135 volts on a Variac and can cause major damage. I know it sounds hot on many amps, but it's looking for trouble.

LEAD DRESS

In an old hand-wired Marshall, the heater leads should be against the chassis. The green grid leads should be up in the air. The purple feedback wires should be kept away from the controls and their leads, except, of course, the presence control to which it's connected. You may move the control and input leads around to obtain minimum noise. Time for another Trainwreck tip: Try doubling the value of feedback resistor in your Marshall. Some amps really open up while others may get sloppy on the bottom. It's worth a try.

ELECTRONIC STUFF

OK, here's a neat little gimmick I came up with back in my days at Ampeg. Originally I called it the "infinity tone control for guitars." These days people call it the Trainwreck "drop out" tone control.

To understand what I've done, first I'll tell you how a tone control works in a typical guitar. The tone control consists of two components. One is a capacitor and the other is a potentiometer. The potentiometer, or pot for short, is a variable resistor. If you put a capacitor of the right value across your pickup, you will short out the high frequencies while the lower frequencies are able to flow from your guitar to your amp. In other words, the capacitor removes the higher frequencies from your signal. To control how much high frequency is removed, a potentiometer is placed in series with the capacitor and this series circuit is placed across, also known as in parallel with, your pickup. As you turn your tone knob, you control resistance of the pot. The lower the resistance, the more high end the capacitor removes from the signal. The trouble is, even up full, the resistance of the pot is still in the circuit and removes a bit of high end.

A classic example of this would be a Strat with a fiveway switch. When you place the selector in the neck and middle pickup 'on' position, both 250K tone pots are in the circuit. The two pots in parallel reduce the total resistance to only 125K. ("K" is electronic shorthand for thousand.) In this position the guitar has its darkest tone. If you switch the selector to bridge pickup only position, the guitar becomes much brighter because all tone controls are out of the circuit at that setting. Now the first idea that comes to mind is to use a higher value pot. A 1000K (1 meg) pot sounds like it would solve the problem because up full its resistance is so high it would have very little effect on the top end. In real life, there's a flaw with this idea. If you used a 1 meg pot, you wouldn't notice any real change in tone until it was turned almost all the way down. This is because you'd have to turn it down until you used up ¾ of its resistance to equal the standard Fender 250K pot, full up. (I'm using Fender as an example, but this would be the same on all guitars that use this type tone circuit, Gibson, Guild, Rick, PRS and so on.) One way around this problem would be to use the desired value of pot with a pull switch to take it out of the circuit. That will work, but who needs to tug at a knob while trying to play guitar? I found my own answer.

THE TRAINWRECK DROP OUT TONE POT

Obtain a pot of the desired value and taper. Make sure it's the kind that can be taken apart by bending up the little metal tabs that hold it together. Also make sure it is the carbon track type which has been the standard for guitars since day one. You will also need an X-acto knife with a No. 11 blade or equivalent. Next, my warning: These knives are very sharp. Be careful taking the control apart, using the knife, putting the control panel back together, and be careful when soldering. On to the task at hand. Carefully bend up the tabs holding the pot together. Remove the back from the pot. On some pots the shaft and copper wiper can be removed from the carbon track. On other pots they remain together. This really doesn't matter, both types can be used. What we are going to do is scrape a bit of the carbon from the carbon track at the high resistance end.

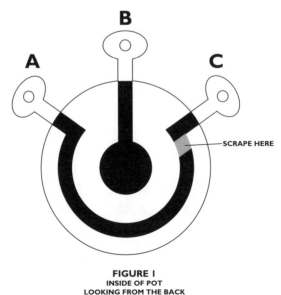

FIGURE I
INSIDE OF POT
LOOKING FROM THE BACK

Tube Amp Talk

Figure 1 shows the carbon track with the shaft and wiper removed. If you have a control that doesn't have a removable shaft and wiper you will still have the same carbon track, but the wiper carrier will partly obscure the carbon track and you will have to work carefully between the wiper carrier and carbon track. You can rotate the shaft to get the best access to the end of the carbon track as shown in Figure 1. Be careful not to position the carbon track wiper over the area we want to scrape. What we want to do is scrape all the carbon from about the last ⅛" of the carbon track using the X-acto knife. You'll want to feather the carbon down from the main part of the track and scrape in a way that will leave a smooth finish. Any roughness here will be felt in the control. Before you put the control back together, use an ohmmeter on a megohm (1000K or greater) range to check that there is no longer a connection between terminals "A" and "C." If you do get a reading, scrape off a little more over the same area. Retest and repeat the above if necessary until you don't get an ohm reading. Reassemble the pot and install it in your guitar just like you would do with a terminal pot.

HOW IT WORKS

When using the pot in a guitar tone control circuit, we only need to use terminals "A" and "B." When used as a volume control we use all the terminals because we use the pot as a "voltage divider." When used in a tone control circuit we do not use the pot as a voltage divider, but as a simple variable resistor. Only two terminals are needed for that use. Point one is that we don't need terminal "C" to be connected to the carbon track because we aren't using it. Point two is that the carbon track is mounted on a nonconductive material. When we scrape the conductive carbon from the end of the track, that spot becomes an insulator, or nonconductive. As you turn the control all the way up, the wiper that normally is always on the conductive carbon will reach the point that we removed the carbon from and will be resting on an insulator. This opens the circuit just like a switch. The real gimmick here is that we've created a tone control off "switch" without adding a "real" switch! And now a word to the guitar companies...I've been using and teaching this trick for over 30 years. Don't go running off to the patent office, it won't stick!

So there you have it. A tone control that works normally until you turn it all the way up. When you hit "10" it automatically removes itself from the circuit and acts as if it wasn't there at all, giving you the last bit of high end possible. In the case of our example using a Strat, all five positions will have the same high end response. You can also put your tone controls at "9," kicking in the carbon track, and the guitar will respond to position switch changes like a stock guitar.

FENDER CHAMP MASTER VOLUME

Many people have called me after reading the Trainwreck pages in *A Desktop Reference of Hip Vintage Guitar Amps*, by Gerald Weber. In that book are four out of more than sixty different master volume circuits I know of. To my surprise, I've been asked several times if I know how to put a master volume control into a Champ. The Champ, or the many similar single-ended small amps, only puts out 3 to 6 watts, one might question the necessity for this control. Maybe I shouldn't be surprised. My own Champ has a Groove Tube #7 6550A output tube, a 10" Celestion speaker, bigger transformers, a hot Marshall style preamp, a presence control, and yes—a master volume! All that is needed is a simple 250K audio taper pot to turn your Champ into a master volume amp. The master volume circuit I'm going to teach you is one of several types that will work in a Champ. This one has the advantage over other types in that when you turn it to "10" ("12" on a Tweed Champ!) the amp is back to its dead stock, nonmaster sound. Before we start I should point out that certain single-ended amps like the Fender Model 5C1 Champ and the Vox AC4 have a single pentode gain stage, no driver or second gain stage, and require a different type of master, in the power stage. I won't cover that here because a power stage master on those amps just doesn't sound very good. All Champs and most other amps that use a 12AX7 preamp tube can use the master that I will describe.

HOW TO INSTALL THE MASTER VOLUME CONTROL

Unplug your amp and discharge the power filters. Locate pin 5 on your output tube socket in the case of a 6V6 tube, and pin 2 on your output tube socket if your amp uses a 6BQ5 or EL84 tube. Connected to this pin will be a resistor, or a wire that runs to a resistor that goes to ground. In a Champ, this resistor is a 220K (red, red, yellow bands). In some old amps it may be a 470K or other value. If it's a 470K, the next most common after the 220K, use a 500K audio taper pot instead of 250K. Connected to the 220K resistor (at the end that goes to pin 5) is a capacitor. Leave the capacitor where it is and remove the 220K resistor. Mount the master volume pot to your chassis. Do not put it in an extra input jack. Electronic feedback will result if you do, making the amp do nasty stuff. Using Figure 1 in our tone control mod as a guide (don't scrape the trace!), remove the wire from the point where the 220K resistor and capacitor had met, and connect it to terminal "B." Leave the other end connected to pin 5 of the output tube socket. (If you have a black or silver face Champ it may have a 330 pf cap between pin 5 and 8. Leave it there.) Connect a wire from the capacitor (at the point where the capacitor and 220K resistor had connected before) to the "C" terminal of the pot. Ground the "A" terminal of the pot and you are done.

HOW IT WORKS

The capacitor feeds the signal to your output tube. The 220K "grid return" resistor provides the bias for the tube by making the grid negative compared to the cathode. It also provides a load that the driver tube, through the capacitor, can supply the signal voltage to. Terminals "A" and "C" of your master volume pot replace the 220K resistor with the 250K resistor which is the pot's carbon element. (The extra 30K makes no difference.) By connecting pin 5 of the output tube to terminal "B" of the master volume pot, you've created an adjustable voltage divider. This lets you adjust the amount of signal that gets to the output tube and therefore controls the power output (volume).

THE UNAUTHORIZED HISTORY OF TRAINWRECK CIRCUITS

FOREWORD

This history will be made up by using both the recollections of Ken Fischer, founder of Trainwreck, and stories told by the voices in his head. Ken is a professional and is guided through the voices by Fireplace (Leona), a Lenape Native American Medicine Woman of the Turtle Tribe. *Do not try this at home.*

THE CHILDHOOD YEARS

SARAH: Ken's mom was a typical housewife during the years he was growing up. He was the first of three children and was born in May of 1945. He has a brother named Scott and a sister named Mona. Their dad worked as both a machinist and a mechanic.

KEN: I remember as a young kid my dad worked for American LaFrance. American LaFrance was to fire engines what Rolls Royce is to automobiles. Part of my dad's job was to drive these fire engines to their new home. Often my dad would take me along on these rides. Back in the 50s a fire engine held great wonder to the boys in Bayonne, New Jersey, where I grew up. Riding on a roaring 12 cylinder monster that was decked out in twenty coats of red lacquer and chromework deep enough to swim in really turned me on to things mechanical. Toys such as Erector Sets didn't hurt this fascination with motors and gears and such to be sure. My best friend Vinny had a fantastic set of Lionel "O" Gauge Electric trains. Here was the mechanical stuff I loved along with electrical wiring.

VG (VOICES OF GIRLS): At age 12 Ken moved to Colonia, NJ with his family. His dad made friends with an electronics engineer from

Lockheed who lived two blocks away. Ken's dad would take him along to this "guru's" house. Pat had radio gear, stereo gear and a Harley in his garage. Pat would give Lenny's boy boxes of electronic parts and instruct him on building all sorts of neat electronic devices to experiment on. At age 15 Ken took on two new interests, guitars and girls! While he was given a guitar and lessons, he wasn't given a girl and lessons. Then he went on to take electronics at the vocational high school. That's where he first met Dennis Kager (Sundown Amplifiers). Years later these two would work together at Ampeg in Linden, New Jersey.

KEN: One summer I got a job at Marcus Transformer company. They were in Avenel, New Jersey along with the R.C.A. Tube Factory. Marcus made transformers for the big electric utility companies. While they made some "small" transformers, they also made power station transformers so large that they had doors and hallways built into them for service and inspection. After Marcus, I took electronics as my vocation and studied the subject for real in New York. Back then, every male had to serve in the military and when my turn came I joined the U.S. Navy.

KLAATU: I came to Earth to stop such nonsense, but Ken was out of the Navy by the time I visited your planet a second time. I found him working in Washington D.C. running a dry cleaning store. I got him a job with Diamond T.V., a sales and repair chain with stores in Virginia, Maryland, and the main shop in D.C. He became a T.V. repairman, on the road in both states and the district. This is hard to believe, but the T.V. repair industry is not always honest. Ken took a very dim view of this fact and soon left the T.V. repair game and Washington D.C. and moved back to New Jersey.

THE MIDDLE PERIOD

KEN: While I worked in Washington D.C., I lived in both Silver Spring and Wheaton, Maryland. Up the road from where I lived was a Yamaha motorcycle dealer. I had a deposit on a Rotary Jet 80. When I made the decision to move back to Jersey, I packed my car, picked up the bike and headed North. That Yamaha was my first motorcycle, but at that point it was just a new toy and I wasn't yet hooked. Back in New Jersey, I moved back into my family home and just hung out and relaxed for a while.

After a while my cash was running low, so I bought a newspaper and looked for a job. Ampeg was hiring assembly line help so I went down to Linden and got that job. After a couple of weeks the people on the assembly line told the higher ups, "Hey, this guy really knows electronics!" I was called into the front office and questioned, and next thing I knew I was the repair tech for the final test room. Later, I was inducted into the engineering department. I worked there a few years but quit on the very day Ampeg was sold and no longer under control of musicians.

Also, by this time I was hooked on motorcycles. While working at Ampeg, I had saved the money I needed to buy a bright red and chrome BSA Lightning Rocket. Everyone wanted Triumph Bonnevilles back then, but the BSA had the American LaFrance look of my childhood. As fate would have it, my car died, so the money was split between buying a 1963 Ford and a Honda Superhawk. The Superhawk wasn't all that super, but I was hooked on motorcycles.

VICTOR FRANKENSTEIN: Motorcycles turned Ken into a real monster. He was in love with the freedom one feels when riding a bike. Unlike me and several of his friends, he never had a mishap on his machines. Unlike myself, he never had to endure being stitched back together. Ken decided to use his skills and became a motorcycle mechanic. After all, every time you tune-up or fix a bike, you take it for a test ride! What more could one ask for? He rode with guys who all had riding names. Ken's riding name was "Trainwreck."

Through the 70s motorcycles were everywhere and he was kept busy. By the 80s, bikes had all but vanished from the scene and times were hard. Ken always did repairs and mods to his friends amplifiers and with some prodding from Steve "Whammy" Hayes, a local guitar builder and repairman, placed an ad in a local musician's newspaper offering his repair and modification services.

TRAINWRECK CIRCUITS

KEN: I started Trainwreck Circuits in 1981. By the time a year had passed, I was doing too much business to remain underground. On March 12, 1982, I registered the Trainwreck Circuits name at the county office for such things. I started a Trainwreck business account at the bank, and then I got my state sales tax number. By the end of

1982, word of mouth about my repairs and mods was so strong, I stopped all advertising. At Ampeg, I had developed my own troubleshooting system. This system works so well, that on average, it takes longer to remove and replace the chassis from a combo or head than it takes to find the problem. With that edge, anyone who wanted to, could wait and watch while I did their amp's repairs. Of course, if I had to order a part like a transformer, the amp would stay until the part came in. This, "fix me while you wait" policy had people driving to me from hundreds of miles away, and I soon set up appointments to handle this work.

I also was doing modifications, but had a real disdain for the typical universal mods being done at the time. I would work to give each amp the best voice it was capable of. That generally means having to engineer a different custom mod for each individual amp. Along about this time I got a call from Dean Farley. At the time he was a Groove Tubes salesman. Dean and Aspen (owner of Groove Tubes) set me up with a GT dealership. That brought in even more amp repairs and mods and I soon became a "tube amp only" shop. From that point, I stopped all transistor amp, PA, and keyboard repair. Groove Tubes were the standard tubes in Trainwreck amps for years. When Aspen switched to the EL-34 from China, I started matching my own tubes. There hasn't been a tube from China to this day that I've ever liked. Groove Tubes always had good EL-84s, and I'd recommend them for any EL-84 equipped amp. Grade #4 for a 'Wreck, and grade #6 for a J.M.I. Vox.

In 1982, I ran across some NOS (new old stock) Ampeg 2x12 cabinets. They had no logos, so my friend Rich Levitch designed and made me the first Trainwreck logos. It's this logo that I've always used on my amps. When Rich designed the logo, I wasn't even thinking of building amps yet. They were made to put on "Trainwreck" Ampeg cabinets! It would take a new friend to persuade me to build him an amp of my own design.

CASPER MCCLOUD: I came over from England to play John Lennon in the original U.S. production of "Beatlemania" at the Winter Garden. Marshall Crenshaw and I alternated, one or the other, playing John in either the early or late show each day. I left my J.M.I. AC-30s over in England, but did bring a J.M.I. AC-50 head. After "Beatlemania," I landed a deal with Atlantic Records. Janis Roeg, who worked for Atlantic, was best friends with Ken's sister Mona, and she arranged for me to visit him at his shop. Ken and I became friends. I told Kenny that I needed a high gain amp for my new album.

Atlantic had purchased a new Boogie for me and whilst it had the gain, the tone wasn't the same as my AC-30s which were still back in England. I finished the album using the Boogie, but I worked with Ken on developing a high gain class "A" EL-84 amp for my personal use. In late 1982, he had a 15 watt prototype built on a gutted Fender chassis. By the first month of 1983, the 30 watt version, built on his own chassis was complete. He named this amp "Ginger" after my wife.

When other players heard my amp, they wanted him to build them one too. I wanted it to be called "The Kenny Amp," but he chose the name "Liverpool 30". He picked this name because the amplifier sounded like a British amp. He didn't foresee that people would connect Liverpool with the Beatles, and of course the Beatles with Vox. For years people thought the Liverpool was a Vox copy, when in fact it is an entirely unique amp completely of Ken's design.

KEN: After I built the first Liverpool, I got many requests from my friends to build them one too. The money from repairs and mods was far greater than the profit on a 'Pool, so I only built them in my spare time. My brother's second ex-wife, Debi, came up with the woodburned faceplates. She's also the one who suggested using hardwood cabinets for the heads. She liked the looks of the hardwood Boogies that I worked on, and hated the looks of all the "ugly black boxes," like Fender, Vox, Marshall and other classics! I decided to give them names instead of serial numbers to keep it on a personal level. In fact, many of my best friends I've met through selling them a Trainwreck Amp.

LES KRYGIER: I saw Ken's ad in a local musician's paper and had heard about his reputation for hot-rodding Marshall Amplifiers, by word of mouth. I always had at least six Marshalls—a Super Lead, Model 1959, 100 watt heads and scores of cabinets on hand. The sound I was going for at the time was the tone on the first Van Halen album.

Ken dialed that one right in, but said he thought he could do better on the solo tones if he modded an amp just for lead work. With six heads I had nothing to lose, and within a few weeks he had developed a one channel, one input preamp and phase inverter for one of my heads. The modified amp had more gain on two than it did stock on ten. It was very thick and very loud, and harmonics literally jumped out of it.

When my band started playing, the Marshalls were too loud. Ken converted them to 6V6 output tubes run on a Variac at 95 volts. This was nice, but I really liked the EL-34 sound. I had played his Liverpool amp and really liked it, but still wanted EL-34s. Working with Ken, he rethought the Liverpool into an class AB amp with EL-34 power. By the first month of 1984, the Express amp was born. I bought two and sold all my Marshall heads except one that had been modded. Later I also bought a Liverpool head and a Strat with single coils.

KEN: When repairing and doing mod work on guitar and bass amps, you soon find out who stands behind their products with easy-to-obtain parts and technical support. I wish I could list all the companies from best to worst, but the worst are just so awful I couldn't tell you about them without being hit with a lawsuit. I can say I never had a parts or technical support problem with Mesa Boogie or Carvin. That's not to say there may have been other companies like those two out there. I worked on several brands that never needed factory parts or technical support. The point is that by the mid-1980s, many companies started saying to me, "We can no longer sell you parts unless you become an authorized warranty service center." Imagine a company that won't sell your favorite tech the parts needed to fix your amp (even after it's out of warranty). Shows how much they care about you once they've made their sale! Sounds like a sad tale, but because they were giving me such a hard time about obtaining their parts, I stopped doing repairs and mods and went to building Trainwreck Amps full time. It looks like I even owe the "Bad Guys" some thanks!

ABOUT MY HEALTH

In 1988 I caught a bad case of the flu. This left me with chronic fatigue syndrome and a vestibular (balance) disorder. In July 1994 I added bleeding ulcers to this list. Kill the rumors, not me; I don't have AIDS or a brain tumor! The CFIDS and vestibular disorders continue. I build amps when I feel up to it, and don't build if I'm not up to it. Who was that doctor who cured Leo?!

Derek Jan: Originally, I'd go to Trainwreck to get my Boogie serviced. The Boogie was a good amp, but I was interested in controlling my sound from the guitar. On a large stage, I didn't want to have to run back to a footswitch to go from clean to dirty.

Ken: Boogie now makes amps you can control from your guitar.

Derek: I bought a Trainwreck Express first. If you ever listened to T.V., radio, or MTV, you've heard my Express! It was used on many major commercials, and I rented it to major rock artists who used it on their hit recordings. It was also a *Guitar Player* magazine product review amp. After a while, I also picked up a Liverpool 30 and really got into the Class "A" EL-34 sound.

Then one day in 1990, I was visiting Ken and saw a Trainwreck amp with four knobs instead of the normal five. When I asked him about it he said, "This one's not for production, I built it for myself." I asked him to please build just one more for my personal use, and Ken agreed to do so. I'm always in major studios and on international tours, and everywhere I went people wanted a "Rocket" like mine. Blame me for starting the demand for the Rocket. It's become my favorite 'Wreck.

Ken: There are lots of people who helped Trainwreck on the way to becoming what it is today, too many to list here. I'd like to thank all the people who have helped me to build the Trainwreck name into what it has become. I'd especially like to thank the people who use Trainwreck Amplifiers as part of their musical creativity. Finally you might ask why I did my own history and interview? *Someone had to do it*!

LAGNIAPPE

INTERACTIVE VOLUME CONTROLS CAN WORK FOR YOU

Interactive volume controls, characteristic to many two channel vintage tube guitar amplifiers, affect your tone to a noticeable degree. Since we already know that they make a difference, the next question is: How can we make this difference work to our advantage?

WHAT'S AN INTERACTIVE VOLUME CONTROL?

Interactive volume controls occur on two channel amplifiers whose channels are coupled by the volume controls. When using an amplifier with interactive volume controls, the volume control of the channel you are not using affects the tone of the channel you are using. Some examples that come to mind are the tweed Deluxe, virtually all early Marshalls, tweed Twin, Gibson BA-15 RV, Gibson GA-77, and the tweed Bassman.

If you would like to test the volume controls on an amp to see if they are interactive, simply plug into one channel and put that channel's volume a third of the way up. Now rotate the volume control on the other channel and listen for a tonal difference. If there is a tonal difference then, by definition, the volume controls under test are interactive. If there is no change in tone, you do not have interactive volume controls.

WHAT CAUSES IT?

When a volume control is in a circuit, it will occur after a coupling capacitor that isolates the low voltage A.C. (signal) from the high voltage D.C. (power supply voltage). If there are two channels, and they are coupled together, some of the A.C. signal from one channel will take another path to ground. Basically what happens is this: Signal comes out of a volume pot and goes backwards through the adjoining volume pot and backwards through the coupling cap, plate resistor, and the filter cap where it is virtually shorted to ground. This is very similar to a tone control circuit in so much as frequency response is altered as the other channel's volume control is turned.

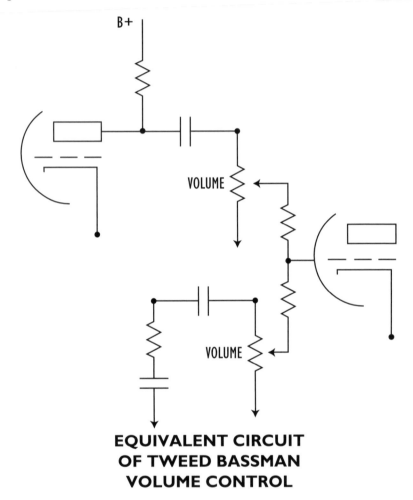

EQUIVALENT CIRCUIT
OF TWEED BASSMAN
VOLUME CONTROL

Take a specific amplifier, let's say the tweed Bassman for instance. As the signal comes out of the first preamp tube, it goes through the .02 coupling cap and to the top of the volume pot. The volume pot is configured as a simple voltage divider whose wiper selects how much voltage. From the wiper, the signal goes through a second voltage divider consisting of two 270K ohm resistors. The junction of these two resistors feeds the next stage. The last of these two resistors does not go directly to ground, but to the wiper of the other volume pot. If this pot is turned all the way down, then the end of the second 270K ohm resistor is at ground potential. If the other pot is not turned off, the resistance between its wiper and ground is added to the second 270K ohm's value resulting in a higher load resistance. This actually increases gain, but wait! The amount of resistance from the second pot's wiper to ground actually has another circuit in parallel with it. From the junction of the second pot's wiper and the last 270K ohm resistor, signal will back up through the top of the pot, through the coupling cap, through the plate resistor, the filter cap and finally to ground. As the second pot is turned up, more high-frequency signal goes the alternative route as opposed to going through the pot directly to ground. While this is occurring, overall gain increase because of the higher load resistance. (Note: The higher load resistance increases the actual gain because the A.C. signal voltage is divided between the 270K ohm resistor and the other 270K ohm resistor plus resistance of the other pot's wiper to ground. Since we are talking about a ratio here, the signal drops less as the other pot's resistance goes higher. We hear that as more gain.) This has the effect of seeming to boost mids. After a certain point, turning the other volume any higher doesn't increase the load resistance enough to add any appreciable gain, and more high-end is rolled off.

Let's look at another example, the tweed Deluxe. This amp is very unique because as the volume controls are not voltage dividers!! Did you ever wonder why all tweed Deluxes seem to have an audio pot with too fast a taper? The volume control in a tweed Deluxe works by "loading down" the signal coming from the plates of the preamp tubes!

If you are plugged into the instrument channel, the signal goes backwards through the microphone channel's volume pot, through the microphone channel's coupling cap, plate resistor, and filter cap to ground. You will get maximum mids in the instrument channel with the microphone channel's volume control turned about half way up. You will get a maximum midrange scoop with the microphone channel's volume control turned full up. This would work the same if you switched channels and were plugged into the microphone channel and adjusting the instrument channel's volume.

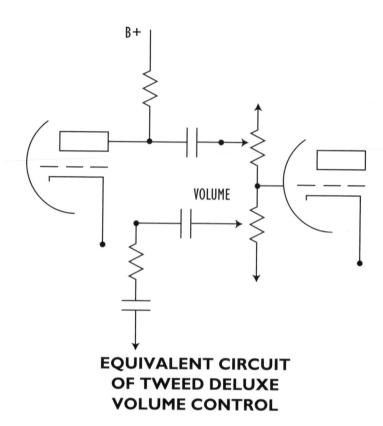

EQUIVALENT CIRCUIT
OF TWEED DELUXE
VOLUME CONTROL

This can be used to your advantage, especially if you have an A/B box. Set the normal channel all the way up and the instrument channel half way up. Use an A/B box to select between microphone and instrument channels.

Here's what will happen: When you select the instrument channel, you will get a fabulous clean tone. Since the instrument volume is turned half way, you are not really overdriving the instrument channel that hard and since the microphone's volume control is turned full up, you are scooping out the mids in the instrument channel. This gives you that "better than blackface Twin" clean tone—to die for.

When you select the microphone channel (which is turned all the way up), you will get a fabulous lead tone. For one thing, you will be overdriving the output stage and because the instrument channel is half way up, you will be boosting mids as much as possible. This results in a thick, creamy, cello-like tone with incredible sustain.

Some readers are probably familiar with Neil Young's stage rig.

There was an article in *Guitar Player* magazine a couple of years ago that explained his stage rig in detail. The heart of his rig is a 1958 tweed Deluxe with servomotors controlling the setting of the volume control on the other channel—the one he is not using, in order to take full advantage of the interactive volume controls.

Of course, there are many other settings between the two extremes described earlier that will give very usable tones. This is where experimentation with your guitar, your playing style and listening will reveal the possibilities.

Tube Amp Talk

GIGGING AMP TONE FOR UNDER A GRAND

So you want to have that cool vintage tube tone, but don't have an extra three grand to get that '59 Twin or tweed Bassman. You need something loud enough for gigging. How do you get the cool tube tone at a price you can afford?

Don't despair. There are still some pretty good buys in used tube amps that can give you a very good tone for under a grand.

THE BOGEN TUBE P.A. HEAD

Don't laugh. Bogen Tube P.A. amplifiers are available in many wattages, the 100 watt and the 50 watt being the most desirable. These amps can be found in pawn shops for a $100 or so. I had one shop owner sell me one for $20 and another for $35. They are wired point-to-point, have paper bobbin transformers (multi-impedance) and use 6L6GCs or 5881s. All used classic Western Electric circuits! Some of the early ones used grid leak bias for the first gain tube. If you are using a high level input signal, the grid leak bias can be easily changed by any competent tech for a few dollars. You might have to buy a ¼" jack and replace the microphone input connector to plug in your guitar cord directly, but you may find the tone very killer. Terry Kath from the band "Chicago" used a 100 watt Bogen P.A. through a two 15" Showman cabinet, and got a great tube tone. He paid $75 for his. Eventually he used two of these with two cabinets and at that time he could have afforded anything he desired. Of course you will need a speaker cabinet, but you will spend less than a grand overall, have great tube tone—with money left over for new tubes and a cap job. An old Bogen will probably need a little service to get tip-top performance.

Make sure you are getting the tube version, because the solidstate version sounds awful. Avoid the models with less than 50 watts, because they will not be loud enough to gig with. Most Bogen P.A.s actually have a master volume control and two or more channels.

SILVERFACE: PRO, BANDMASTER, SHOWMAN, BASSMAN, SUPER REVERB

Since silverface Fenders were the least desirable Fender amps in the late 60s and early 70s, they can be had for considerably less money than the blackface counterparts. A silver to blackface modification can be performed (around $75) and *voila!* A cool-sounding, giggable amp for under a grand! This mod is described in detail in my first book *A Desktop Reference of Hip Vintage Guitar Amps.* I have actually heard some of these silverface amps, converted to blackface, that sounded as good as or better than the blackface (after the blackface mod was performed).

Avoid amps with rust on the transformers. If needed, the rest of the amp can be overhauled for a couple of hundred dollars. These amps are all handwired, point-to-point, with no printed circuit boards. Most early ones have paper bobbin transformers. Avoid those from the late 70s or later because extensive mods may be needed to get a good tone. Avoid, like the plague, any that have a bias balance or hum balance control on the rear panel. These are inherently hard sounding and you probably will be disappointed in these. Try to stick with late 60s to early 70s for best results.

FOUR OUTPUT TUBE SUPRO COMBO AMP

I've seen these at shows for $150 to $350. Hendrix used one before he used the Marshalls, and of course he did pretty well with it (his was the two output tube version.) They have two 12" speakers and they come in a couple of different guitar models and a Bass model. The Bass model has more umph. Avoid the two output tube models because some of these are not all that loud.

Supro amps were very well-made amps—meticulously handwired and laid out perfectly. Many had dual chassis to isolate the power supply from the preamp (reduces hum by keeping A.C. away from the early gain stages). Paper bobbin transformers were standard and all components were over-designed. Cabinets were solid pine. Any amp built this well, in our modern times, would definitely sell for big bucks.

FOUR 6V6 TRAYNOR HEAD

Traynor made a tube head that was an exact '59 Bassman circuit but with four 6V6s. When used with the right speakers, it had great tone and dynamic response, with very smooth break-up. A guy came in my shop the other day with one that he had paid only $75 at a local pawn shop. It needed a cap job and tubes, but it was entirely handwired, laid out correctly and had some really good sounding transformers. It probably put out about 50 watts.

Traynor amps were never very well known, but they are unsung heroes. Almost all used classic Western Electric circuitry, point-to-point wiring and decent layout. They also built an amp that was a '59 Bassman circuit but with EL34s. Sound familiar? Jim Marshall would have thought so.

TUBE RECTIFIERS

Most of us already know that virtually all vintage amplifiers use tube rectifiers, but let's dig a little deeper and discover what makes one type of rectifier different from another. How does all this affect tone? Which rectifiers could be direct replacements for other types with a resulting difference in tone?

Let's start with a basic concept of what a rectifier does. All tube amplifiers need a high voltage direct current (D.C.) power source in order to operate the power tubes and preamp tubes. Your wall electricity is 120 volt alternating current (A.C.). The amplifier has a power supply that changes 120 volt A.C. to several hundred volts D.C. in order to handle this requirement. A power transformer (this is a step-up transformer) increases the voltage, and the rectifier tube changes it from A.C. (which alternates back and forth) to D.C. (which all goes in the same direction.) Anytime alternating current is changed to direct current, it is said to have been "rectified," hence the name "rectifier tube."

HOW DOES IT WORK?

Basically there are two different styles of rectifier tubes used in guitar amps—directly heated or indirectly heated. In a directly heated rectifier tube there are two plates and a filament heater, inside a glass vacuum tube. When the amp is turned on, the filament heater is heated up inside the tube. This heat excites the electrons on it. (The heater is the part of the tube that lights up and typically works on very low voltage and high current—typically 5 volts at 2 or 3 amps of current.) Each of the two plates are connected to opposite ends of a high voltage step-up transformer. (700 volt range is typical, 350 volts positive and 350 volts negative.) One of these plates will be positive while the other is negative. (These polarities of positive and negative will change back and forth 60 times per second because that's how fast the wall current alternates.) Excited electrons from the heater are attracted to whichever plate happens to be positive at the time. (One or the other will always be positive at any given moment; therefore, current will flow at all times from heater to one plate.) Popular types of directly heated rectifier tubes are: 5AS4, 5AU4, 5AW4, 5AX4, 5AZ4, 5R4, 5T4, 5U4, 5V3, 5W4, 5Y3, 5Z3, 5X4, 5Y4, 80, 82, 83 and 5931. Because the filament heater is itself acting as a type of cathode, this type of tube is called a directly heated cathode-type rectifier (even though it has no separate cathode.)

To summarize: The current enters the tube through the heater (this heater acts as a cathode) and leaves through one of the plates, whichever one happens to be positive at the time.

Indirectly heated cathode-type rectifiers work exactly the same way, except there is another element inside called a cathode. The heater is situated next to the cathode and heats it up. The electrons on the cathode get excited as they are heated and they are attracted to one of the plates, depending on which one happens to be positive at the time. Why don't electrons from the heater travel to the plate? Because the cathode is in between the heater and the plates, it actually hides the heater. The cathode does not produce its own heat but gets it from the heater and is therefore said to be indirectly heated. In the indirectly heated cathode rectifier, having a separate cathode allows for closer proximity to the plates; therefore, the electrons do not have to travel as far to reach the plates. Some examples of indirectly heated cathode type rectifiers would include 5AR4/GZ34, 5AT4, 5CG4, 5V4, 83V, 5Z4, 6087, 6106, 84/6Z4, 35Z5, 45Z5, 35W4.

To restate and summarize: The current enters the tube at the cathode and leaves through one of the plates (whichever one is positive at the time).

WHAT AFFECTS TONE?

There is a natural type of compression that occurs with a rectifier tube that is related to its internal resistance. The indirectly heated type rectifiers have less resistance as a rule since the distance from the cathode to the plate can be kept very small and the electrons just don't have as far to go. Also the cathode surface area itself can be made larger, thus reducing the resistance of the tube. Directly heated types generally have more resistance, because the distance from heater to plate is much greater than in indirectly heated types, and the surface area of the heater is not that large.

HOW IS TONE AFFECTED?

All electricity that is used for the amplifier must pass through the rectifier tube (except bias voltage and filament heater voltage). The resistance of the rectifier acts as a bottleneck when a great deal of current tries to get through. Before any signal, the high voltage on the power tubes is as high as it gets. Now you play a note. Current flows through the power tubes to the rectifier. The internal resistance of the rectifier causes a voltage drop across it. This voltage drop is subtracted from the voltage that was originally on the power tubes. So now the voltage on the power tubes is momentarily less. As voltage on the the power tubes drop, the output level drops, the tone becomes browner (less high-end definition). All this was happening as the string was first played, but as the string begins to decay, there is less actual signal. With less signal, the current going to the rectifier is less and therefore less voltage is dropped across the rectifier. Lesser voltage drop means more plate voltage on the power tubes which also means higher output and more definition.

To restate and summarize: The rectifier tube's internal resistance causes the overall volume of the amp to decrease as the guitar signal is pushed harder and causes the overall amp volume to increase as the notes begin to decay. This cause an envelope of attack, decay, sustain which is a type of natural tube compression. This is sometimes referred to as sag or envelope. This contributes to an amp's ability to breathe.

The rectifier tube is not the only component that affects sag. Sag is caused by power supply resistance. Although the resistance in the rectifier tube makes most of the difference, all other power supply resistances (such as the resistance of the internal power transformer winding) do affect sag and natural compression.

WHAT'S THE DIFFERENCE FROM ONE RECTIFIER TO ANOTHER?

With rectifiers that have greater internal resistances, this sag is enhanced (all other factors being equal). Operating volume will be slightly less and dynamic response will not be as quick and punchy.

Other than directly or indirectly heated cathodes and internal resistances, there are seven other differences from one rectifier to another. These differences include: pin out, heater filament voltage, heater filament current, maximum plate current (D.C. output), maximum peak inverse voltage, RMS supply voltage, and maximum peak current per plate.

We have covered the basic concept of what a rectifier actually does, how the tube rectifier affects tone, and the two styles of cathode heating. Now we will look at the other seven differences from one rectifier to another, talk about differences in tone, and look at a chart with popular rectifier tubes and their characteristics.

The first and probably most obvious difference is the pin configuration. Most common rectifier tubes for guitar amps use the familiar 5DA, 5L or 5T base connections which all are interchangeable. Pins #4 and #6 are connected to the high voltage winding on the power transformer and pins #2 and #8 are connected to the 5 volt filament heater; the cathode is internally connected to pin #8 (5DA, 5L) or it is directly heated and pin #8 is used as the cathode connection (5T). However, there are other pin configurations which are not interchangeable.

Next is the heater filament voltage and current. Almost all guitar rectifiers are 5 volt or 6.3 volt. The first number of the tube type will tell the voltage requirement. (Example: 5AR4 would require 5 volts or a 6X4 would require 6.3 volts.)

Heater current is different for each rectifier depending on type. Check the chart for exact values. The current we are talking about here is how much current the filament heater draws. If you have a tube that normally draws 1 amp, it is not advisable to replace it with a different type that draws 3 amps because the transformer will overheat and the voltage will drop considerably.

On the other hand, many companies over-design their transformers and the heater winding may be wound for two or three times the current of the stock rectifier. In this case, changing to a different type rectifier that draws twice the current may be no problem.

How much potential difference between the negative voltage swing and the positive cathode voltage before arcing occurs is called the peak inverse voltage. The peak inverse voltage, or PIV as it is sometimes called, should be at least triple your actual rectifier output voltage. For instance, if your rectifier output voltage is 450 volts it is best to use a tube with a peak inverse voltage of 1350 volts or better.

Maximum RMS supply voltage is the amount of supply voltage that can be placed on the plates without causing a malfunction. Maximum supply voltage on most 5 volt rectifier tubes is well within the parameters of most guitar amp power transformers.

Maximum peak current is the amount of peak current that can pass through the tube without causing a malfunction. Again, although this is usually not a problem, it is something to be considered if you are trying a different type of rectifier than stock. For instance, if you are using a 5Y3 in a 5AR4 circuit, the circuit may draw so much current (during peaks) that the 5Y3 arcs and becomes history. Have mercy!

Maximum plate current is the amount of current that can pass through the plate on a regular basis without causing a malfunction. Like maximum peak current, this is something to consider when using other than stock rectifier types.

Let's talk tone. The attack/decay envelope, compression, sustain, low-end-definition (or lack of it), volume, and punch are all affected by choice of rectifier tube. Certain rectifiers will give more or less voltage than other rectifiers. The ones that give the higher voltages (these are the ones with the least drop on the chart) will make the amp have more headroom, volume, and tighter bottom. Conversely, the ones that give lower voltages will sound creamier with smoother top end and faster breakup. You can check the output voltage of the rectifier and see what voltage it is putting out. In the same exact circuit, different rectifier tubes will produce different results.

For instance, if you have an amp that is too tight and doesn't break up fast enough, perhaps changing the rectifier could be just the ticket to loosen things up a bit. On the other hand, you could have an amp that just doesn't have enough clarity in the bottom-end. A more powerful rectifier tube could add some definition to that bottom-end when you play that open E string.

The circuit that the rectifier operates within makes a huge difference in how the tube sags (envelope). For instance, if a tube that is rated for 125 mA max plate current is placed in a circuit that idles at 100 mA, the amp will have a spongy envelope and considerable compression and sag, especially when the amp is being driven hard.

It will not have very much punch. This may be very desirable for recording and yet a punchier tube with more headroom may be more suitable for playing live gigs. There is no one best rectifier. To sum it up, a rectifier will sound slightly different from one amp to another because it may be asked to work a little harder due to circuit variations. There is no substitute for a trial and error listening test. Most of the 5 volt rectifier tubes are interchangeable.

Tube Type	Classification by Construction	Base Connections	Outline Dwg	Filament Volts	Max Plate Watts		Other Characteristics
5AR4	Full-Wave High-Vacuum Rectifier	5DA	T-X	5.0	1.9	—	Max D.C. output current=250 mA; max peak inverse voltage=1,500 volts; rms supply voltage per plate=450; max peak current per plate=750mA
5AS4-A	Full-Wave High-Vacuum Rectifier	5T	12-15	5.0	3.0	—	Max D.C. output current=275 mA; max peak inverse voltage=1,550 volts; max rms supply voltage per plate=450; max peak current per plate=1,000 mA. Tube Voltage Drop: ♣ 50 volts at 275 mA D.C.
5AV4	Full-Wave High-Vacuum Rectifier	5T	T-X	5.0	3.7	—	Max D.C. output current=250 mA; max peak inverse voltage=1,550 volts; rms supply voltage per plate=450; max peak current per plate=750 mA. Tube Voltage Drop: ♣ 46 volts at 250 mA D.C.
5AX4-GT	Full-Wave High-Vacuum Rectifier	5T	9-13	5.0	2.5	—	Max D.C. output current=175 mA; max peak inverse voltage=1,400 volts; rms supply voltage per plate=350; max peak current per plate=525 mA. Tube Voltage Drop: ♣ 65 volts at 175 mA D.C.
5AZ3 ■	Full-Wave High-Vacuum Rectifier	12BR	12-62	5.0	3.0	—	Max D.C. output current ✪ =275 mA; max peak inverse voltage ✪ =1,700 volts; rms supply voltage per plate ✝ =600; max peak current per plate ✪ =1,000 mA. Tube Voltage Drop: ♣ 44 volts at 225 mA D.C.
5AZ4	Full-Wave High-Vacuum Rectifier	5T	9-31	5.0	2.0	—	Max D.C. output current =125 mA; max peak inverse voltage =1,400 volts; rms supply voltage per plate =350; max peak current per plate =375 mA. Tube Voltage Drop: ♣ 60 volts at 125 mA D.C.
5R4-G 5R4-GY	Full-Wave High-Vacuum Rectifier	5T	16-3 16-3	5.0	2.0	—	Max D.C. output current =250 mA; max peak inverse voltage =2,800 volts; rms supply voltage per plate =750; max peak current per plate =650 mA. Tube Voltage Drop: ♣ 67 volts at 250 mA D.C.
5R4-GYA	Full-Wave High-Vacuum Rectifier	5T	T-X	5.0	2.0	—	Max D.C. output current =250 mA; max peak inverse voltage =2,800 volts; rms supply voltage per plate =750; max peak current per plate =650 mA. Tube Voltage Drop: ♣ 50 volts at 275 mA D.C.

■ Compactron   ♣ Per section   ✪ Design maximum rating.

Tube Type	Classification by Construction	Base Connections	Outline Dwg	Filament Volts	Max Plate Watts		Tube Voltage Drop	Other Characteristics
5R4-GYB	Full-Wave High-Vacuum Rectifier	5T	12-15	5.0	2.0	—	♠ Tube Voltage Drop: 63 volts at 250 mA D.C.	Max D.C. output current ✪ = 250 mA; max peak inverse voltage ✪ =3,100 volts; rms supply voltage per plate ✪ =900; max peak current per plate ✪ =715 mA
5T4	Full-Wave High-Vacuum Rectifier	5T	10-1	5.0	2.0	—	♠ Tube Voltage Drop: 45 volts at 225 mA D.C.	Max D.C. output current=225 mA; max peak inverse voltage=1,550 volts; max rms supply voltage per plate=450; max peak current per plate=675 mA
5U4-G	Full-Wave High-Vacuum Rectifier	5T	16-3	5.0	3.0	—	♠ Tube Voltage Drop: 44 volts at 225 mA D.C.	Max D.C. output current=225 mA; max peak inverse voltage=1,550 volts; rms supply voltage per plate=450; max peak current per plate=800 mA
5U4-GA	Full-Wave High-Vacuum Rectifier	5T	T-X	5.0	3.0	—	♠ Tube Voltage Drop: 44 volts at 225 mA D.C.	Max D.C. output current=250 mA; max peak inverse voltage=1,550 volts; rms supply voltage per plate=450; max peak current per plate=900 mA
5U4-GB	Full-Wave High-Vacuum Rectifier	5T	12-16	5.0	3.0	—	♠ Tube Voltage Drop: 50 volts at 275 mA D.C.	Max D.C. output current=275 mA; max peak inverse voltage=1,550 volts; rms supply voltage per plate=450; max peak current per plate=1000 mA
5V3	Full-Wave High-Vacuum Rectifier	5T	12-16	5.0	3.8	—	♠ Tube Voltage Drop: 47 volts at 350 mA D.C.	Max D.C. output current =350 mA; max peak inverse voltage =1,400 volts; rms supply voltage per plate =425; max peak current per plate =1,200 mA
5V3-A	Full-Wave High-Vacuum Rectifier	5T	12-16	5.0	3.0	—	♠ Tube Voltage Drop: 44 volts at 350 mA D.C.	Max D.C. output current ✪= 415 mA; max peak inverse voltage ✪ =1,550 volts; rms supply voltage per plate ✪ =550; max peak current per plate =1,400 mA
5V4-G 5V4-GA	Full-Wave High-Vacuum Rectifier	5L	14-3 12-14	5.0	2.0	—	♠ Tube Voltage Drop: 25 volts at 175 mA D.C.	Max D.C. output current =175 mA; max peak inverse voltage =1,400 volts; rms supply voltage per plate =375; max peak current per plate =525 mA

■ Compactron   ♠ Per section   ✪ Design maximum rating.

Tube Type	Classification by Construction	Base Connections	Outline Dwg	Filament Volts	Max Plate Watts		Other Characteristics	
5W4 5W4-GT	Full-Wave High-Vacuum Rectifier	5T	8-6 9-13	5.0	1.5	—	Tube Voltage Drop: ♠ 45 volts at 100 mA D.C.	Max D.C. output current=100 mA; max peak inverse voltage=1,400 volts; max rms supply voltage per plate=350; max peak current per plate=300 mA
5X4-G	Full-Wave High-Vacuum Rectifier	5Q	16-3	5.0	3.0	—	Tube Voltage Drop: ♠ 58 volts at 225 mA D.C.	Max D.C. output current=225 mA; max peak inverse voltage=1,550 volts; max rms supply voltage per plate=450; max peak current per plate=675 mA
5X4-GA	Full-Wave High-Vacuum Rectifier	5Q	12-16	5.0	3.0	—	Tube Voltage Drop: ♠ 47 volts at 250 mA D.C.	Max D.C. output current=250 mA; max peak inverse voltage=1,550 volts; rms supply voltage per plate=450; max peak current per plate=900 mA
5Y3-G	Full-Wave High-Vacuum Rectifier	5T	14-3	5.0	2.0	—	Tube Voltage Drop: ♠ 60 volts at 125 mA D.C.	Max D.C. output current=125 mA; max peak inverse voltage=1,400 volts; rms supply voltage per plate=350; max peak current per plate=375 mA
5Y3-GA	Full-Wave High-Vacuum Rectifier	5T	12-16 9-13	5.0	2.0	—	Tube Voltage Drop: ♠ 60 volts at 125 mA D.C.	Max D.C. output current=125 mA; max peak inverse voltage=1,400 volts; rms supply voltage per plate=350; max peak current per plate=440 mA
5Y3-GT	Full-Wave High-Vacuum Rectifier	5T	9-13 or 9-42	5.0	2.0	—	Tube Voltage Drop: ♠ 50 volts at 125 mA D.C.	Max D.C. output current =125 mA; max peak inverse voltage =1,400 volts; rms supply voltage per plate =350; max peak current per plate =440 mA
5Y4-G	Full-Wave High-Vacuum Rectifier	5Q	14-3	5.0	2.0	—	Tube Voltage Drop: ♠ 60 volts at 125 mA D.C.	Max D.C. output current =125 mA; max peak inverse voltage =1,400 volts; rms supply voltage per plate =350; max peak current per plate =375 mA
5Y4-GA 5Y4-GT	Full-Wave High-Vacuum Rectifier	5Q	12-16 9-13 9-42	5.0	2.0	—	Tube Voltage Drop: ♠ 60 volts at 125 mA D.C.	Max D.C. output current =125 mA; max peak inverse voltage =1,400 volts; rms supply voltage per plate =350; max peak current per plate =400 mA

■ Compactron   ♠ Per section   ✪ Design maximum rating.

Tube Type	Classification by Construction	Base Connections	Outline Dwg	Filament Volts	Max Plate Watts		Other Characteristics
5Z3	Full-Wave High-Vacuum Rectifier	4C	16-1	5.0	3.0	—	Tube Voltage Drop: ♠ 58 volts at 225 mA D.C.	Max D.C. output current=225 mA; max peak inverse voltage=1,550 volts; max rms supply voltage per plate=450; max peak current per plate=675 mA
5Z4 5Z4-GT	Full-Wave High-Vacuum Rectifier	5L 5L	8-6 9-11	5.0 5.0	2.0 2.0	—	Tube Voltage Drop: ♠ 20 volts at 125 mA D.C.	Max D.C. output current=125 mA; max peak inverse voltage=1,400 volts; max rms supply voltage per plate=350; max peak current per plate=375 mA

■ Compactron   ♠ Per section   ✪ Design maximum rating.

DE-MYSTIFYING THE FILAMENT CIRCUIT IN YOUR AMP

If you play a vintage tube amplifier, chances are that it hums more than you'd like. There are basically two types of hum that could come from an amplifier, namely, 60 cycle hum and 120 cycle hum. The 120 cycle hum comes from the D.C. power supply while the 60 cycle usually comes from the filament supply. Filament supplies are almost always A.C. and since the A.C. from your wall outlet has 60 cycle hum, so does the filament supply. Early amp designers, never dreaming anyone would "crank it up," were not testing amps at a loud enough volume to perceive the hum as a problem. As well, many vintage amps were designed to run on 110 volt wall A.C. and today we have 120 volts (and in some areas even higher)! This causes "higher than original" voltages in the amp which results in us hearing more hum than the original designer heard. Because hum wasn't a problem in the 40s and 50s, many amp designers used a non-hum-canceling filament supply simply because it was the easy way to go. This dog won't hunt! Especially for today's players that want to "turn it up" and make those output tubes work.

What type of supply do I have?

There are basically six styles of running a filament supply.

1. Series
2. D.C.
3. Daisy chain (non-hum-canceling)
4. Hum-canceling with centertapped ground
5. Hum-canceling with artificial ground (either fixed or adjustable)
6. Positive biased; all of which have various advantages and disadvantages.

You will rarely see a series filament circuit on a vintage amp, although some Dan Electro amps (amp in the guitar case) were wired that way. The disadvantage of the series-style circuit is that if any tube filament burns out, all the filaments turn off and you have a big problem figuring out which tube is burned out. Also, they are not hum-canceling.

Even rarer is the D.C. filament supply. In fact, I can't think of any particular amp that used this style. I used D.C. filament circuits when I first started building amps because it was the ultimate in terms of zero hum. I quickly found out the problems of working with D.C. filaments. For one thing, there are many parts in this circuit, all of which will be reduced to ashes if an output tube shorts plate to heater. Also, the D.C. supply uses a solidstate device to convert A.C. to D.C. These devices are very heat-sensitive, that is to say that if a 10 amp rectifier is hot and you run 3 amps of current through it, it will probably blow from the heat. I was using 25 watt rated rectifiers on a 3 amp circuit. Another major disadvantage is voltage regulation. The least little change in heat causes a change in resistance of the semiconductor device, resulting in a difference in actual filament voltage. D.C. filament supplies also use a very large capacitor that is bulky, expensive and nearly impossible to find. These caps are also heat-sensitive and prone to failure. Voltage regulators big enough to handle the type of current in a filament circuit are difficult or impossible to find and also prone to failure—again from the heat of a tube amp. I would not recommend D.C. filament supplies.

Let's talk about the four most common filament styles starting with the daisy chain (non-hum-canceling) supply. Every tube has two filament leads. In the daisy chain style, a single wire runs from one filament lead on one tube to one filament lead of the next tube to one filament lead of the next tube, etc. Ultimately it connects to one end of the transformer filament winding. This would leave one more filament lead for each tube. These are all grounded to the chassis, as is the other end of the transformer filament winding. This style of wiring was used on many amps and is the easiest to wire and the cheapest to build. Unfortunately, it hums the most of the six filament styles. It uses a non-center-tapped filament transformer. Some examples of this would include the tweed Deluxe (5E3 and earlier) and tweed Champ (5F1 and earlier) and 5E7 Bandmaster.

Hum-canceling (with center-tapped ground) style circuits were rarely used until the late 50s. In this style of wiring, a twisted pair of wires goes from tube to tube connecting all of the filaments in parallel. Neither wire is grounded. The transformer filament winding is center-tapped and grounded. This balances the A.C. to reduce hum and the twisted pair adds some inductance which helps. The advantage of this style is it is hum-canceling. Disadvantages include the time and difficulty in wiring. Also, if an output tube shorts plate to heater, you will be running the entire power of that tube through the filament winding. This could, in some rare instances, smoke the filament winding—certainly a disadvantage! You will find this type of circuit in the 5F6 and 5F6A Bassman amps as well as most Gibson, Magnatones, Marshall and blackface Fenders.

Hum-canceling (with artificial ground) style filament circuits are wired identically as the hum-canceling (with center-tapped ground) except there is no center-tap on the transformer filament winding. Instead there are two resistors (usually 100 ohms) that go from each side of the filament winding to the chassis ground. This tricks the circuit into thinking that the filament winding is center-tapped because the ground reference is midway between the two winding ends. The advantage is that if a tube shorts plate to heater and the entire output of that tube is shorted to the heater, all the current goes through the resistors and blows the resistors. This acts like a fuse and "disconnects the ground reference" thus not damaging the transformer winding. Some examples of this style would include the blackface Super Reverb (AB568 and later) and the Bandmaster (AA1069 and later). An alteration of this same style is to make it adjustable by using a pot in place of the resistors. One end of a 100 ohm or 250 ohm pot goes to one end of the transformer filament winding and the other end of the pot to the other filament winding. The wiper of the pot gets grounded. The pot can be adjusted for the least amount of hum. This is especially nice, if your output stage is a little off balance, because you can dial in the exact point of least hum. Examples of this type of circuit are found in the Ampeg V2, V3, V4, and SVT.

Positive biased filament windings, in my opinion, are the best overall for the least amount of hum and are very dependable. In this style, the tubes are wired exactly the same as in other hum-canceling styles but instead of the transformer filament center-tap or artificial center-tap being grounded, it is connected to a positive bias supply of usually 30 to 40 volts. Ampeg used this circuit on many cathode-biased amps and used the top of the power tube cathode resistor as the positive bias supply. Very clever indeed. Some amps used this and actually had a separate positive bias supply. Some examples that come to mind are the Marshall 4104 and 4010 (JCM 800 Lead—both 50 and 100 watt). I've used this circuit myself to eliminate hum in single-ended amps (such as a Fender Champ). Single-ended amps have an especially big problem with hum because the output stage itself is not hum-canceling as in a push-pull type amp.

THE EVOLUTION OF VINTAGE REVERB CIRCUITS

The first time I remember hearing an amp with reverb was as a 12-year-old window shopper at Swicegood Music on 9th Avenue in Port Arthur, Texas. The store had just received an amp called a "Super Reverb" amp. I was sufficiently impressed as the salesman demonstrated the feature. The year was 1964 and I can remember thinking at the time that my novice guitar playing would not have to be *that* good in order to sound real good with this new feature.

Although I had never noticed reverb before that time, it had been a feature offered on amps much before then. Some manufacturers called it "Echo," some called it "Delay," and some called it "Reverb." Regardless of what you called it, it was a feature that was "hip" and here to stay. Some of these circuits were alike in certain ways and different in others. This article will distinguish the different types, how they sound differently, and what they have in common.

All vintage reverb circuits have two things in common. For one thing they are all tube driven, and they all use a Hammond style reverb tank. This tank or pan, as it is sometimes called, consists of a housing with two or three springs stretched between a pair of transducers. One transducer is like a speaker in that it vibrates the springs and the other is like a microphone in that it picks up the signal on the other end of the spring. The differences lie in how the tanks are driven, how they are coupled, where the reverb signal is reproduced, and ultimately how they sound.

SPEAKER DRIVEN REVERB

Perhaps the oldest setup of reverb was to take a little dry signal off the speaker (by using a voltage divider) and use that to drive a reverb tank. Usually about 1 or 2 watts of power were used for this. The output side of the tank was then sent to a preamp tube for voltage gain and then sent to a power tube (and output transformer) that beefed up the reverb signal powerful enough to drive a separate speaker. The output tube/tubes of the reverb circuit were usually less powerful than the original output section, and had their own separate speaker. In the end, the amp would have one speaker (or set of speakers) with the dry sound and a separate speaker (or set of speakers) with the reverb sound. Some manufacturers carried this to the extreme and had the reverb speakers in a separate extension cabinet that could be setup on the other side of the stage for better stereo imaging. This was high tech stuff for back then.

Some examples of this type reverb would include the Ampeg Echo Twin (Models ET1 and ET1-B) and the Ampeg Super Echo Twin (Models ET2 and ET2-B). The Ampeg B12X and B12XY also used this setup. I have seen Gretsch amps use this but I can't remember the model numbers. The Leslie 122RV organ speaker uses this type of reverb, as do most other organ amplifiers.

Speaker driven reverbs sounded very good **at low volume** and especially good with the separate satellite speakers. However, they were expensive to build because they required an extra output stage—complete with speaker, output transformer, power tube/tubes and preamp section. One pitfall was the fact that when cranking up the amp, output tube distortion could feed the reverb pan. This distortion would be reverbed and the distortion was further amplified upon recovery. In the end, it could get muddy, and would ultimately show up as distorted reverb. There had to be a better way that would sound better and not cost so much to build.

CAPACITANCE COUPLED REVERB

Although my least favorite type of reverb, the capacitance coupled reverb was used on a great number of amps. This system used a coupling capacitor to take signal off of a preamp tube (usually a 12AU7, 6CG7 or 6U10) to drive the reverb tank. A special tank with much higher input impedance was used. The return signal was amplified by a preamp tube and injected back into the signal path. This is perhaps the most inexpensive way to make a reverb circuit and, in my opinion, the least desirable. There was usually a gain pot to determine how much of the return signal was sent back into the circuit. This control was called different names depending on the manufacturer. Magnatone called it "Reverb Depth," Ampeg called it "Dimension," but they all worked the same.

To me, this setup sounds weak and thin. For one thing, the capacitor coupled reverb does not let much low-end through the capacitor and therefore the reverb is at best midrangey with an abundance of high-end. Why did manufacturers use this setup? Capacitors cost less than transformers.

Some examples of amps with this type of reverb would include most Ampegs (Models B22X, B42X, G12, G15, G20, GS15R, GU12, GV15, GV22, AC12, etc.), most Magnatones (Models 431, M4, M8, M10, M10A, M13), all Danelectros, and many others. Fender and Gibson never used this type of reverb.

TRANSFORMER COUPLED

There had to be a better way to make a reverb circuit that had the fullness and the lushness of the "speaker driven" style reverb, but without the muddiness or the expense. Fender, Gibson, Marshall, and some Magnatones (Models MP-1, MP-3, MP-5) used a high-powered preamp tube (usually a 12AT7 or 12AU7—we are talking power, not gain here) coupled to a small output transformer to achieve this. A 12AT7 or 12AU7 properly coupled to a small output transformer can deliver one or two watts of power. Not only that, but we are talking about a very low impedance output. This is perfect for driving a reverb tank but without the output tube distortion of the "speaker driven" style reverb. The cost was only a little more than the "capacitance coupled" reverb. Of course, the return signal went to a preamp tube that re-injected the reverb sound back into the amp circuit. This worked very well. It had the lushness because the transformer did not roll off the low-end like the "capacitor coupled" style reverb and at the same time it didn't have the muddiness of the "speaker driven" style reverb.

POWER TUBE TRANSFORMER COUPLED

Even though the "transformer coupled" style was much better than the other types, it still had its drawbacks. For one thing, it was still driven by a preamp tube and there was no way to actually alter the way the reverb sounded. You could have more reverb or less reverb, but you were stuck with the reverb sound being only the way that particular reverb sounded.

Enter the Fender 6G15 Reverb unit. For the first time, players could actually "dial-in" the reverb sound they wanted. If you consider that reverb is a natural phenomenon that occurs naturally without electronics, you will see that there are certain natural parameters that alter the way reverb occurs.

For one thing, natural reverb always occurs in a "room." (I am using the word "room" in a very broad sense; the Grand Canyon is a "room".) The reverb sound changes as the room size is altered. The 6G15 had a control called "Dwell" that could alter how hard the reverb tank was driven. This would sound like a larger room at higher settings and a smaller room at lower settings. Very cool.

In natural reverb, the resonance of the walls alter the way reverb occurs. Steel walls sound "wetter" than wooden walls. 6G15s had a tone control that only affected the tone of the reverb signal and not the fundamental dry signal. This could be set so that the reverb part of the signal was more trebly than the dry part, thus giving that "wet" tone. Dick Dale, the Beach Boys, and the Ventures are some examples of this setting. Also, by setting the tone less trebly than the fundamental, tones akin to playing in a wooden barn could be available to the player. This worked great for blues.

In natural occurring reverb, where the listener is "listening from" makes a difference in how the reverb occurs. If you had your head in the back of an amp, you would hear mostly dry signal and very little reverb. Conversely, if you were standing against a far wall of a large room, you would hear mostly reverb signal and very little dry. The 6G15 had a true "mix" control that when turned one way would give all dry signal and turned to the other extreme would give all reverb signal. Of course the player could dial in the exact "mix" he desired. This is similar to selecting the appropriate place in a room to listen from!

The tanks were driven with a transformer. The transformer was not driven with a preamp tube, but used a 6K6 power tube instead. (This tube is very similar to a 6V6 but is not rated for as high a wattage. The 6V6 will work in a 6K6 socket.) The 6G15 unit was placed between the guitar and the amp and responded well to variations in pick attack making it the most musical of all reverbs. It is no wonder that a blond or brown 6G15 in excellent condition could bring over $1,000 on the resale market.

This unit too had a few drawbacks. For one thing, it was designed for use with a single coil pickup and therefore would distort if using a humbucking pickup. Also, Fender left off a grid-return resistor on the reverb recovery circuit. If you unplugged the pan or had a bad return cable, the 12AX7 would go into runaway and burn up. Another flaw was the absence of a screen grid resistor on the 6K6 power tube to limit screen current. Of course, 6G15 reverb units can easily be modified to correct such flaws.

EVERYTHING YOU ALWAYS WANTED TO KNOW ABOUT TRANSFORMERS (BUT WERE AFRAID TO ASK)

Because of the vast number of calls I receive daily concerning transformers, I am certain that there is a lot of confusion about transformers. We will cover all types of transformers that are used in vintage tube amplifiers in this chapter. There is a little math involved in some of the examples given, but nothing more complicated than multiplication, division and square roots. You might not be familiar with some terms in this chapter, so I recommend reading this in a quiet environment when you can read slowly, pay attention, and re-read a sentence if you don't comprehend it the first time through. Although transformers are simple devices, it could take years of study to really get a grasp on everything that is happening inside a transformer. This is my attempt to simplify a not-so-simple topic. I write for musicians, not engineers, and I have tried to keep this as basic and simple as possible. However, there are some mathematical relationships in transformers that must be addressed.

WHAT IS AN OUTPUT TRANSFORMER?

An output transformer is actually two coils of wire wrapped around a metal core. One coil of wire connects to the output tube/tubes and power supply, and the other coil connects to the speaker. Neither coil is directly attached to the other, but the magnetic field that is expanding and collapsing from the one coil induces electrical current in the other coil. Actually the output transformer is what couples the output tubes to the speaker. The output transformer has two jobs: 1. to match the impedance of the tubes to the impedance of the speaker, 2. to block D.C. voltage so that no D.C. appears on the speaker.

Tubes operate at very high impedances and speakers operate at very low impedances. Generally speaking, low impedance circuits are high current and low voltage while high impedance circuits are low current and high voltage. It takes a lot of current to operate a speaker and tubes do not really produce that much current. The output transformer makes it possibly to transform high voltage low current into low voltage high current. Tubes need D.C. voltage on the plates in order to function but speakers cannot tolerate D.C. Since the two coils of wire are coupled magnetically and not directly, there will be no D.C. voltage on the secondary regardless of how much D.C. is on the primary. Only A.C. appears on the secondary winding.

HOW DOES IT WORK?

Let's say we wrap wire around an iron core 100 times (or turns). Now we take a second wire and wrap it 50 turns over the first winding. The first winding we will call the primary and the second winding we will call the secondary. If we hook the primary up to a 100 volt A.C. source, and hook the secondary to an A.C. voltmeter, we will get 50 volts A.C. across the secondary (on the meter). In other words there will be the same ratio of voltage as there is ratio of turns. Stated another way, the number of A.C. volts per turn on the primary (input winding) will be the same for the secondary. Example: 100 volts A.C. on a primary winding of 100 turns equals 1 volt per turn. The secondary will have also 1 volt per turn—in this case 50 turns gives 50 volts. If we had only 30 turns on the secondary, we would get 30 volts on the secondary.

There are actually three transformations going on in every transformer—voltage transformation (as in the example above), current transformation, and impedance transformation. The factor that makes a power transformer different from an output transformer is simply that in its design, one aspect of transformation has been selected for emphasis over others.

OUTPUT TRANSFORMER DESIGN

Suppose you have a pair of 5881 power tubes that require 4200 ohms plate to plate primary impedance and you want to drive an 8 ohm speaker. How do you know what turns ratio you need to make the thing work? The turns ratio of primary to secondary will be the square root of (4200 divided by 8). That's right; you simply divide the primary load impedance by the secondary load impedance and then find the square root of that quotient. The answer is 22.92 and would be expressed as 22.92 to 1 ratio. This means for every 1 turn on the secondary, there needs to be 22.92 turns on the primary. Of course, the transformer will have to be wound with large enough wire to accommodate the current and there are other factors, but this is how you figure turns ratio to match impedances.

Let's say you are using four 5881 tubes and therefore need to use a 2100 ohm plate to plate primary impedance and let's say you are going to connect this to a 4 ohm speaker load. The formula is the same: divide 2100 (primary load impedance) by 4 (secondary load impedance) and then find the square root of the quotient. 22.92 to 1 is the ratio again. Does this mean that you could use a Fender Twin output transformer in a two output tube amp if you ran it at 8 ohms? You betcha. In fact, there have been many Vibroverb re-issues that have been retrofitted with a single 15" speaker and a Twin output transformer. This improves the tone dramatically, especially if you use a NOS Twin transformer that was wound on a paper bobbin. More on this later.

OTHER DESIGN CONSIDERATIONS

While you could use a Twin output transformer in a 40 watt amp running at 8 ohms, you could not use a blackface Vibroverb output transformer on a Twin. Why? Because the thickness of the wire has to be thick enough to accommodate the current and the Twin is running at approximately twice the current of the Vibroverb. A stock Vibroverb transformer would burn up if you put it in the Twin, but the Twin output transformer can easily take the current put out by the Vibroverb. Incidentally, for those of you who like to remove the two outside power tubes from your Twin for gigging at a lower volume, you can see now how important it is to disconnect one of the 12" speakers in order to keep the tubes running into the correct impedance. If you continue to run it at 4 ohms (both 12" speakers), the tubes will think they are running into 2100 ohms and this will shorten tube life dramatically. Why, that's halfway between what they should be running and a dead short!!!!!

There is a lot more to output transformers than turns ratio. For example, the core material (iron) alloy makes a huge difference in the tone, as does the bobbin material and physical location of the windings, core laminate thickness, core stack size, primary incremental inductance, capacitive losses, copper losses, hysteresis loss, etc.; however, I realize that I'm writing this for a bunch of guitarists and not for transformer design engineers.

HOW ARE OUTPUT TRANSFORMERS WOUND?

Perhaps the two most important aspects of transformer design are the materials that are used and the physical location of the windings. If you wind all of one winding and then all of the other winding, the transformer is said to be "straight wound." This is different than "interleaved." In an interleaved design, a little bit of the primary is wound and then a little bit of the secondary then a little bit of the primary and then a little secondary, etc. This is a superior way to go because there is less loss in an interleaved design. A good example of an interleaved transformer is the 1959 Bassman which has a total of seven windings. Four windings are for the primary and three are for the secondary. A good example of a straight wound transformer is the tweed Deluxe. It has all of the secondary and then all of the primary.

Most modern output transformers are straight wound on a plastic bobbin which does not sound as good as an interleaved design wound on a paper bobbin. Paper is thinner than plastic and therefore the windings in a paper bobbin are closer to the core. Less loss is the result. Interleaving improves coupling and the result here is also less loss. If that's the case, why do major manufacturers use plastic bobbin straight wind transformers? Because they are cheaper and those manufacturers can sell their products for less money, thus creating the illusion of better value. (Perhaps it is a better value if the amp is purchased just because it's cheap and tone is of little consideration.) However, there are players that want every advantage they can get, regardless of cost, even if the difference is only 10 to 20% improvement in tone!

WHAT'S A CENTER-TAP?

Almost all amplifiers that use two or more power tubes operate in push-pull. Operating in push-pull requires a center-tapped primary.

There is a lead that is connected to the exact center of the winding. This is not necessarily the center of the length of wire used to wind the primary, but rather the center of turns. From the center-tap to the start of the winding there would be exactly the same number of turns as from the center-tap to the finish of the winding. The center-tap actually connects to the high voltage power supply; the start connects to the plate of one tube while the finish connects to the plate of the other tube. Additional pairs of tubes can be added one at each end to get a four tube, six tube, or eight tube output stage configuration. Stated another way, the B plus is connected to the center-tap, the plates of half of the output tubes are connected to the start of the primary and the plates of the other half of the output tubes are connected to the finish of the primary.

COMMON MISCONCEPTIONS

I get so many calls from technicians that think they have a bad output transformer when they don't really. In fact, my company sells transformers and the one I get the most calls on is the tweed Deluxe. This transformer couples two 6V6s to an 8 ohm speaker load. Here's the misconception. A technician will measure the D.C. resistance from one end of the primary to the center-tap and measure the D.C. resistance of the other end of the primary to the center-tap. The end of the winding to the center-tap will always have more D.C. resistance. The tech will conclude that some of the beginning to center-tap windings must be shorted because of the resistance difference. This is a common mistake. Transformers are not wound by resistance, they are wound by turns and anyone that has ever run track knows that the outside lane is longer. In other words, as I wind a couple of hundred turns of wire around a bobbin, the bobbin diameter is constantly increasing and the last turn may take twice as much wire as the first turn. On a 200 turn primary, the transformer is tapped between the 100th and the 101st turn, the actual D.C. resistance of the wire will be more from center-tap to the ending (called the finish) than it will be from center-tap to beginning (called start).

ANOTHER COMMON MISCONCEPTION

The second most common misconception is checking the plate voltage on both power tubes, finding one plate voltage higher than the other and concluding something must be wrong with the transformer. If the tubes are not matched and one is drawing more current than the other, there will be a larger voltage drop on that side and therefore less plate voltage available. On transformers wound with fairly small wire (tweed Deluxe) the plate voltage of the output tubes will almost never match, even if the tubes are matched. We already talked about D.C. resistance, and if one side has more resistance, it will have a larger voltage drop even if the tubes are matched. I've seen this confuse many otherwise knowledgeable technicians.

ADVANCED TROUBLESHOOTING

To check a push-pull output transformer for primary shorts, disconnect all the leads. Put a 1000Hz A.C. signal on the secondary. Now measure the A.C. voltage across the primary from start to center-tap. Measure from finish to center-tap. These measurements should be identical within perhaps ¼ volt. If either section of the primary is shorted, its voltage will read lower than the other section. Please note how this is different from measuring D.C. resistance. (Actually D.C. resistance could be considered a measurement for "copper loss" and is not directly related to turns ratio.)

MORE ADVANCED TROUBLE SHOOTING TECHNIQUES

I would like for you to know one correct way to check an output transformer turns ratio and consequently the impedances. This information can be very useful if you have a transformer and want to know the secondary load impedance (what ohm speaker load to use).

Remove the transformer from the amp. Get a signal generator and put a 1000Hz signal on the secondary. Get an A.C. voltmeter and see how many volts you are putting into the secondary (I like to adjust my signal generator to ½ volt). Connect the voltmeter to the primary leads, both start and finish. If the transformer is center-tapped, the center-tap is not connected. Now divide the secondary signal generator voltage into the primary voltage and you will have a quotient. This quotient is the turns ratio (quotient : 1) of primary to secondary. In other words whenever you have this many volts on the primary, you will have 1 volt on the secondary.

This is pretty cool because you can actually square the turns ratio and multiply by what impedance you would like to run as a speaker load and this product will give you the primary impedance. Or you could square the turns ratio and divide this into what you would like to use for a primary impedance to get the secondary load impedance (speaker impedance). Example: You are at an electronics surplus sale and you notice a nice looking output transformer for sale and nobody knows what it is. You see that it has five wires (three smaller gauge wires and two very large wires, one of the smaller wires is red and therefore usually the center-tap.) It is for sale for $2 but here again nobody knows what it goes with. Since it is not rusted (rust is bad because it causes eddy currents in adjacent laminates, thus zapping efficiency), you decide to take a chance and purchase it.

When you get it home, you put a signal generator across the secondary at ½ volt A.C. When you measure the primary from start to finish (the center-tap is not connected), you read 14.35 volts. This would make the turns ratio 28.7 to 1. (I got that by taking the 14.35 and dividing by ½.) You may also put voltage across the primary and raise it until you see ½ volt on the secondary. This may be necessary if the secondary draws too much current.

Next you square the 28.7 and get 823.69. You quickly reason that if you multiply this number times your 8 ohm speaker load you get 6589.52 ohms. Knowing that a pair 6V6 tubes usually are run into approximately 6600 ohms, you surmise that this transformer is for a pair of 6V6s and an 8 ohm speaker load.

You could also have figured it another way. Let's suppose you squared the turns ratio and got 823.69. At that point, you could choose what kind of primary load impedance you want and then determine what speaker load you would need. For instance, let's say you wanted to run a pair of 6V6s at 6600 ohms. You would divide the 6600 by 823.69 and get 8.01 ohms for a secondary load (speaker load).

There is a little bit of guesswork involved here because you want to know that the transformer can take the current. If you have a very small transformer that looks like it came out of a Princeton, you would not want to hook it up to a 100 watt Marshall, because the size is so small it couldn't possibly have been made with large enough diameter wire to handle that much current. You need to use a little bit of educated guesswork.

Also, if you square the turns ratio to get 823.69 and wanted to run it at 4 ohms, you multiply the 4 ohms times the 823.69 and get 3292 ohms. This is about the right primary impedance for a quartet of 6V6s but when you look at the transformer, you don't think it is large enough to handle 50 or 60 watts of power. At the same time, 3292 ohms is too small an impedance to run pairs of 6L6s, EL34s, or 6550s. From this, you reason that it must be for a pair of 6V6s into 8 ohms. Gotta use that gray noodle.

Here's another application. You purchase a cool Showman head at a garage sale for $150. You're not quite sure if it was originally sold as a Showman or a Dual Showman (the Showman was 8 ohms output and the Dual Showman was 4 ohms output). Knowing that Fender usually ran a quartet of 6L6s at 2100 ohms, you disconnect the output transformer leads and put a ½ volt 1000 Hz signal on the secondary. Measuring the primary for A.C. voltage, your meter says 8.1 volts. You divide this by the ½ volt you are putting on the secondary and come up with the turns ratio of 16.2 : 1. Now you square the turns ratio (16.2 X 16.2) and get 262.44. Divide the primary load impedance (2100 ohms) by this number and you get 8 ohms. You know now that it was a "single 15" Showman because it was running at 8 ohms!

WHAT'S A POWER TRANSFORMER?

A power transformer is a device that is actually two or more (usually several) coils of wire wrapped around a metal core. One coil of wire is called a primary and connects to your wall outlet 120 volts A.C. Each of the other coils of wire connect to some part of the amplifier to supply certain amounts of voltage. None of the coils of wire are directly attached to each other, but the magnetic field that is expanding and collapsing in the primary induces electrical current in each of the other coils of wire that are wrapped around the same iron core.

WHY DO I NEED A POWER TRANSFORMER?

Consider that your guitar amplifier needs many different voltages for many different functions. For instance, most vacuum tubes used in guitar amps need 6.3 volts A.C. to heat the filament inside the tube. Most common tube rectifiers need 5 volts A.C. to heat their filaments. The plates of the output tubes need 350 to 500 volts D.C.

to make them work and a bias supply voltage of minus 10 to minus 60 volts D.C. to make them work. The preamp tubes need 100 to 250 volts D.C. for them to work. Despite all this, your wall outlet provides approximately 120 A.C.

In the earliest stages of tube amplification they just used a lot of batteries. The batteries that heated up the filaments were called the "A" supply batteries. The high voltage batteries for the plates of the tubes were called the "B" batteries and if a bias battery was used, it was called the "C" battery. They didn't need a battery for a rectifier filament, because all the batteries were already D.C., so they didn't need a rectifier. Many of these battery terms carried on even when the batteries were no longer used. For example, the high voltage D.C. supply voltage that is used for the plates of the tubes is often referred to as "B+" voltage. It was always positive and high voltage. I've heard many old timers refer to the bias supply voltage as the "C-" supply. Of course this voltage was always D.C. and minus. I've even heard folks speak of the "A" supply—referring of course to the 6.3 volt filament supply for the tubes. This voltage could be A.C. or D.C.

HOW DOES IT WORK?

Transformers are much more complicated than this explanation will make it seem. There are certain losses that have to be figured into the equations, but this is the simplified (let's neglect the losses) explanation of how it works. Let's say we wrap wire around an iron core 120 times (or turns). This is what we will call our primary, and we will attach it to our 120 volt A.C. wall outlet. Since there are 120 turns of wire and we are connecting this to 120 volt wall outlet, we will have exactly 1 volt per turn of wire. Now if we want to wind a coil of wire for our B+ winding of, let's say, 700 volts center-tapped, we will wind 350 turns, solder a center-tap lead and wind 350 more turns. The 1 volt per turn that was in the primary will now induce 1 volt per turn in this B+ winding. So the B+ winding is only one secondary. Now we will want a 5 volt winding for our rectifier tube filament, so that winding will need to be 5 turns because it too will pick up the 1 volt per turn from the primary. Now we will need an "A" supply so we wind 6 and one-third turns to get our 6.3 volt winding. Now we've got ourselves a power transformer with multiple secondaries—each secondary supplying the necessary voltages for various different functions within the amp. To get lesser B+ voltages for the preamp tubes, and screen supply, we will just use some dropping resistors from our main B+ supply.

There are actually three transformations going on in every transformer—voltage transformation (as in the example above), current transformation, and impedance transformation. The factor that makes a power transformer different from an output transformer is simply that in its design, one aspect of transformation has been selected for emphasis over others.

For instance, the "A" supply would have to be wound with very large wire because even though it is only 6.3 volts, it is running at considerable current. The B+ winding would use much smaller wire because it is not using nearly as much current, and the primary would have to be wound using big enough wire to handle all the wattage of all the secondaries combined, plus a little extra for losses and even a little extra for safety.

OTHER DESIGN CONSIDERATIONS

The thickness of wire used for the B+ winding has some resistance, like any length of wire. Because it is the high voltage winding, it will always have the greatest number of turns of any of the secondary windings. Simply stated, the smaller the wire, the more the resistance and the thicker the wire, the less resistance. Whatever the resistance happens to be is actually in series with the B+ power supply. The more resistance, the more sag and the slower the recovery rate. Conversely, the less resistance, the faster the recovery rate and the less sag. Put a Twin power transformer in a blackface Vibrolux and you will hear the difference. The Twin has very little resistance (internal resistance of the actual winding). Jim Marshall used this trick on the Park 50 watt amps—he used a 100 watt power transformer and reduced the power supply sag. Another example of this was the very first Bedrock amps, the ones handbuilt by Brad Jeters. Brad used a 1 amp B+ winding in his 50 watt amps. This was virtually unheard of. This is about five times thicker wire than the B+ winding of a 45 watt Super Reverb. No wonder those original Bedrocks had huge bottom end punch!

SIMPLE TROUBLESHOOTING

Sometimes a couple of primary turns could be shorted and the transformer would still work, but run a little hot and the voltages on the secondaries would be high. If a few turns of the primary are shorted, you would then have slightly more volts per turn than normal and all of the secondaries would also have slightly more volts per turn than normal. This would account for the slightly high secondaries. An easy way to check is to measure your 6.3 volt filament leads (all of the tubes must be installed and the amp placed in the play mode). If the filaments read anywhere from 6.2 to 6.7 volts, you're probably OK. Remember that most vintage amps were actually designed for 110 volts and we now have 117 or 120 volts depending on your particular area. But let's say you monitored the A.C. input voltage with a Variac and set it to exactly 110 volts and the heater filament is running 6.9 volts. My guess is that you have a few primary turns shorted.

If you think you have a bad power transformer, disconnect all the leads except the primary. If it blows a fuse, it's definitely a bad transformer.

We talked about transformer basics and we went into depth about output transformers and advanced troubleshooting techniques. We discussed power transformers and troubleshooting techniques. Now we will look at interstage inversion transformers, reverb driver transformers, filter chokes (inductors), and troubleshooting techniques.

WHAT'S AN INTERSTAGE INVERSION TRANSFORMER?

Although most amplifiers are designed with capacitor/resistor coupling from one stage to the next, it is possible to couple stages with a transformer. This is very expensive (relative to the cost of a resistor and a capacitor) and is almost never used in guitar amplifiers. The resistor/capacitor coupling almost always sounds better. Occasionally, however, a push-pull transformer is used as a phase inverter. For example, many Gibson amps such as the EH150, GA35RVT, and the GA20 used this design. It had an advantage in that it would add gain and achieve phase inversion and thus the cost could be justified.

HOW DOES IT WORK?

The primary has one lead connected to the power supply and the other end to the plate of the preamp tube of the previous stage. On the secondary, there is a center tap which goes to ground if the amp is cathode-biased or it goes to the negative bias supply if the amp is fixed-bias. One end of the secondary goes to the grid of one output tube and the other end goes to the grid of the other output tube. Since the secondary is center-tapped, the ends are exactly 180 degrees out of phase and since the turns ratio is less than one, there is a step-up action taking place that actually provides gain. A transformer with a turns ratio of about one-half and rated at 2 watts will work nicely for most of these type inversion transformers. Of course, it must have a center-tap on the secondary for inversion to be achieved and the primary is operated single-ended.

TROUBLESHOOTING TECHNIQUES

In an interstage transformer, the primary can be checked for continuity by using an ohmmeter. Make sure and disconnect all leads before troubleshooting. The ohmmeter can also be used to make sure the secondary does not have continuity with the primary. You could check for secondary shorts to secondary by putting a signal generator on the primary and then measure one end of the secondary to the center-tap with a voltmeter. Now measure the other end of the secondary to center-tap. The two voltage readings should match. If one measurement is much lower than the other, then there is a short in some of the secondary turns.

REVERB DRIVER TRANSFORMERS

Many amps use a small transformer to drive a reverb pan. This is basically the same idea as a single-ended (no center-tap) output transformer with a very high turns ratio. The reverb pan is 8 ohms and the tube is much higher, depending on what kind of tube is being used. The idea is to take a high impedance signal which is low current/high voltage and transform it to a high current/low voltage signal. For instance, the reverb driver transformer in a Fender Super Reverb is rated for about 3 watts and has a turns ratio of 56 to 1. You may have noticed that this is very close to the turns ratio in a Champ, and the Champ transformer could be substituted with very satisfactory results.

TROUBLESHOOTING

Just like the interstage inversion transformer, you would disconnect the leads from the circuit and check continuity of the primary, check continuity of the secondary and check to make sure there is no continuity between primary and secondary. Sometimes the primary winding will short and everything will seem to be OK on a meter check. Substituting a known good transformer for the suspected bad one is probably the quickest way to determine if there is a problem with the reverb driver.

WHAT'S A FILTER CHOKE?

It's not technically a transformer, but we will cover it here since it is very similar to a transformer, it is made by transformer companies, and they look like a transformer. It is actually a transformer with only one winding on it. The one winding is used to resist any changes in current. How this works is very interesting. Since there is only one winding, the magnetic field that is generated from one turn of winding actually is collapsing and expanding on every other turn of the winding. This induces current in the other windings. This means that the current does not want to stop and will resist any changes. These chokes are usually used in the power supply, especially in Marshall, Vox and Fender amps. Not all amps use these.

TROUBLESHOOTING

They almost never go bad. If some of the windings short, it will still work, but you might notice a little more hum in the amp. Sometimes these will become open, in which case a ohmmeter check will show no continuity.

SUMMARY

Two or more coils of wire wrapped around an iron core constitute an iron core transformer. Transformers are useful in guitar amps because they can transform voltages, impedances and current. Since the two or more coils of wire are not directly connected to each other, they will block D.C. and only pass A.C. voltage.

FENDER TRANSFORMER REFERENCE AND CONVERSIONS

Will a power transformer for a Super Reverb work in a Pro? My amp looks like it may have had the transformer replaced because it just doesn't look original. What is the correct replacement part for a tweed Princeton? Is it true that the 2x10 Brown Super used a '59 Bassman power transformer? Can I replace the burned-out output transformer in my blackface Twin with one from a Showman?

If you ever needed answers to some of these transformer questions and didn't know where to get them, here's a Fender Amp transformer cross reference and conversion/substitution chart. The cross reference chart shows what transformer was used in what amp and the conversion chart shows later part numbers that can be substituted for the originals. Enjoy.

Fender Amp Transformer Cross References

Name	Production	Power	Choke	Output
Bandmaster	6G71	67233	125C1A	45217
	AA763	125P7D	125C1A	12A6A
	AC568	125P7D	125C1A	12A6A
Bandmaster Reverb	AA270	125P5D	125C1A	125A6A
	AA763	125P5D	125C1A	125A6A
	AC568	125P5D	125C1A	125A6A
Bassman	5F6	8087	14684	45249
	5F6A	8087	14684	45249
	6G6	125P5A	125C1A	125A5A
	6G6A	125P7A	125C1A	125A13A
	AA270	125P7D	125C1A	125A13A
	AA864	125P7D	125C1A	125A13A
	AB165	125P7D	125C1A	125A13A
	AC568	125P7D	125C1A	125A13A
	Bass 70	013895		013897
	Bass 135	013692		013691
Champ	AA764	125P1B		125A35A
Concert	6G12	67233	125C1A	45249
	AB763	125P7D	125C1A	125A9A
Deluxe / Reverb	6G3	125P2A		125A1A
	AA763	125P23B	125C3A	125A1A
	AB868	125P23B	125C3A	125A1A
Harvard	6G10	125P1A		125A2A
Princeton	5F2A	66079		265
	AA1164	125P1B		125A10B
	B1270	125P1B		125A20B
Pro / Reverb	5C5	6516		1846
	5D5	6516		1846
	5E5 (A)	6516		1846
	6G5 (A)	125P7D	125C1A	125A7A
	AB763	125P5D	125C1A	125A7A
	AA270	125P5D	125C1A	125A6A
	AA1069	125P5D	125C1A	125A6A
Reverb Unit	6G15	125P24A		125A12A

Showman	6G14	67233	125C1A	45216
	AA763	125P34B	125C1A	125A30A
	AB763 (Dual)	125P34B	125C1A	125A29A
Super / Reverb	6G4	8087	125C1A	45216
	6G4A	125P6A	125C1A	125A6A
	AB563	125P5D	125C1A	125A9A
	AA763	125P5D	125C1A	125A9A
	AB763	125P5D	125C1A	125A9A
	AA270	125P5D	125C1A	125A9A
Tremolux	5G9	8160	14684	108
	6G9	125P6A	125C1A	45217
	6G9A	125P6A	125C1A	45217
	6G9B	68409	125C3A	125A6A
	AB763	125P26A	125C3A	125A6A
Twin / Reverb	5G8 (A)	7993	14684	45268
	6G8	67233	125C1A	45548
	AB763	125P34A	125C1A	125A29A
(*7591 tubes*)	AB763	125P19A	125C1A	125A18A
	AC568	125P34A	125C1A	125A29A
	AA769	125P34A	125C1A	125A29A
	AA270	125P34A	125C1A	125A29A
Vibrolux / Reverb	6G11	125P6A	125C3A	45217
	6G11A	68409	125C3A	125A7A
	AB763	125P26A	125C3A	125A7A
	AA965	125P26A	125C3A	125A6A
	AB568	125P26A	125C3A	125A6A
Vibroverb	6G16	125P6A	125C1A	45217
	AA763	125P5D	125C1A	125A7A

Please note that reverb driver transformers in all amps use #022921

Fender Conversion / Substitution Chart

ORIGINAL PART	60s PART	90s PART
125A1A	022640	022640
125A6A	022848	036968
125A7A	036968	018343
125A9A	022855	022855
125A10B	022913	022913
125A12A	022990	
125A13A	022871	018343
125A20A	022921	022921
125A29A	022889	022889
125A30A	022897	022897
125A35A	022905	022905
125C1A	022699	022699
125C3A	022707	022707
125P1B	022772	022772
125P5D	022798	022798
125P6A	036483	036483
125P7D	022814	022814
125P23B	025130	
125P24A	012671	
125P26A	022723	036483
125P34A	022756	022756

Tube Amp Talk

COOL TOOLS
FOR AMPS

When visiting my shop, many have remarked on the large number of tools and test devices that are simple and homemade. In this chapter I will show you how to build a simple tube matcher, a signal tracer for trouble shooting, and a current limiter for trouble shooting.

WHY DO I NEED A TUBE MATCHER?

Although many new old stock tubes are available that sound much better than modern production "Music Store" tubes, these NOS tubes are almost useless without the ability to test and match pairs and quartets for amps. "Drug Store" type tube testers do not really give an accurate reading, because the test voltages they use are so much lower than those used in a guitar amp.

A SIMPLE TUBE TESTER

This device can be built in about an hour for under $20 and can be used to match any output tube whose cathode is on pin #8. The 6L6, 6V6, 5881, 7581A, EL34, EL37, KT66, KT77, KT88, 6550 tubes are all in this category. This passive device plugs into both an output tube socket of an amp (that uses the kind of tubes you are testing), and a good quality digital volt meter. What makes this device really cool is that since you are using a guitar amp to test the tubes, you will be testing at the high operating voltages found in guitar amp circuits and not the typical 150 volts used in commercial tube testers. Don't be fooled by the simplicity of this device; it's amazingly accurate.

THEORY OF OPERATION

Since the current going into a tube equals the current going out of the tube, the cathode current is equal to plate current. (For all practical purposes, grid current is negligible.) Since current running through a resistance produces a voltage, we know that we could run the cathode current through a small resistor and develop a voltage across the resistor. In choosing a value for this resistor, we will select a 1 ohm ½ watt resistor. By using 1 ohm, measuring the voltage across the resistor in volts will convert directly to amps. This voltage is very small, therefore you will need a very accurate meter. A typical reading might be .035 volts or .042 volts which converts directly to .035 amps and .042 amps respectively. This is the same as 35 milliamps and 42 milliamps respectively. We are actually measuring the current that flows through the tube when the tube is idling in a high voltage guitar amp circuit.

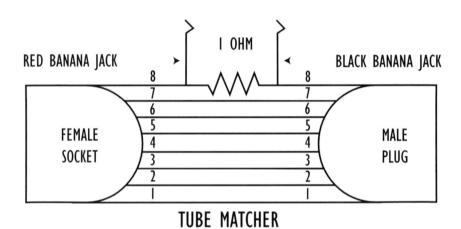

TUBE MATCHER

LET'S START BUILDING

You will need:

1. Approximately 35 feet of hook-up wire
2. One eight-pin tube socket (female)
3. One eight-pin plug (male)
4. A 1 ohm ½ watt resistor
5. Two banana jacks (one black and one red)
6. A small enclosure.

Mount the eight-pin tube socket (female) in the enclosure. Mount the two banana jacks in the enclosure. Mount the 1 ohm resistor between the two banana jacks, one end of the resistor to the black jack and one end to the red jack. Don't solder it yet.

Hook up a wire from pin #8 on the socket to the red banana jack and solder both pin #8 of the socket and and the red banana jack. Cut the remaining hook-up wire into eight equal lengths (about 4 feet long each). Run a wire from the black banana jack to pin #8 of the male eight pin plug. Solder both ends of this wire.

Take the remaining hook-up wire and connect the socket leads to the male plug leads. Match pin numbers on all pins except pin #8, which is already hooked up. For example, pin #1 from the socket goes to pin #1 of the plug, pin #2 of the socket goes to pin #2 of the plug, pin #3 socket to pin #3 plug, etc. All connections are soldered.

LET'S TEST SOME TUBES

Get an amp that uses the same kind of tubes we are going to test. Unplug the output tubes of the amp. If it uses more than one output tube, remove them all. Now plug the eight-pin male plug into any output tube socket on the amp. With your digital voltmeter plugged into the banana jacks and set for the 1 volt range, insert a tube to be tested into the tube socket on the tube matcher. Warm the tube up first, using the "standby" function on the amp. All volume and tone knobs on the amp should be turned all the way off. Put the amp in the "play" mode and watch the meter. The voltage reading will drift around for a few minutes and then stabilize. Record the reading, put the amp in the "standby" mode and remove the tube. Be careful not to burn your hands. Now let's test another tube.

If, when testing a tube, there is a loud hum, turn the amp off immediately (to avoid blowing a fuse). The tube is shorted and drawing mega-current. In a moment I will show you how to build a current limiter that will interface with the tube matcher's amp to guard against this condition.

Another nice advantage to this tester/matcher is that if a tube is noisy, you will hear the noise coming out of the speaker. Commercial testers can't check for noise!

After you have tested a few tubes, you will notice different readings. Anything that is matched within .005 volts can be used as a matched pair. There are more sophisticated ways of matching tubes, but this way will work fine for guitar amps, and is easy and inexpensive to build.

The power supply of the individual amp will make a difference in the reading you get on your matcher; therefore, to be consistent, always use the same amp for the particular type of tube you are matching.

WHAT'S A CURRENT LIMITER?

A current limiter is a device that restricts current. It works because it is actually a series resistance (in series with an amplifier) so that when something in the amp is shorted, the device will provide enough added resistance to limit the current that is drawn by the A.C. (120 volt) power line. By limiting the current, you eliminate blowing up transformers, fuses, diodes, or anything else that too much current could ruin.

Have you ever had an amp that was blowing fuses? Have you ever replaced a power transformer in an amp with a used or not exactly stock part? Have you ever built an experimental circuit and been afraid to plug it in for fear of blowing something up? Or what about when you put a bad tube in the matcher built earlier and it blew a fuse? A current limiter is a useful device that can help solve all of the problems listed above.

A SIMPLE CURRENT LIMITER

This device can be built in about an hour for under $20. The idea was originally introduced to me by an old transformer engineer and I modified it for use in amplifier repair. What really makes this device cool is the speed at which an amp can be repaired and the money saved on blown fuses, transformers, diodes or tubes.

THEORY OF OPERATION

We will add series resistance to the A.C. line power supply. With the resistance in series with the amp, if the amp shorts, there will still be the resistance in the circuit to limit current. Rather than use a conventional resistor, we will use a 100 watt light bulb as a resistor. This has the added value of lighting up to its full brightness when it is drawing full current. Conversely it will be dim when the amp is not shorting. Once you become accustomed to the light bulb's intensity, you will be able to judge (by the brightness/dimness of the 100 watt light) how much current is being drawn.

LET'S START BUILDING

You will need:

1. An A.C. line cord a few feet long with a 3-prong male end connector
2. A small aluminum chassis box
3. Two single 120 volt female outlets (do not use a double outlet)
4. One utility light fixture with a 100 watt light bulb and a line cord
5. A few inches of 18-gauge hook-up wire

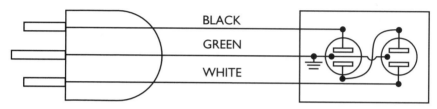

TWO SINGLE AC OUTLETS ARE WIRED IN SERIES TO MAKE THE CURRENT LIMITER,
A DOUBLE OUTLET WILL NOT WORK BECAUSE IT IS PREWIRED IN PARALLEL.

Install both 120 volt female outlets in the small chassis box. These will be wired in series so you will **not** be able to use a double female outlet. The double outlets are internally wired in parallel and you want these outlets to be wired in series. Here's how it is done. You will notice that there are three leads on the female outlets. The ground lead (round), the hot lead (slotted) and the neutral lead (other slotted). Please note that if your outlets have screw terminals, the copper screw is hot and the aluminum screw is neutral. Using the hook-up wire, connect the hot lead of one female outlet to the neutral lead of the other female outlet. Now look at your A.C. line cord. It should have a green wire, a black wire, and a white wire. The white goes to the neutral lead on the first female outlet, the black goes to the hot lead of the second female outlet, and the green goes to each ground terminal on the female outlets. Ground this connection to the chassis as well.

Check your utility light to see if it has a switch. If it does, make sure the switch is in the on position. Now plug your light into one of the female outlets. Your current limiter is complete.

LET'S TROUBLESHOOT SOME AMPS

Let's say that you have an amp that is blowing fuses. Here's what to do. Replace the fuse with the correct value. Plug the amp you are troubleshooting into the remaining female outlet on the current limiter chassis and put the amp in the standby mode. When the amp is warmed up a little, move the standby switch to the play mode.

The 100 watt light bulb will light up its full brightness. (Assuming, of course, that there is some kind of short in the amplifier.) Remove the power tubes one at a time. When the shorted tube is removed, the light will go very dim. You now know which tube was shorted.

But what if the tubes are not shorting? Or what if a new power tube is installed and the 100 watt light is still bright?

1. Then remove **both** power tubes. If the light dims considerably, you probably have a problem with your bias supply voltage. If the light is still glowing brightly with both power tubes removed, you could have a bad rectifier, shorted power transformer, or shorted filter capacitor.

2. Put the amp back into the standby mode. If the light goes out, the problem is in the filter caps and if the light stays on, the transformer is shorted, or the rectifier is bad.

3. Remove the rectifier. If it is a tube rectifier simply take it out of the socket, if it is a solidstate rectifier, simply clip the two red wires that go to the rectifier from the power transformer. If the light still glows brightly, your power transformer has a short. If the light goes dim, it is the rectifier. If your rectifier is solidstate and not tube, it is very common for these to blow. They are cheap to replace and easily available.

LET'S USE IT WITH THE TUBE MATCHER

You've just gotten some new old stock 6V6 tubes and want to match them. You plug in your tube matcher but you suspect that some of these tubes are shorted and don't want to spend three dollars in fuses to find out which ones. Your matcher is plugged into your blackface Deluxe. Here's what to do. Plug the Deluxe into the current limiter and put a tube in the matcher to warm up. Now take the amp off of standby and put in the play mode. If the light in the limiter glows brightly, the tube is shorted. If the light is dim, take the Deluxe out of the limiter and plug it into regular power before taking your tube reading for matching. You will not get an accurate reading if the amp is in the limiter.

There are many other uses around the shop for the limiter. At our shop, we installed an SPST toggle switch across the hot and neutral leads that go to the outlet feeding the 100 watt light. With the switch off, we have current limiting. But with the switch on, the light is shorted and we have regular power. We use this because we are constantly switching from limited current to regular current, and this eliminates constantly unplugging between the limited and full power.

WHAT'S A SIGNAL TRACER?

Have you ever had an amp that didn't pass signal? The tubes were lit up, everything seemed OK, it wasn't blowing fuses, but still the thing would not make a sound. The signal path was interrupted some-where—the question was where? A signal tracer is a device that can check the signal path to see if there is signal on the part of the circuit you are checking. You could start off by checking the input jack for signal and follow the signal path until you find a place in the signal path where signal is lost. When you locate the part of the circuit where signal is lost, you will have a good idea which joint, wire, or component is inter-rupting the signal path and you will be able to repair the amp.

A SIMPLE SIGNAL TRACER

A signal tracer can be made in minutes from a few dollars worth of parts and a small battery-powered amplifier. A probe is fashioned from an axial capacitor taped to a wooden chopstick. A lead then goes from the probe to the input of a small battery-powered amp. Of course you will also have a ground lead coming from the ground side of the input jack (on the battery-powered amp) to an alligator clip. This clip is attached to the chassis of the amplifier under test to complete the input circuit.

THEORY OF OPERATION

Here's how it works. A signal is plugged into the input of an amp to be tested. The technician tests various places in the amp to see if the signal is present in that part of the circuit. He tests with a probe. The probe is a blocking capacitor with a handle. The blocking ca-pacitor passes signal but no D.C.; that gets the signal to the input of the battery-powered amp. The battery-powered amp amplifies the signal for the technician to hear. The technician then knows if there is signal present at the particular part of the circuit being tested.

LET'S START BUILDING

You will need:

1. One .05 microfarad 630 or 800 volt axial capacitor
2. Two 5-foot lengths of 18-gauge stranded wire with 630 volt insulation
3. Electrical tape and possibly some shrink tubing
4. A wooden chopstick or other convenient non-conductor handle
5. A small battery-powered amplifier with internal speaker (hereafter called the tracer amp), anywhere from 1 to 10 watts will do nicely. Input impedance should be fairly high—the higher the better. (Radio Shack has at least one battery-powered amp available that will work well.)
6. Either a ¼" male plug or whatever type of input connector is needed for the tracer amp
7. An alligator clip
8. A signal source. If you don't have a signal generator, a cassette player that has an earphone output jack works nicely. Usually this type of player will kill its internal speaker when the earphone jack is used. You will need a patch cord with the earphone plug on one end and a ¼" male plug on the other end. This connects the output of the cassette player to the input jack of the amplifier under test.

I will describe a way to make this signal tracer. You may want to improvise a little. Here are the two basic steps:

MAKING THE TEST PROBE

Start by soldering one end of a 5-foot length of 18-gauge wire to one lead on the .05 microfarad 630 volt capacitor. Insulate this connection with shrink tubing or electrical tape.

Next place the capacitor next to the chopstick in such a way that the non-soldered lead hangs over the small end of the chopstick by ½". (See diagram.) Using electrical tape, wrap the capacitor and the entire length of the chopstick leaving only the ½" capacitor lead exposed. When you finish, you should have a chopstick wrapped in tape; under the tape will be a capacitor soldered to a wire—the wire hangs out the big end of the chopstick approximately four and a

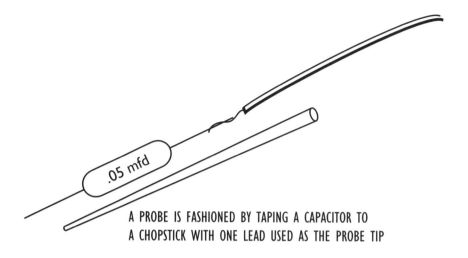

A PROBE IS FASHIONED BY TAPING A CAPACITOR TO
A CHOPSTICK WITH ONE LEAD USED AS THE PROBE TIP

half feet. The other capacitor lead hangs out ½" from the small end of the chopstick. Of course, this lead is left bare.

WIRE THE INPUT PLUG

Take the other end of the 18-gauge wire coming from your test probe and solder it to the hot lead of the connector mentioned in item #6 of the parts list above. Now take the other 5-foot length of 18-gauge wire and solder one end to the ground lead of the connector mentioned above. The other end of this wire gets soldered to an alligator clip.

LET'S TROUBLESHOOT SOME AMPS

Plug the connector from the probe into the tracer amp. Attach the alligator clip to any convenient ground on the amp under test. Connect a signal source to the input of the amp under test. Take your probe and touch the bare capacitor lead to the hot lead on the input jack. You should have signal and will hear it through your tracer amp. Proceed through the amp until you find the spot where the signal is lost.

TIME-SAVING TIPS

Since you have a blocking capacitor on your probe, you can touch any single point in the amp without fear of damaging anything. Almost all amps are laid out so that the signal starts on one side of the amp and progresses towards the output tubes. Any gain stage ultimately goes through a tube and the output of the tube is almost always the plate. (The exception being the cathode follower circuit in which case the output is the cathode. Pins #3 and #8 are the cathodes on a 12AX7 style tube.) If your amp uses 12AX7 style tubes, the plate of each triode is pin #1 and pin #6. Quickly checking each of these can be done in less than a minute and will reveal the approximate vicinity of the problem.

Another thing to save time—check the phase inverter tube first. (This is the preamp tube next to your power tubes.) If you are getting signal here the problem is in the output stage. You should check the plates of the output tubes (pin#3 on 6L6 or 6V6) for signal and continue with the speaker, etc. On the other hand, if there is no signal at the phase inverter, the problem is in the preamp or the phase inverter itself.

The phase inverter is approximately halfway through the amp's circuitry; by checking there first, you can quickly determine in which half of the amp the problem lies.

Tube Amp Talk

TESTING CAPACITORS FOR LEAKAGE

Necessity is truly the mother of invention. When testing a Kendrick ABC Box, I noticed a scratchy pot on my guitar that would indicate D.C. voltage on the pot or a bad pot. I tried two more boxes with the same result. Then I tried the guitar directly into an amp. The scratchiness went away indicating D.C. coming from the ABC Box.

In the ABC Box there is an electrolytic capacitor isolating the input from the supply voltage. This capacitor was obviously leaking D.C. I needed to devise a tester, so I put a 9 volt battery in series with the capacitor and a volt meter to see how much it was leaking. Would you believe 8 volts? A new Sprague capacitor with 85% leakage?!? What's up with that? It seems the supplier had these caps for a while before they were sold.

I talked to a friend who works for a company that manufactures caps and he informed me that all electrolytic caps begin to crystallize after one year. Attention collectors: don't just look at your amps, play them at least once or twice a year, if for no other reason than to keep the caps from crystallizing!

The way my friend described it, the inside of a cap is filled with a liquid electrolyte. Just like honey crystallizes inside the honey jar, the electrolyte does the same thing inside the cap. He informed me that his company tracked their inventory and baked any cap that was not sold within a year. This baking causes the electrolyte to re-liquefy — just like heating crystallized honey changes the honey back into a liquid.

I asked him how I could bake caps and he suggested placing the caps to be baked in a shoe box and heating them with a hair dryer until it was quite hot to the touch (approximately 130 to 140 degrees Fahrenheit). You could also make a small oven with a metal box and a light bulb, experimenting with different wattage bulbs and a thermometer to get the right amount of heat. Of course you don't want to get so much heat that the plastic insulating shrink wrapper (on the outside of the cap) is destroyed. He assured me that this would reform the electrolytic. According to my friend, if a cap is several years old and has not been reformed every year, it can never be reformed!

I also remember reading a differing opinion in an antique radio restoration publication about a method to bring back an electrolytic cap that has been crystallized for years. I have never tried this before, but for the benefit of those that may like to experiment with this idea, here it is.

Take the cap to be reformed and hook it up to a D.C. voltage supply that is equal to the cap's full rated voltage, but put a large value (several meg) resistor in series with the cap. Be sure to observe correct polarity (the positive of the cap goes to the positive of the circuit). Let this remain connected for several days. The resistor limits the current so much, that the cap cannot blow and the voltage will eventually charge the cap to the full D.C. working voltage. The electrolyte is supposed to re-liquefy and become reformed in the process. If anyone out there tries this, please drop me a line and let me know your research results. The author of that particular article swore by the process, but like I said, I have never tried this myself and pass it on only for those that would like to experiment.

WHAT IS D.C. LEAKAGE?

Capacitors are not supposed to conduct D.C. They are only supposed to conduct A.C. When a cap is crystallized internally, it will conduct some D.C. This is bad. There are four types of caps in an amplifier, namely coupling caps, tone shaping caps, bypass caps, and filter caps. The bypass caps and filter caps are usually electrolytic type and the coupling and tone caps are usually not electrolytic. Even non-electrolytic caps can leak D.C. with really bad consequences.

WHAT ARE THE CONSEQUENCES?

When tone caps or coupling caps leak, D.C. is forced onto the grid of the next stage of amplification. This throws off the bias of the next tube and can result in improper gain, unwanted distortion, and ultimately bad tone. If a potentiometer is in the circuit, it will become scratchy — even if there is no problem with the potentiometer. When bypass caps leak, the bias of that particular stage is thrown off and the consequences are the same. When a filter cap leaks, unnatural harmonics, improper plate voltage, and excessive hum are the consequences. In addition to that, the extra current can cause the power transformer to overheat and in some instances even burn out.

Although electrolytics are more prone to D.C. leakage than non-electrolytics, any cap can leak D.C.

What are we going to do?

We are going to build a tester to test capacitors for D.C. leakage. You will need a small enclosure, two banana jacks (female—one red and one black), two alligator clips (one red and one black), a couple of feet of hook-up wire, a 9 volt battery, a 9 volt battery connector and a momentary push button SPST switch. You will also need a voltmeter, preferably a high quality digital meter.

LET'S BUILD A TESTER

Start off by mounting the banana jacks and momentary switch to the enclosure. Hook a wire from one jack to one side of the switch and from the other jack to the other side of the switch. (When we are finished, your voltmeter will plug into the banana jacks and the switch will short out the meter so the cap under test can charge.)

Next we are going to hook a wire to the red banana jack, with the other end going to a black alligator clip. The red alligator clip goes to the red wire (positive) of the 9 volt battery clip. The black (negative) wire from the battery clip goes to the black banana jack. Plug in the 9 volt battery and you are done.

LET'S TEST SOME CAPS

For starters, plug the positive lead from your meter into the red banana jack and the negative lead into the black banana jack. Set your meter for low voltage setting (10 or 20 volts). Connect the two alligator clips together and look at your meter. This will measure how much voltage your 9 volt battery is really putting out. Make a note of this amount because it is your applied voltage and will be used later to calculate the percentage D.C. leakage.

If you are using polarized caps (all electrolytics are polarized), the red alligator clip goes to the positive side of the cap and the black alligator clip goes to the negative side of the cap. Correct polarity is very important, because if you get the hook-up backwards, even a good electrolytic will leak 100% of the applied voltage! If you are testing a non-polarized cap, it doesn't matter which alligator clip goes to which lead of the capacitor.

Press the momentary switch and hold for a few seconds. This will charge the cap. Release the switch and watch the meter. You will see it go from zero up. When the reading stabilizes, divide this number by the applied voltage (measured earlier) and you will have the percentage D.C. leakage.

Depending on where the cap is used in an amp, D.C. leakage of even 1 percent can cause dire consequences in an amp. For instance, a coupling cap in a 300 volt circuit that is leaking 3 volts will absolutely ruin your tone and throw the following stage's bias completely off. On the other hand, a filter cap that is leaking 3 volts will not cause any noticeable effect on tone.

OTHER OPTION FOR COUPLING CAPS

I like to test coupling caps in the actual circuit they are in. I do this by disconnecting the end of the coupling cap that does not go to the plate of the tube. I connect the positive end of a voltmeter to this end and the negative side of the meter to the chassis ground. If it leaks more than a tenth of a volt, it gets replaced.

TEXAS
TONE DEL MAXIMO

UNDEFINABLE TEXAS TONE DEL MAXIMO—DEFINED

When I perform, people expect more from me in the way of killer guitar tone. I am not sure if it's because I'm from Texas or because I own an amplifier company, or a combination of the two. For the last several years, I have been totally happy with my tone. The tone I get on stage actually "turns me on" so much that when I'm performing, I become inspired to play riffs, figures, and phrases that I feel certain could never have happened otherwise.

Don't expect me to explain every detail—my purpose here is to share the basics and then it's up to you to personalize your sound to your own ear. You must have the tone that will satisfy you! The kind of tone I'm talking about is really so subtly multi-timbred and layered that it defies definition in words, although I can give you part of the answer.

DIFFERENCES IN AMPLIFIER TONE

Consider the fact that vacuum tube amps all sound differently depending on circuitry, power supply, tubes, and speakers, etc. For instance, a blackface Showman with its four 6L6 output tubes, single 15" speaker, fixed-bias circuitry, and solidstate rectifier will have a completely different tone from a 1954 tweed Super that uses two 6L6Gs, a pair of 10" speakers, cathode bias circuitry, and tube rectifier. Both tones are valid, but how are they different?

They are different are in the following ways: the overall envelope of attack, the compression, overall sustain, clarity, frequency responses, distortion characteristics, and harmonic content. So an amplifier has a voice of its own. If that's the case, why not use more than one voice? If three voices could sing perfectly in tune and phrase together so perfectly that it sounded like one multi-timbred voice, wouldn't that sound better than one voice? There are some other

Texans that figured this out long ago. Stevie Ray Vaughan, Jimmy Vaughan, Billy Gibbons, and Eric Johnson have all used multi-amp set-ups for both live and recorded performances.

USING A MULTI-AMP SETUP

I will explain to you my personal setup and then you're on your own to develop your own rig. I use three amplifiers, each chosen for a distinct voice. My idea was to balance these three voices so that the sound was both clean and dirty, while sounding compressed and dynamic. Does this sound like a paradox? Read on!

The first amp/voice was selected for clear, clean bottom-end without much top-end. I wanted very little compression with a tight bottom-end. The second amp/voice was selected to have a nice big sag to the envelope with a not-so-tight bottom-end. This amp would have enough smooth distortion to cream-out the tone and would have a full-range frequency response. The third amp/voice was selected for airiness and presence. This one would not have much bottom-end and would have a very small floating baffleboard for maximum airiness. I want more distortion and grind in this amp but without the compression on the front of the note. This amp must have good sustain.

HERE'S WHAT I USE

The first of these is a Kendrick 4000 head that has two output tubes removed. This changes it from 80 watts 4 ohms to 40 watts 8 ohms. This takes advantage of the 80 watt power supply used on a 40 watt circuit to tighten up the bottom. I use 5881s with a solidstate rectifier. The amp is fixed-biased and the speaker cabinet is a Kendrick 1x15 cabinet (floating baffleboard) loaded with a JBL D140 bass speaker. The controls are set for maximum clean. If you heard this amp by itself, you'd probably laugh and then decide you hate it. The bass speaker will not produce highs at all, but the bottom is full, rich and tight. There isn't much sustain. I set the cabinet on the floor for a bigger bass sound.

The second amp is a Kendrick Spindletop 100 watt Texas Crude Head. This amp uses EL34s and the bias selector switch is in the cathode-bias mode. It pushes a Kendrick 2x12 cabinet loaded with a pair of Brownframe 12" Kendrick speakers. This gives me a full range sound with lots of sag to the envelope. The mids bark and the highs are creamy.

Of course those EL34s have a certain grind to the distortion that is very smooth. The cathode bias mode allows singing compression and sustain but without much attack on the front-end of the note. It also has tons of bottom-end, but this bottom-end is much looser than the first amp.

The third amp is a Kendrick Do-Awl Combo. This amp uses 5881s and is fixed-bias. The cabinet is a V-front style like an old Dual Professional. This cabinet style is extremely airy without much bottom-end. Two 10" Blackframe speakers are in this amp. The circuitry of this amp includes a special tube that generates harmonics and rounds out the waveform. Since it is fixed-bias, the attack is fast. I use a Real Tube pedal in front of this amp which adds yet another color to the distortion characteristics. This amp is not on the floor! I set it where the speakers are at guitar string level for three important reasons. It maximizes airiness, minimizes bottom-end, and interacts with the strings on my instrument so that I can obtain harmonic feedback at will. Actually, I can get both fundamental and harmonic feedback depending on where I'm standing! An added benefit is that when I get feedback from this one amp, the strings on my guitar are resonating, so I still get fundamental notes coming out of the other two amps, even though the 1x15 cabinet doesn't normally have much sustain.

FINE TUNING THE SET-UP

Of critical importance is the actual balance of volume of these three amps. When I am playing, I cannot actually hear the sound coming out of the 1x15 cabinet, but if I turn it off or down, I can really feel the difference. I like the beef to be coming out of the 2x12 cabinet, but with enough of the 2x10s to cut through and define the front of the notes.

You need only experiment to find the voices that you want for your personalized sound. And you don't have to turn the volume up to get great tone. The first time that I set up three amps at rehearsal, I saw a worried look on my drummer's face. He assumed I was using three amps to get three times louder. I actually set the amps at a very low volume and it sounded great. His look of worry changed to one of surprise when he heard it. If you are concerned about carrying around this much equipment, get a roadie, a stronger girlfriend, or just accept the fact that you can't get the Texas Tone del Maximo any other way.

FINE TUNING THE TEXAS TONE DEL MAXIMO

Since the article, "Undefinable Texas Tone Del Maximo—Defined" appeared in the January 1996 *Vintage Guitar* magazine, I have gotten mucho el requestos for more information on how to "fine tune" a three amp set-up. There are considerations concerning grounding, A.C. wall voltage polarity, speaker polarity, and basic hook-up considerations that will make a huge difference in tone, noise and hum.

To refresh your memory, we talked about connecting three amplifiers together that all have valid but different tones. The idea is to get a layered, multi-timbre tone that contains many facets of killer tone—all at the same time. For instance, it can sound compressed yet dynamic—clean yet distorted—clear yet grindy—all at the same time!!! Many Texan players have known this for years (Stevie Ray, Billy, Jimmy, Eric, etc.) We also pointed out that the tones obtainable by "Texas Tone Del Maximo" are simply not available when using only one amp.

The original article only gave the basics and many who tried running three amps succeeded in producing "amplifier ground loop hum del maximo" and in some cases phase cancelled tone, which sounds pretty bad. The purpose of this chapter is to clear up the how's and why's and what's concerning "Texas Tone Del Maximo" done right—the way we do it in Texas.

ELIMINATING GROUND LOOP HUM

TEXAS TONE DEL MAXIMO AXIOM #1: When using a multi-amp set-up, remember to have only one piece of equipment earth grounded (via the A.C. three prong wall plug) and lift the A.C. plug earth ground on all other pieces of equipment! This is a very basic rule whether interconnecting amplifiers, recording gear, home stereo, or any audio device that is connected to another audio device. The reason is simple. The earth ground on your amplifier is the third prong on an ordinary three prong A.C. (male) wall plug. It is actually connected internally to the device's metal chassis. The wall outlet has that connection going to an earth ground. If everything in the set-up is earth grounded through this third prong, it seems like that would be correct—but it is not. When hooking more than one audio device together, the interconnecting cable shielding makes a connection to the chassis through the sleeve of the quarter inch phone jack. That is to say—when two amplifiers' inputs are connected, their chassis' are also connected together. If each has its own earth ground via the three prong wall plug, then you will have everything double grounded which will result in a ground loop. The loop would actually start at one amp's chassis, go through the shielding of the inter-connecting cable to the next amp's chassis, from that chassis to the third prong of the second amp's A.C. plug, through the wiring of the house, back through the third prong of the first amp and back to the first amp's chassis. Now we're talking "Hum el Grande." With only one piece of equipment directly earth grounded (via the three prong wall plug), every device in the set-up has only one path to earth ground and the loop is broken. If the interconnecting cable is in place, everything in the set-up will "see" an earth ground (via the interconnecting cable), even though several devices may not have a direct connection to earth ground via the three prong wall plug!

A word of caution—make sure to connect all audio cables before connecting the A.C. wall voltage to any device. If you mistakenly plug amplifiers into the wall A.C. before signal wires are connected, you will be dealing with ungrounded equipment. This could result in electrical shock because you will undoubtedly touch two pieces of equipment with different chassis potentials. Remember the grounds are "made" through the audio hook-up cabling.

ELIMINATING POLARITY HUM

TEXAS TONE DEL MAXIMO AXIOM #2: When using a multi amp set-up, correct A.C. wall polarities must be proper, or excessive hum will occur. To understand this, consider that not all A.C. wall outlets are wired correctly. Of the two A.C. leads, one is supposed to be hot and the other neutral. Many amplifiers have a "polarity," or "reverse" switch to change the A.C. wall voltage polarity. (Note: On some of the earlier vintage amps, the polarity switch is labeled "ground" and the A.C. male connector is two prong). Turn the "polarity switch" one way and the hum will increase; turn it the other way and the hum will decrease. I would recommend using only one amp and finding the setting that particular amp likes and then adding another amp. Through trial and error, check both polarity settings on the second amp and choose the setting where hum is at a minimum. Add yet another amp and continue in the same manner.

Depending on what you are using to connect the inputs of all these amps, you may need to adjust the polarity of that device as well, but only if it uses A.C. wall current instead of batteries. In my personal three amp set-up, I use the Kendrick ABC box which uses wall current. I must add that to the first amp and adjust polarity before adding the second amp. If you are using a Morley ABY box to connect the amps' inputs, you will not need to perform this test, because the Morley uses batteries (D.C.) and therefore is not affected by A.C. polarity.

What do you do if the amps you are using do not have polarity switches? It really depends on the type of A.C. male plug they use. If it is a vintage two prong male plug, simply turn it over in the wall outlet to reverse polarity. If it is three prong plug and you are using a ground lift adapter, you may need to do a slight modification. The three prong to two prong A.C. ground lift adapter assumes that the wall outlet is wired to correct polarity and therefore has one prong slightly bigger than the other. This assures that the plug can only go in one way. The bigger prong of the ground lift adapter can be filed down to match the size of the smaller prong and the plug can then be reversed in the wall outlet. Another possible suggestion: Get a six way outlet strip and experiment to find how everything plugs in with the least amount of hum. Using different color markers, color code each groundlift adapter/male plug and each female outlet on the strip. Leave the ground lift adapters permanently installed on the line cords. Every time you set up your amplifiers, plug them up the same way and you will reduce your set-up time because you will not need to do any testing. If your amp has a "polarity" or "reverse" switch, you might also want to mark the switch settings with post-it tape to eliminate any guesswork here.

CORRECTING FOR PROPER SPEAKER PHASING

Earlier articles have addressed proper speaker phasing, but here's the TEXAS TONE DEL MAXIMO AXIOM #3: All the speakers must move the same way at the same time or phase cancellation occurs thus ruining your tone.

How do I correct phasing? The simple way is to listen to two amps together. Reverse the speaker leads on one of the amplifiers and listen again. The hookup combination that sounded like Texas Tone Del Maximo is correct phasing. The other way which sounded like Minimus Tonus del Minimo was the incorrect hookup combination. With the first two amps connected properly, with respect to phase, add a third amplifier and listen. If it sounds better than the first two amps by themselves, you are hooked up correctly. If, when adding the third amp, you notice it sounding not as good as just the two amps by themselves, then you know the third amp is hooked up in-correctly. To remedy this situation, simply reverse the speaker leads of the third amp and every speaker will be moving the same way. Important note: When listening to these amplifiers, make sure the speaker cabinets are all pointing the same direction. Properly phased amplifiers will be out of phase if they are facing each other!

PROPER PHASING FOR YOUR AMPLIFIERS

Anytime you are using more than one speaker or more than one amplifier, proper phasing is absolutely critical to your tone. If any speaker is out of phase with any other speaker within an amp or if one amp is out of phase with any other amp, a phenomenon known as "phase cancellation" will occur and will ruin your sound.

WHAT IS PHASE?

Sound is a vibration or wave. With a guitar, the vibration or sound wave occurs as a string moves back and forth. This is then amplified and drives a speaker that also vibrates back and forth. Actually the speaker's cone moves forward then backward very rapidly. There are two possible ways it could begin its vibrations. It could go forward **first**, then backward, or it could go backward **first**, then forward. If it goes forward first, then backward, it is said to be "in phase." If it goes backward first, it is said to be "out of phase." There is a little more to it, but for now we will call forward "in phase" and backward "out of phase."

WHAT IS PHASE CANCELLATION?

A speaker could be considered to be an air pump. When the speaker cone pushes forward in its vibration cycle, the air around it is also pushed forward. When it moves backward in its vibration cycle, the air around it also moves backward. Of course, this happens very quickly; a cone could move back and forth several hundred, even several thousand, times per second. When you have two or more speakers in a cabinet, and they are all going forward at the same time, they will also be going backward at the same time. In this condition, the speakers are said to be wired "in phase" with each other. If one speaker's cone is going forward while another one is going backward, the speakers are said to be wired "out of phase" with each other. When two speakers are wired "out of phase" with each other, the air in front of the speaker cone being moved forward rushes to fill the partial vacuum created by the other speaker cone that is moving backward. Likewise, in back of the speakers, the air moving backwards from one speaker cone rushes to fill the partial vacuum created by the other speaker cone moving forward.

This phenomenon is called "phase cancellation." The vibrations from one speaker are actually canceling the vibrations of the other. The resulting sound can be summarized as follows: extremely thin tone, absolute zero bottom-end (because low-end frequencies cancel each other more easily since the air is not moving/vibrating as quickly), no dynamics (when you attempt to accent a note, the speakers simply phase-cancel these dynamics), very little or no sustain.

HOW DO I CORRECT THE PHASE OF MY SPEAKERS?

When two speakers are in an amp and they are wired "out of phase" with each other, all you need to do is reverse the leads on one of the speakers. This will reverse the phase of the one speaker and now they will go the same direction at the same time.

Theoretically, you should be able to place a 9 volt battery on the speaker leads with the positive terminal of the battery going to the positive terminal of the speaker and the speaker cone should move forward. (Do not leave it connected for very long, because speakers don't tolerate D.C. voltage very well!) In fact, this is the test that is used to determine which speaker lead is positive and which is negative. Simply touch the battery to the leads to make the cone move forward. If it doesn't move forward, reverse the battery leads and the cone should go forward. Once you have determined which way the battery terminals must be connected in order to make the cone move forward, note which terminal of the speaker is touching the positive lead of the battery. This is your positive speaker terminal. Do not trust any markings on a vintage speaker because before there was any standardization on speaker markings, some manufacturers, including Jensen, marked speakers exactly the opposite from what has become today's standards. If you had a pair of these speakers and one speaker had been reconed using a modern coil, chances are that these speakers would not be marked consistently on the terminals. Hook them up using the terminal markings as a guide, and you might have two speakers canceling each other. That is why you must always check speaker phase with a battery.

AMPLIFIERS HAVE PHASE TOO!

Just as speakers must operate "in phase" with each other, so must amplifiers. If the positive portion of a signal appears at the input of an amplifier at the same time the positive portion of the amplified signal appears at the output, the amp is said to play "in phase." If the speakers are connected to the amp so that the negative lead of the speaker is connected to ground and the positive lead goes to the "hot" lead of the amplifiers output (standard speaker wiring), the amp is said to play "forward."

Conversely, let's suppose that the positive portion of a signal appears at the input at the same time the negative portion of the amplified signal appears at the output. This amplifier is said to play "out of phase." It is said to "invert" the signal. If the speakers are connected to this amplifier with negative to ground and positive to "hot"(standard wiring), the amplifier is said to play "backward."

ARE ALL CHANNELS IN PHASE?

Some amplifiers have two channels that are not "in phase" with each other. A good example of this is a blackface Twin. In this amp, as well as many blackface Fenders with reverb, the normal channel plays "in phase" while the reverb channel plays "out of phase" (inverts the signal). If you try to bridge the channels on this type of amp, you will get "phase cancellation" in the amplifier itself! This will, of course, sound terrible.

PHASE WHEN USING A MULTI-AMP SETUP

Let's say that you are going for Texas tone del Maximo and you want to use a multi-amp setup. It is imperative that every speaker in every amp move forward at the same time. This cannot be accomplished without considering which amps are playing "in phase" and which ones "invert" the signal. Speaker wiring is usually done with the negative terminal of your speaker going to ground and the positive terminal of your speaker going to the amplifier's hot lead. Although this is general practice, the wiring can be reversed to "re-invert" the signal of an amp that normally plays "backward," so that it will be "in phase" with the others.

In other words, you can take an amp that "inverts" the signal and therefore plays "backward," and make it play "forward" by simply reversing the speaker leads connecting the amp to the speaker.

I have double-checked every amp I own so that every amp plays "forward" whether the amp chassis inverts the signal or not. How can you tell? One way is to listen to two amps on at the same time. If it sounds extremely thin, with no bottom-end, the amps are fighting each other (out of phase). After you reverse the speaker leads on one of the amps, the amps should play "in phase" with each other. Now add another amp and listen to it. Here again, if the tone doesn't inspire you, you may need to swap phase on the speaker, too. Proceed in a like manner until all amplifiers are sounding good with each other. Get out your guitar and get inspired!

Important note: When listening to these amplifiers, make sure the speaker cabinets are all pointing the in the same direction. Properly phased amplifiers/speakers will be out of phase if they are facing each other.

Q&A

FENDER

Can anything be done to reduce the pulsating noise of Fender amp vibrato?

Yes and no. It depends on where the noise is coming from, and what type of vibrato circuit you are using. Fender used four different types of vibrato circuits. Actually three of these are tremolo and only one type is true vibrato, even though Fender did not make the distinction. They are:

1. PREAMP TUBE BIAS MODULATION. This was only used on a very few Fender amps. The black Vibrochamp and the tweed 5E9A Tremolux are really about the only two I can think of that used this circuit.

2. OUTPUT TUBE BIAS MODULATION. Though rarely used, this circuit was found on the brown Deluxe, blackface Princeton, tweed Vibrolux (model 5F11 and 6G11) and perhaps the most famous — the brown Vibroverb.

3. PITCH VIBRATO. This was used on most brown amps such as the Bandmaster, Concert, Pro, Showman, Twin, and Super. You will never find this circuit on a blackface or tweed amp. It uses two or three tubes and actually alters the pitch of the notes.

4. OPTOCOUPLER GROUNDING OF THE SIGNAL. This was used on almost all blackface amps except the Princeton and the Vibrochamp. You will never find this circuit in a tweed or brown amp.

Now we must look at the source of the noise. If the noise is coming from the preamp tubes, changing to quieter preamp tubes will help dramatically. Because the vibrato circuit is actually making the gain of the amp increase and then decrease on all of the circuits mentioned except type #3 above, noise coming from preamp tubes or just gain hiss will be more noticeable with the vibrato circuit on. Circuit type #3 above will have a swirling sound if the preamp tubes are noisy.

Sometimes noise will be coming from the plate load resistors. (These are the resistors connected to pin 1 and 6 of the 12AX7 and 12AT7 tubes.) These are made from compressed carbon and can have small pockets of air inside. Electrons will sometimes arc across the small pockets and cause noise. Changing the resistor with the same value resistance, but upgrading to twice the wattage rating of the original, will cure this problem. Sometimes humidity will be absorbed into the carbon, in which case the noise will be more of a popping sound. This can sometimes be remedied simply by leaving the amp on and in the play mode for three or four hours.

Pitch vibrato has a beating that is not caused by resistors, humidity or tube noise. It is caused by the oscillator clipping. Using a weaker 12AX7 in the oscillator socket (usually the third socket from the end on the opposite side of power tubes) will help. Try a Sovtek 12AX7WA; these have slightly less gain. Voltage tweaking may also help, but I have never seen one that didn't beat a little.

I own a Showman (blackface) head. How do I know if it is 4 ohm or 8 ohm? Didn't Fender make them two different ways, depending on what kind of speaker it was sold with?

You are correct. Fender did make the Showmans in two styles, the difference being in the output transformer. The 8 ohm version was for the single speaker Showman and the 4 ohm was for the Dual Showman. An easy way to tell which you have is to check the transformer number. The 4 ohm output transformer will have either numbers 022889 or 125A29A; the 8 ohm output transformer will be marked either 125A30A or 022897.

Wires from tubes and components are dressed differently on my Fenders. Pre-CBS has wires twisted together dressed flush to chassis. CBS has hook-up wire coiled around other wires (like shielding) and grounded at both ends. Now I see a transition silverface dressed the same as my CBS but with shielded wire (it looks like microphone cable) grounded only at one end. Could you revive the art of lead dress and layout and explain what is going on here? With modern shielded wire going from the preamp to the front panel controls, why do we ground the shielding only at one end, usually at the grounding strip on the inside front panel?

Lead dress and layout are very important things to consider when designing an amp. If done correctly, the amp will be stable and not prone to parasitic oscillations. If done incorrectly, the amp may have

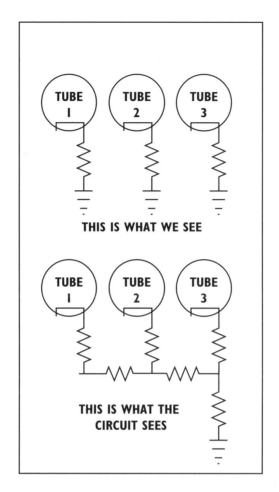

THIS IS WHAT WE SEE

THIS IS WHAT THE CIRCUIT SEES

intermittent problems or may not work at all. As a matter of fact, many silverface amps were unstable and would not work without corrective circuitry to bleed off high-end. (I'm talking about those 2200 pf disc caps going from the grid of the output tubes to ground. This was a real tone killer, but necessary because the rest of the lead dress was bad.) Lead dress and layout also make a difference in hum levels of the amp and frequency response. As gain increases, lead dress and layout become even more critical.

Layout and lead dress rules are far too complex to fully explain here, but I'll give you the basics in a nutshell.

1. Inputs should be as far away as possible from the outputs, the signal should move along gradually from input towards output and not cross back over on itself.

2. The power supply should be far away from inputs and the grounds should occur physically in the same order as they occur schematically (looking from the power supply towards the input.)

3. Grid wires should be short; plate wires should be long.

4. A.C. filament heater wires should be twisted in pairs and close to the chassis with some kind of ground reference. (The best ground reference would be a center tap on the transformer, but there are three other ways this could be done without a center tap.)

5. When using shielded wire for low level signals connected to the grids of preamp tubes, only one end should be grounded. If both ends are grounded, a ground loop could occur where signal or hum is picked up by the shielding and moves to one end of the shielding, through the chassis and back to the other chassis ground on the shielding and back through the shielding to the other chassis ground. Round and round she goes.

6. To reduce the loading effect of the resistance of the chassis itself, a grounding buss made of a very low resistance conductor that is easily soldered (such as copper) should be used. When you consider that all preamp tube cathodes are going to ground (via the cathode resistors), you can see that a resistance in the chassis, between that particular cathode and the power supply ground, would be seen as an extra cathode load resistor (with signal developed across it) and attached to another cathode load resistor (actually a resistance in the chassis) going to the next stage. See diagram above.

Fender switched from 400 volt to 600 volt caps in the late 60s. Should I be afraid to use 400 volt caps in a silverface Pro or Twin? Have you ever found, given the same brand and type of cap, the voltage rating to affect sound?

I assume you are talking about the coupling capacitors. Don't be afraid to use 400 volt caps in your amp! In actual construction of a capacitor, to make a 600 volt cap, the distance between the two conductors must be increased. This would cause the capacitance to drop, so in order to keep the capacitance the same, the surface area of the conductors must be increased. Therefore a 600 volt cap will have more space between the conductors, but with greater surface area. This would make a difference in the sound. There are many other things that would make a bigger difference—for instance, what dielectric the capacitor uses. Some caps that (in theory) should sound great, actually sound horrible. Good tone is not theory and math. My advice is to experiment and trust you ears.

I have two 1953 Deluxes, both are model 5C3, and both are stamped CK. One of them has for tubes: 1-5Y3, 2-6V6 and 2-6SC7, the other has 1-5Y3, 2-6V6, 1-6SC7 and 1-12AY7. The one with the 12AY7 has the same tube chart as the other one and the socket screws look like they have never been taken off. Did Fender have both of these setups? Could both amps be stock, with one of the amps with the wrong tube chart?

Both of them could be stock. These amps were built at a transitional time when Fender was changing from the 8 pin preamp tubes to the 9 pin miniature. Often they would simply cross out the tube on the chart and write in the correct tube. I have seen many chassis of this era that were punched for the 8 pin socket, but had adapters with the 9 pin miniature socket installed.

What is the correct preamp tube for the Fender Princeton 5B2? Is it 6SC7 or 6SL7? I've seen tube charts with the 6SC7 crossed out (on the chart that's in the amp) and 6SL7 written in its place. Do you know which one is right?

Though both are twin triodes and both have an amplification factor of 70, the 6SC7 and the 6SL7 have different base connections and are therefore not interchangeable. The 6SC7 has only one cathode that is used for both triodes and has no connection on pin #1. The 6SL7 has separate cathodes for each section. Pin #1 is the extra cathode base connection. Check pin #1 and see if it goes to a cathode resistor. If it does, my guess is that you need a 6SL7. If there is no connection on pin #1 of the socket, you will need a 6SC7. At this time in Fender's history, they may have been experimenting with different tube types and didn't want to have special labels printed; so they simply crossed it out on the tube chart and wrote the tube type by hand.

I have a stock '59 Bassman re-issue. Other than installing a rectifier tube, what can I do to make it honk like an old one? César

Diaz told me there are a couple of resistors I can change. What tubes can I use for the first preamp tube? I use a '52 Telecaster and play through the normal channel.

Although the '59 Bassman re-issue is, in my opinion, the best sounding amp Fender has made in decades, you don't really want to know what you would have to do to get your re-issue to sound like an old one. Since you asked, I'll tell you; then I'll tell you what you could do short of that to make it sound better than it sounds now.

For starters, get rid of the cabinet and have a new cabinet made out of solid pine. The solid pine has a woody resonance whereas the stock plywood sounds flat and lifeless. Next, replace both the power and output transformers. Unlike an original Bassman, the stock re-issue power transformer voltage is too high on the B plus and will always sound harsh and brittle. The stock re-issue output transformer must be replaced. Next, get rid of the printed circuit board, gut the inside and re-wire the entire amp point to point. Make sure and use Allen Bradley carbon composition resistors and not the cheap carbon film resistors that are in the re-issue. Use the resistor wattage ratings that were used in the original '59 (most are twice the wattage rating used on the re-issue). Next replace the 10" speakers, everyone knows what kind to use. Now replace the tubes— all of them—use Tung-Sol 5881s and American made 12AX7s and a 12AY7 for the first gain stage. Use a Mullard GZ34 rectifier. All of this work could be done for somewhere under $1600.00—not bad when you consider how good it will sound.

Here's what you could do short of that to make it sound much better than it sounds now. Replace the first and second preamp tube (counting from the input side wall) with 12AY7s. This will draw more current and ultimately reduce the preamp tube's plate voltages (and preamp harshness). Replace the output tubes with Tung Sol 5881 power tubes and bias them to draw 35 mA a side. Try both a GZ34 and a 5U4 rectifier tube and listen for the envelope characteristics. Choose the best sounding one for your music. Replace the speakers with four Kendrick 10" speakers. These are made with real 3KSP paper cones pressed on the same original dye as Jensens. Put the re-issue speakers in the attic. They will probably sound pretty good in about 25 years when the Alnico 5 starts to mellow.

Where can I get an exact replacement fuse holder for my '64 Black Tolex Champ?

That part is still being made today. It is Littelfuse part #342012A and it can be purchased at any electronics distributor that handles Littelfuse products.

I've got an early 60s Bandmaster Combo that came to me with two Radio Shack speakers, wired for 4 ohms. It is basically a nice sounding amp, but it is anemic with those speakers. If I hook it up to my Super Reverb, it's dramatic—it sounds great. The best vibrato ever. What speaker load do I want? If I ask amp guys around here, they don't believe it came with three 10s, but it did. In just the right light you can see 3 faded circles in the grille cloth. The transformer is 125P7A.

Certainly it came with three 10s, that's the only way Bandmasters were made until '61 when the piggy-back design was introduced. The amp requires a 2.66 ohm speaker load. That transformer number you gave is a power transformer that was used on the Bassman, Pro, Super (2x10), and Bandmaster.

My '61 Super amp (brown Tolex, 2x10s, model 6G4A) has lots of power for its size, but nothing I've tried so far can make its tone approach the transparency, sparkle, and just generally groovy sound (at any volume) of my Bassman or AC30. The Bassman and Super have the same year and model Jensens in them. I've tried all sorts of coupling cap mods, etc., to try and de-muddify the Super, but it still sounds dull and lifeless next to the Bassman. Any suggestions? (I know apples can't be oranges.) All voltages check out per schematic, etc. Is the tremolo circuit eating my signal? I can do without it.

What you have here are two different amps. Although the output stage and speakers are basically identical, the preamp circuitries are entirely different. The Super does have a tone shaping cap (across the second gain stage plate load resistor) that bleeds off high end. Perhaps removal of this cap (.003 mfd) will give you the clarity you desire. There is also a .003 mfd on the treble pot to ground. I suggest experimenting with removal of it as well.

This amp could definitely be modified to do what you like; however, fine tuning an amp for tone is a very personal thing that should be done by ear. Trial and error listening tests while doing modifications are the best way to fine tune the tone.

If you have not had the amp overhauled recently, I suggest that you do so before trying any modifications. Oftentimes, faulty worn-out components can adversely affect the tone of an amp and therefore modifications may not be necessary. If the amp has bad components in it and you try to modify it, you might be correcting the symptom instead of the problem.

The voltages on early Fenders should read higher than the schematic (6% to 12%) because the amps were originally built for 110 volts and today's wall juice is more like 120 volts (or more).

The vibrato circuit is probably not eating signal. Later blackface vibrato circuits had that problem. I would recommend leaving the vibrato alone. It is a cool circuit (three tube pitch vibrato) and probably the best sounding vibrato circuit that was ever put on a Fender amp.

I have heard that in the tweed era, the Champ was the only Fender amp that had a 4 ohm speaker. But I have checked the ohms on three tweed Princetons that I believe are completely original and the speakers were all 4 ohm. Can you set the record straight? Also will an 8 ohm speaker hurt an amp that should have a 4 ohm and vice-versa?

This sounds logical and I am not surprised. Early tweed Princetons were actually a Champ circuit, but with the addition of a tone control. The transformers were the same and therefore a 4 ohm speaker would be appropriate.

Most amps are very tolerable of mismatched impedances, especially mismatches of 100% or less. Certain amps, especially British amps (generally because of the lower primary impedances on the output transformers) are more sensitive to impedance mismatches. Maximum signal transfer occurs when the impedances are matched; therefore, any mismatch will hurt efficiency. This means that an 8 ohm speaker on a 4 ohm amp will result in less power transfer and a higher reflected impedance to the tubes. This means the output tubes will not be working quite as hard because of the current limiting effect of the higher reflected impedance. The tubes should last longer. Using a smaller impedance speaker, say a 4 ohm with an 8 ohm amplifier will also result in less power, but a lower reflected impedance to the tubes. This means the tubes will work extra hard because the lower reflected impedance tends to encourage higher than normal current flow. The output tubes in this circuit will not last as long, all other factors being equal.

My '56 Champ has a speaker coded 395606. I examined a '54 Champ with a 6" speaker that also had the 395 prefix. The '54 looked original. I cannot find this prefix listed anywhere as a manufacturer's code, and it seems like the chances of both of these Champs having the same prefix code on the speaker is too remote for this to be a coincidence. If you could tell me what kind of a speaker this is, it would solve the mystery. Thanks!

Although it is not listed in the manufacturer's source codes, the "395" prefix was used by a company named "Permaflux." As far as I know, the company is no longer in business. I can not be certain that they made the Champ speaker, but I know they did sell aftermarket voice coils.

I have a transformer question. My '53 tweed Deluxe has a T538, which I believe is original. Would this work in a '54 Deluxe? What is the correct transformer for a '54?

The tweed Deluxe transformers are all interchangeable. The number you gave (T538) does not sound familiar. The '54 Deluxe used an 0246 output transformer. These were also used on the 5E4A Super Amp.

How about a diagram for changing a '69 Twin balance adjust over to a straight bias adjust (and why convert anyway?) My '69 Twin doesn't have an overseas voltage selector switch, but the tube chart says "For Export only-117 volt." What gives?

Check out the silverface to blackface conversion chapter in my other book, *A Desktop Reference of Hip Vintage Guitar Amps*. There is a detailed description (with graphics) of how to change the balance bias to adjustable straight bias.

Why change? Well, for one thing, the bias balance adjustment will not **set** bias, but can only balance what is already set. An adjustable straight bias (blackface style) will allow real adjustment of bias without having to alter fixed resistors in the circuitry.

Export transformers usually sound better because they are built to operate at 50 cycles and this means extra core material. When they are run at 60 cycles, there is less loss. Export transformers have several extra wires for the primary taps. You can look at the power transformer and see if there are several extra wires that are just taped off and not being used. This will let you know if you have an export or domestic transformer. Not all export models used a selector switch, but all export Fender transformers have the extra unused wires.

Can I use 6L6 power tubes in my Super 5E4 which calls for 6V6. My power transformer is a Peerless 480 and I read 410 volts on the power tubes. I notice that the schematic of the 5F4 Super (which uses 6L6 tubes) reads 415 volts, so I'm assuming I can do this by changing the resistor that controls bias (6800 in 5F4 and 18K in 5E4.) If possible, what should the bias setting be? Do I need to change anything else? You recently said that the selenium rectifier used on my Super was used to change A.C. to pulsating D.C. Can I still get that part or is there a different way to achieve this with different parts?

Yes, with proper modifications, you can use 6L6s in your Super. The non-original Peerless transformer may not be hefty enough. I really don't know, because I am not familiar with that particular model number transformer. You could try it and see if the transformer overheats and check the A.C. filament voltage. If it doesn't overheat and the A.C. filament voltage is 6.3 volts or better, it will work. You will also need to make other modifications.

Among these other mods: Change the output transformer to one that is specifically for a pair of 6L6s to 4 ohm speaker load. The original transformer is actually wound for about 15 watts and the 6L6s will burn it up. Also, you are on the right track about the bias circuit, but your thinking is a wee bit off. You will need to set bias, but since you are using a non-original transformer, I recommend using the shunt/current method to find the exact resistor values. Do not assume anything about the 18K resistor in the 5F4 schematic. That resistor value just happened to be the value needed with original transformers and 110 volt line and a new selenium rectifier. None of those conditions exist now.

Selenium rectifiers will often leak current when they get old. They can also become unstable and are affected by heat. I recommend replacing a selenium rectifier with a silicone rectifier. I use a 1N4003.

This rectifier circuit does change A.C. bias voltage to pulsating D.C. and the capacitors in the bias circuit change the pulsating D.C. to pure D.C. Do not confuse this bias voltage rectifier with the rectifier for the power supply.

I am considering buying either a Fender Princeton/Reverb or a Deluxe/Reverb. I want to use it in a band situation and for possible club work. Which one will be best for getting that great cranked up Fender sound (turned up to 10) without being too loud for the situation? Would a power attenuator be the answer? Also, I have a Fender Bronco amp. The transformer is dated 1971. Is it anything rare? Some people have said they are, some say they aren't, but regardless it sounds killer.

I would hardly think you would need a power attenuator with a 20 watt amp unless your next gig is the midnight show at the old folks' home or your band rehearses in the lobby of the municipal library. I assume that you play with a drummer.

Either amp would be a good choice if it were overhauled electronically and had a good sounding speaker. They are both 20 watts and both sound good cranked. The Princeton has a 10" speaker while the Deluxe has a 12". The circuitry is very different, but they both sound Fendery.

What is important here is whether you get the reverb version of the Princeton or Deluxe. You will not be happy with either the Deluxe or the Princeton if you get it in a non-reverb version. Even if you never use the reverb, the one with the reverb sounds much, much better. This is due to the fact that it has an extra stage of gain that makes all the difference in terms of dynamics, sustain, natural compression and overdrive.

Your Bronco is rare. Not many of those were built and that makes them rare.

I have a '65 Fender Deluxe amp. The amp appears to be all original; however, I'm not certain about the speaker, as it has the

"Jensen Special Design" sticker and no number on the frame. I've got another 1965 amp that I bought new from the store and the case is the same, a sticker, but no stamp. The part that confuses me is that the speaker has a metal dustcover that you can see through the grille. Since I'm new at this, I do not know if this type of speaker was available back in '65. It really doesn't matter as it sounds fine, but could it be that it's the original speaker?

Also, this amp lacks power. I know it's only rated at 20 watts, but my 70s Champ is louder. I've swapped out tubes, replaced the rectifier and hooked up other speakers and I have the same results—good sound, low noise, not much power. The only things different about this amp are the rectifier and the fact that it is an export model with switchable voltages and channel one appears to be louder than channel two. What's your opinion?

The metal dustcap did not come on the Jensen speaker. My guess is that someone installed the metal dustcap to increase top end or the speaker was re-coned and the metal dustcap was installed at that time.

I think that your Deluxe has some faulty parts in it that must be replaced. The first thing I would check for would be leaky coupling caps. If they leak a small amount of D.C., let's say 1 or 2 volts, then the grid bias for the next stage is altered dramatically. Believe it or not, this tiny bit of leakage can cut the power of an amp to practically nothing. I think that it's time for a complete check-up on the amp. (Voltages checked, tubes re-biased,etc.)

I have a 4 ohm amp (e.g. Fender Showman) and would like to build a 4x10 speaker cabinet. Problem is, the best sounding speakers are 8 ohm, not 16 ohm. (Comments?) Is there some way to wire together four 8 ohm speakers to produce a 4 ohm load? Will I get a better sound building my 4x10 using mediocre 16 ohm speakers with proper load, or cool vintage 8 ohm speakers?

There is no way to wire four 8 ohm speakers to get a 4 ohm load. You have plenty of options. One option would be to wire the four speakers for 2 ohms and just run it at 2 ohms. Fender sold them with an extra speaker jack and an extra speaker. When you run them that way, it's 2 ohms anyway. As a matter of fact, in the 60s, my band used a Dual Showman for P. A. and we actually used two 4 ohm cabinets in parallel (2 ohms) for many years with no problems.

Another option might be to use six speakers wired in parallel series to get 5⅓ ohms. When you figure how much power that Showman can deliver, you might feel better running it through six speakers instead of four.

Another possibility would be to use two cabinets, each with four 10" speakers and each cabinet wired for 8 ohms. When both cabinets are plugged in to the head, the nominal impedance will be 4 ohms.

Still another possibility would be to wire four 10" speakers for 8 ohms and just run it with 100% mismatch. There are many players that actually prefer this setup. You will lose a couple of watts, but your tubes will last longer and the high-end will be smoother.

I've heard that if I replace the 6L6s in my blackface Pro Reverb (or any amp for that matter) with 6550 power tubes, I could boost my power from 40 watts to around 60 watts, with no fear of damaging the amp. Is it true? Do I need to re-bias? Also, how long should power tubes last in an amp used at least 12-15 hours a week? In a Fender Twin Reverb, what actually happens when you use it with only two of the power tubes? Also, in reference to my power tube life span question, is fuzzy low-end a symptom or could it be my stock speakers are losing it after 30 years?

It is possible to run your Pro Reverb Amp on 6550s. However, I would not recommend it without adding an extra filament heater transformer (6.3 volt). The problem is that each 6L6 draws only .9 amps, while each 6550 draws 1.8 amps of filament current. This means that a pair of 6550s will draw double the filament heater current of the 6L6s. That's enough to possibly overheat a power transformer and ultimately burn it up. You could use an auxiliary 3 amp / 6.3 volt transformer. It is best to switch your preamp tube heaters to the auxiliary transformer and leave the 6550s connected to the power transformer. I also recommend changing the output transformer as well. If you run 60 watts of power through a 40 watt transformer for a length of time, you are asking for trouble. Not only should you re-bias, you might need to mod the preamp gain to drive the 6550s hard enough to get 60 watts.

The life of a power tube is contingent upon four things:

1. The brand of tube.
2. The type of amp it is being used in.
3. How hard it is worked.
4. How many hours it is played. In general, the average tube should last somewhere around 1000 to 2000 hours.

When you use only two of the output tubes in a Twin, you are reducing the output by half, and you are cutting the load impedance in half. This will wear your tubes out prematurely, unless you do something to correct the load impedance. An easy way to correct the load impedance is to unplug one of the speakers or to use an 8 ohm speaker load.

Fuzzy low-end could be caused by speakers or output tubes. Simple substitution will tell the story on this. Try a different speaker and different tubes.

I purchased a rebuilt '59 Bassman. It has a new finger-jointed cabinet, but the baffle is mahogany plywood, possibly ¼" thick. What is the proper thickness and material for the baffleboard? Also, there are four bolts per speaker. Shouldn't there be a bolt for each mounting hole? How are these bolts attached to the baffleboard? There is a buzzing sound when I turn it up. I'm hoping a thicker baffleboard and more mounting bolts will solve this.

The original Bassman used ⁵⁄₁₆" plywood. This is not available anymore. I use ⅜" Medium Density Overlay (also called MDO). This plywood actually measures ⁵⁄₁₆". There are only four mounting screws per speaker. The actual screw is a 8/32 1½" flathead Phillips screw. These screws are countersunk from the front and a flat washer and nut are used on the back of the speaker. These speakers are not spaced exactly even.

There are many things that could cause a buzz. Check the nameplate. These sometimes buzz and they are not suspected as the problem. Dropped voice coil speakers can sound good at low volumes and buzz like a bee when cranked.

I own a very early Fender Tremolux amplifier. It was owned by Motown Records in the 60s. It is blond Tolex with an oxblood grille and has the standard size cabinet that was used for 10" speakers.

My bottom came from the factory with two 12" speakers. It was a special order and you can tell it was done that way. My problem is nobody believes me that it came from the factory with two 12s. Have you ever heard of anything similar, special ordering speakers from Fender back then? Any help would be appreciated.

I can not certify that your amp came from Fender with 12s and I have never seen one that fits your description. However, I have seen many special orders amps that were different from what was normally sold. I also know from personal experience that an amplifier company is capable of giving a customer what the customer requests, especially something as simple as speakers. I would suggest that you code date the amp and speakers to see if the speakers were made before the amp. If they were, it will strengthen your case about the amp being a factory special order.

I recently purchased a 1957 Bandmaster. The top speaker does not have a cover/cap over the magnet like the bottom two speakers. I was wondering if this was done on purpose by Fender, because it appears the chassis and transformer would not fit if it did have a cover/cap. All three speakers have matching numbers. I have read that you can lower power by removing the outside power tubes of a four output tube amp. Does the same thing stand true of an amp with two power tubes and removing one?

Yes, you are correct. The top speaker has had its cover/cap removed during manufacture in order to fit the speakers into the cabinet.

You can lower the output of a four output tube amp by removing the outside pair of power tubes and doubling the speaker impedance; however, that is not quite the same as removing one tube from a two output tube amp.

Please note that in a four output tube amp with two tubes removed, you are still operating push-pull and in Class "AB." If you remove one tube from a two output tube amp, you are running single-ended and all single-ended amps must be run Class "A" or serious non-musical distortion may occur. Most two tube amps are Class "AB," but will actually operate in Class "A" at low volumes. Therefore, you could probably remove one tube if you were going to run it at low volume. The exception would be if you were trying this with a cathode-biased amp, in which case the bias resistor would not develop enough bias voltage because only half as much current would be flowing through the cathode resistor.

You could try it and see, but I recommend that you constantly watch the output tube to make sure that the plates don't go cherry red on you. If they do, simply turn off the amp quickly. There are many lower gain amps that would work nicely at low volumes when using this trick.

I like a solidstate rectifier and 6L6s in my Fender Deluxe Reverb—now the amp is loud enough to gig with. Some say this mod is safe, others say damage to the power transformer may occur due to increased heater current. What's your opinion? Would pulling the unused tubes free up enough current to make a difference? Otherwise, if an auxiliary 2 amp transformer is recommended, where do we get them and how about giving us a hook-up diagram? Also, how many milliamps of current should we bias the 6L6s in this amp? The power transformer runs pretty hot—how do we tell when it's too hot? Will this mod hurt the output transformer?

Although heater current can be a problem with EL34s in a Deluxe, the transformer is sufficiently over-designed to handle a pair of 6L6GCs in most cases. Heater current on a 6V6 is .45 amps while a 6L6 is .9 amps. Therefore a pair of 6L6s would draw an extra .9 amps of current and the winding is over-designed by more than that.

A good way to tell if the transformer likes the 6L6s is to check the A.C. heater voltage. If it is dropping below 6.3 volts, the transformer doesn't like the extra current. If the voltage is 6.3 volts or a little higher, it will tolerate the extra current well.

If your particular amp doesn't like the 6L6s, you could pull some unused preamp tubes to make up most of the difference. Assuming you are only using the vibrato channel, taking out the normal channel preamp tube (first preamp tube) will free up an extra .3 amps of heater current; however, removing this tube will affect the preamp bias of the vibrato channel second gain stage (more gain). Changing the cathode resistor from an 820 ohm to a 1500 ohm (pin 8 of the second preamp tube) will bring it back to stock gain characteristics. You might like the extra gain characteristics that will occur when the normal channel preamp tube is removed; so make sure and try it first, before changing the cathode resistor.

If you remove your vibrato tube, you will free up yet another .3 amps of heater current. If you are going to take this tube out anyway, you might want to disconnect the yellow wire on the right terminal of the intensity pot (looking at the back of the pot). This will dramatically improve the sound of the amp, but will disable your vibrato circuit. If you are going to take the tube out anyway and not use vibrato, you might as well disconnect the yellow wire and improve fidelity at the same time.

I would bias the 6L6s to idle at 30 to 35 milliamps for the best tone. The weak link is going to be your output transformer. The one you have will run pretty hot and chances are that it will eventually go south if you keep running the 6L6s. Move up to a heavier duty output transformer. I would go to a 50 watt transformer that matches 6L6s to 8 ohms. Marshall, Boogie, and Kendrick have these type of transformers. The Kendrick is interleaved and sounds like it was made in Texas, the Marshall sounds like it was made in England, and the Boogie sounds like it was made in California. Kendrick also offers a hot rod, an 11 winding interleaved output transformer made for Deluxes and Princetons that will make pair of 6V6s (not 6L6s) sound about as loud as a 50 watt Marshall.

The power transformer running hot may not be a problem. Again you should also check your plate voltage (pin 3 of either output tube) to tell if the transformer likes being run on 6L6s. If you get about the same voltages as with 6V6s, you have nothing to worry about. If the voltages drop 20 volts or more, you can either bias the tubes cooler (less idle current and therefore less voltage dropped across the transformer's internal resistance) or go to a beefier transformer. Of course you must rebias when changing from 6V6s to 6L6s and vice-versa.

I recently purchased a '62 Bassman head and would like to know more about it. My questions are: How does it differ from the original tweed Bassman? How many watts is it? What impedance speakers should be used? Also, I will be keeping this amp in my room which is located in a finished basement. The room is

part of the house's central air system, and a dehumidifier runs during the summer. In a previous column you stated that paper insulated output transformers can absorb moisture, causing them to arc, if stored in a basement environment. Will my amp be safe in my room under the conditions I mention?

I'll answer your last question first and say you will have no problem keeping the amp in your basement if it has a dehumidifier and is part of the house's central air system. The key here is humidity, and it sounds like humidity isn't a problem in your basement.

That amp is approximately 45 watts and was actually made three different ways. The model 6G6 Bassman had a tube rectifier, the model 6G6-A had a solidstate rectifier, and the model 6G6-B also had a solidstate rectifier with some minor circuit differences to darken the tone of the bass channel.

These amps used the Schumacher transformers unlike the tweed Bassman that used Triad transformers. The tweed Bassman only had one set of tone controls that served both channels. The brown Tolex Bassman that you have has separate tone controls for each channel.

Your amp is a piggy-back style, designed to push 4 ohms (two 8 ohm 12" speakers in a closed-back cabinet) while the tweed Bassman is a combo amp designed to push 2 ohms (four 8 ohm 10" speakers in an open back cabinet.)

The preamp circuitry of each amp is completely different, although both style Bassmans have a very desirable and rich tone.

I'm restoring a 1966 blackface Fender Vibrolux Reverb and have to make a couple of decisions:
1. The amp had been loaded with one 12" J.B.L. E-120 and a different baffleboard. I want to go back to 2x10s, but need to know more about whose speaker replacements sound good?
2. The amp needs to be re-tubed and re-biased. I have a guy here to help with the bias setting, but want to know more about which tubes would be good for replacement. I am looking for the cleanest sound I can get. I'm playing a '74 Guild Artist Award archtop. I've also got a '51 Epiphone Zephyr Regent Deluxe. I'd like better definition on the low-end and something on the high-end that's not as brittle sounding as the J.B.L. 12".

Your basic big-box jazz sound is pretty much it. Please let me know what you think. Any help will be appreciated!

Before I answer your questions concerning your amp, let me answer the question concerning the tone you desire. Try plugging your guitar in the Vibrato channel number 2 input and rolling the volume on your guitar down a couple of numbers. This will reduce the signal going into the first tube and make a much cleaner sound with more punch. Use the volume control on your amp to get your loudness. Make sure you have a real GZ34 rectifier tube, preferably a Mullard. There are some 5Y3s that are being sold as GZ34s and will not give you the headroom of a real GZ34. The real ones have five pins, the phony ones have four pins. The Mullards are made in Great Britain. A solid-state rectifier will give you a little better headroom, but you will lose some sustain/natural tube compression and your envelope will be lost.

Let's talk about that baffleboard. The originals had a baffleboard made of ½" plywood with a 1" wide strip of ¼" plywood around the outside edge. This made the total thickness around the edge ¾".

This will sound really good for archtop, however if you want a chunkier sound with less airy midrange, make the baffleboard out of ¾" plywood and still use the one inch wide strip of ¼" plywood around the edge. This would make the total thickness 1" around the edge.

However, if you want a more open and airy midrange with a little more cut (are you playing with a live drummer?), you may want to move to a ⅜" baffleboard and use a 1" wide strip of ⅜" plywood around the edges. This would have the effect of keeping the edge thickness at ¾" (for ease of mounting the baffleboard to the cabinet), but having a overall thinner baffleboard (easier for the speaker to resonate). Do not use a baffleboard any thinner than ⅜" because the speakers will rip through the plywood of anything any thinner.

Output tubes are critical to your sound. I would recommend you trying a matched pair of 7581A output tubes made by R.C.A. or G.E. This tube is a massive overkill 20% upgrade to a 6L6GC. It will give you the headroom you desire with a strong bold bottom, transparent midrange and smooth top end. Also, since these tubes can take much more current than a 6L6GC, your amp can be biased with a little extra current running through the output tubes at idle. This will ultimately have the output tubes (that are running in Class AB) biased closer to Class A thus giving you smoother rounder tone but without sacrificing the dynamics of running pure Class A.

You are not the only archtop player that is faced with the dilemma of a vintage-style speaker having too much break-up and a modern speaker being too brittle for the tone you are after. A full range Bass Guitar speaker might be a good place to start. These speakers will generally have a much larger magnet and therefore can give the headroom you need for a clean sound, especially on the bottom end. Celestion, J.B.L., and Kendrick make such a speaker.

Will I damage my blackface Vibrolux if I play a six-string bass through it?

You will not damage your amplifier if you use it to play bass. You might damage your speakers, depending on what kind of speakers you have and how loud you play. Since impedance gets lower as frequency gets lower, you will be running fairly high current through the speakers when playing a bass loud. This could overheat the voice coils in the speaker and blow them.

On the other hand, there are bass players who use and like guitar speakers. One person in particular who has always used guitar speakers is Dusty Hill of ZZ Top. The important thing is to use guitar speakers that have a high enough wattage rating so that they can take the heat without blowing.

Again the speaker wattages are rated at 1000 hertz. Bass players generally always play at lower frequencies than that and lower frequencies mean lower impedance (higher current and more heat). Therefore you should use highly over-rated guitar speakers (at least double the wattage of the amp) when using them for bass.

I just bought a blackface Pro Reverb amp and need to know what model it is so I can put in the correct rectifier tube. The number inside is 1201067 and stamped on the back is A06004.

Although the model number will most likely be on the tube chart, I would recommend trying two different rectifiers and see which one sounds best to you.

One type to try is the 5U4GB and the other would be a GZ34 (a.k.a. 5AR4). The GZ34 will give you a little more headroom and volume with slightly less envelope (power supply sag). Although the 5U4GB will not be as loud, you might like its spongier feel and natural compression better. My guess is that you will prefer the 5U4GB for recording and the GZ34 for live gigs, but let your ears decide. The one that sounds the best for you is the correct one to use. Tone is in the ear of the listener.

Did any of the 2x10 Supers have reverb?

Fender never advertised a Super Reverb with only two speakers. It is possible that someone out there has one that was actually made that way. I have never seen or heard of one, but there are some "one-of-a-kinds" out there that were made by special request. Fender did make a 2x10 amp with reverb; it was called a Vibroverb.

I have a 1964 Fender Princeton Reverb amp which I got cheap, due to a blown speaker. I have since replaced it with a 25 watt, 10" Celestion speaker. What would be the most economical way to get more of a crunch sound, since I have never actually had this amp to a tech? Should I check into bigger tubes? Or just keep using a fuzz box? Any ideas would be appreciated.

The most economical way to get what you want is to keep using your fuzz box (assuming it is paid for). However, the best thing you could do for your amp is to have it serviced by a professional. Your amp probably needs a cap job and possibly some new tubes, but not bigger tubes—better tubes. Electrolytic capacitors in the power supply of the amp will wear out causing weak, mushy response and, in extreme cases, can cause your power transformer to overheat and possibly burn up. Tubes become cathode striped after long periods of usage and lose their sparkle, power and dynamics.

When it comes to dealing with a tube amp, there are going to be times that you will have to dig deep into your pocket and part with some of that hard-earned American green, if you want your amp to sound its best. I have found that the best way to look at these things is on the long term. For instance, if a cap job costing you a hundred dollars is done once every ten years, then your actual cost is only $10 a year. Spend a hundred and a half on very high quality tubes and use them a couple thousand hours, and your actual cost is pennies per gig.

I own a white Tolex Tremolux model 6G9-A, production I MA. This unit also has a separate jack for an extension speaker cabinet in addition to the standard speaker output. What is the proper impedance to use with this amplifier, first with one cabinet and then with two?

If the output transformer is original, the output impedance is 8 ohms. If you add another speaker, make sure it is an 8 ohm speaker. This, combined with the first speaker, will actually give you a total nominal speaker load of 4 ohms, and the power will be evenly divided between the two speakers. Should you wish to use a 16 ohm extension speaker, this will work fine and give you a nominal impedance of 5.33 ohms; however, the 8 ohm speaker will get two-thirds of the power and the 16 ohm speaker will only get a third of the power. On vintage Fender amplifiers, unlike many other brands, there is no internal switching of transformer secondary taps, when an extra speaker is plugged in (these units are very tolerable of a 100% mismatch). You could also use two 16 ohm speakers, which would make a combined total nominal impedance of 8 ohms and, of course, your output transformer is designed for 8 ohms.

I recently purchased a 1961 Fender Bandmaster with a 1x15 cab from (allegedly) the original owner. My question concerns the covering. I thought all Fender amps of this vintage were either covered in white or brown Tolex, but this one is black with a silver grille cloth. The previous owner states that it has never been recovered or painted. The head has a brown faceplate with white knobs (except the presence knob, which is brown) and both the head and the cabinet have flat Fender logos. Have you ever seen this configuration before? Also, could you suggest a good replacement 15" speaker?

I have seen this color configuration before, although only a couple of times. My guess is that Fender was just getting into trying the black Tolex and didn't have knobs or faceplates to match. The single 15" cabinet seems a little suspect. My guess is that the cabinet was either custom ordered, or the dealer sold a Showman 1x15 cabinet with the Bandmaster. A cool way to tell—the Showman cabinet would be 36" from side to side and the Bandmaster cabinet would be 30" from side to side. Although Fender also made a 1x15" speaker cabinet, it is much more square shaped than the picture you sent to me.

Funny thing you should ask about the 15" speaker. There are not any new replacement 15" speakers that I would recommend for guitar amps. That is why I am currently designing one that will probably be out in a few months. In the meantime, depending on what kind of sound you are after, you might try a used C.T.S. 15". These were used in the Leslie tone cabinets and can probably be found through an organ repair service. I think they sound very good for guitar.

I recently acquired a Fender 4x10 Bassman re-issue. When I first got the amp, I re-tubed using a GZ34 rectifier tube and a 12AY7 for the first tube. (I understand that this was the correct tube complement of the original Bassman.) This worked OK for a while, and the amp did not sound quite as harsh, but lately I have been getting these horrible intermittent crackling noises, making the amp completely unusable. What should I do?

Don't despair, the problem you describe is easy to correct. Locate the two 100K resistors that go to pins #1 and #6 of your first preamp tube. Replace these with ½ watt 100K ohm resistors and your problem will disappear. It may take a couple of hours to replace these, because the entire board will have to be removed (due to inaccessibility of the printed circuit board for soldering.) When Fender re-issued the 4x10 Bassman, they used ¼ watt plate load resistors instead of ½ watt. Although this never made sense to me because ½ watt resistors only cost a couple of pennies more than ¼ watt, it would work with the Chinese 12AX7 tube that the factory put in that socket; however, when you changed to the 12AY7 (like an original), the power being used by the tube is slightly over a ¼ watt. The resistor is slowly burning up and that is most likely what's causing the noise. Also, you might try going to a 5U4GB rectifier tube. It has a larger voltage drop than a GZ34 and since the re-issue was designed with way too much voltage, this will help compensate to eliminate more of the harshness.

I've noticed that the 6G15 Fender Reverb schematic looks a lot like a glorified Champ. Am I crazy to think that I could rewire

a Vibro Champ into an outboard reverb? How would the voltages and transformers match up?

Surely the Vibro Champ could be gutted and made into a Fender 6G15 Reverb Unit. The output transformer will actually sound better than an original because of the extra core material. You will have to pad down the B plus voltage with a couple of pi filters, or you might have to go to a half wave power supply, like an original 6G15 Fender Reverb, in order to duplicate the voltages. The only real problem you may encounter is fitting a reverb pan into the bottom of that small cabinet. Will it fit?

Be careful to lay it out exactly like an original, otherwise parasitic oscillations could ruin your day.

I just got a late 70s Pro Reverb with a "hum balance" adjustment pot in the back which connects to the heater voltage pins of the second preamp tube. How does it work? Also, the output jack says 70 watts—aren't these supposed to be 45 watts? Can they really get 70 watts out of two 6L6s? I read 505 volts on pin #3 of the output tubes.

Let's answer the second question first. It is physically impossible to get 70 watts out of a pair of 6L6GCs. When you consider that the plate watt dissipation of a 6L6GC is 30 watts, it doesn't take a rocket scientist to figure out that two of them are incapable of more than 60 watts.

Let's get on with the first question concerning the "hum balance" pot. In a push-pull output stage such as the Pro Reverb, the output tubes are wired out of phase and therefore are hum cancelling, provided the output tubes are matched! Also the preamp tubes are wired humbucking and therefore are hum cancelling, provided the gain of each half of the tube is matched. But wait a minute—what does a manufacturer do to avoid the expense of matching tubes, and the time necessary to select preamp tubes that are quiet in the actual circuit? They take out the bias control and put a bias balance control in its place and then take out the filament center-tap (ground) and put an artificial movable ground (hum balance pot) to balance out hum. With this set-up, they can put any pair of output tubes and any preamp tubes in it and turn the bias balance control (there is no actual bias voltage adjustment pot) to where it makes the least amount of hum, then turn the hum balance control until you find the spot that makes the least amount of hum. This saved a lot of valuable time that could be used to build more amps quickly and it worked very well—for the manufacturer!

I have a 1971 Fender Vibrolux. The output transformer is hooked up with the green and black output wires to the 4 ohm speaker jack. Brought back into the chassis through the same bushing as the green and black are a green/yellow and a white wire, which are clipped off inside the chassis. Could these unused wires be the 8 ohm and 16 ohm speaker taps? The transformer is a Fender 01834, EIA606..., but some of the numbers are illegible. The transformer that you describe sounds like a 018343 which is a multiple tap, plastic bobbin, catch-all replacement transformer. The taps are 2 ohm, 4 ohm, and 8 ohm. The black wire is the bottom of the winding. If you want to see what the taps are, measure from each secondary wire to the black wire with an ohmmeter. The tap with the most resistance is the 8 ohm tap, the least resistance is the 2 ohm tap and the only other remaining lead will be your 4 ohm tap.

On my Pro Reverb, I have noticed that the Reverb is not as strong and deep as on my other Fender Reverb amps. If I turn the Reverb up, I get more Reverb but it sounds thin. I have tried different reverb tubes, cables and pan with no change. I tried a 7025 tube as the driver, but it tends to distort. The inside of the amp looks normal—nothing burnt or broken. I checked what I could with my ohmmeter. If you could shed some light on my problem I would be very grateful. This is a great sounding amp otherwise. For starters, I would recommend checking the plate voltages (pin #1 and pin #6) against the schematic on both the reverb driver tube (12AT7) and the reverb recovery tube (7025). If either of these voltages are low, that could be your problem. Assuming these voltages are good, and considering you have already checked the cables, pan and tubes with known good parts, I would begin suspecting the reverb driver output transformer. This is the small transformer mounted by the reverb "in" and "out" jacks. A replacement transformer is very cheap and easy to replace. If you haven't done so already, try substituting the reverb recovery tube with a known good-sounding 7025 or 12AX7. Use a 12AT7 for the driver and a 12AX7/7025 for the reverb recovery.

Reverbs are very sensitive and the least little thing can make a huge difference in how it sounds. What about the first gain stage tube for the vibrato channel (second preamp tube)? If this tube sounds bad or has low gain, it will dramatically affect the overall reverb sound. After all, the reverb driver is driven by the vibrato channel preamp tube. Try substituting a known good 7025.

I have a blackface Pro Reverb with no speakers. How can I tell what the speaker impedance should be, and how can I change it to use one 8 ohm speaker?

If your amp has the original output transformer, it is 4 ohms. You can hook it up to an 8 ohm speaker load without any damage to the amp. Vintage Fenders are very tolerant of impedance mismatches. You may lose a few watts of power, but your tubes will probably last longer due to the higher reflected impedance. As a matter of fact, I know some highly respected musicians who actually prefer a deliberate mismatch of having the speaker twice the ohms of the output transformer.

The higher reflected impedance makes the tubes think they are operating into a higher impedance primary. This has somewhat of a current limiting effect that makes the tone a little bit more compressed and not quite as loud.

I have a blackface Princeton Reverb that I play with the volume on three and connect it to a small Polytone amp (via the Ext. speaker jack) with the volume at one. Am I damaging either amp by not making a line out?

You are for sure not damaging the Princeton because the input impedance of the Polytone is so high (as compared to the 8 ohm speaker load of the Princeton in parallel with it) that it is negligible. On the other hand, I wouldn't be so sure about the safety of the Polytone. The rule is: Never connect anything rated in watts to the input of another amp. But must we always go by the rules? Since the input impedance of the Polytone is high, it is not really taking the high current from the Princeton. The Princeton's speaker is taking the current. The only thing that could be a problem is the signal voltage and, if you run the Princeton at a low enough volume setting, you won't be generating much signal voltage anyway. I say if you like the sound and it's working (not blowing anything up) go for it!

On the other hand, you could turn your Princeton to any volume and be perfectly safe, if you made a line out. You could even make a special cord with the resistors wired inside the plug and not have to mod your amp! Or you could make a clip-on line out jack to clip on to your speaker leads!

I have a '63 Vibroverb re-issue amplifier and I would like to make the tone warmer, especially at lower volumes (3 or less). I understand that some of the harshness that I am hearing may be due to the solidstate rectifier and/or low efficiency speakers (Oxford). What modification do you recommend to reduce the harshness. I'm hoping that simple "bolt on" mods, like different tubes and/or speakers will be the answer.

You're going in the right direction with your idea about tubes and speakers. The three things that affect tone the most are tubes, speakers, and transformers. Unfortunately these are the three things that cost the most on an amplifier. (Most manufacturers skimp on these items for the same reason.)

For starters, the easiest mod to perform is to replace the output tubes with either Tung-Sol 5881s or R.C.A. 6L6GCs. This alone will make a dramatic improvement.

Next you need to make a decision on the speaker configuration. If you keep the 2x10" configuration, use Kendrick Blackframes; they will be more defined and warmer than the Oxfords. Another possibility is to change to a 1x15" configuration. A new baffleboard and either a JBL 15" speaker or a Jensen P15N would be a cool way to go. The JBL will be clearer on the bottom end, but the Jensen will be warmer at low volumes. Changing to any of these recommended speakers will be magical.

Next, you're going to need an output transformer with some iron in it! If you keep the 2x10" configuration you will need a 4 ohm output transformer and I would recommend either a 4 ohm Bassman transformer (125A13A) or a Kendrick 2210A. The Kendrick is interleaved, but either of these will improve the tone dramatically. If you use a 1x15" configuration I would suggest either a blackface Twin, a Kendrick 4212A or a Kendrick 2210A-8. All of these have the correct turns ratio to run two output tubes into 8 ohms. The 4212A will be the loudest and most defined but it is huge (2.5 inches of core material). The Twin will also sound very good and have a lot of headroom, and the 2210A-8 (interleaved) will have slightly more break-up than the 4212A or the Twin but will be the richest—harmonically speaking. If you're planning on playing the amp softly, the 2210A-8 is the way to go.

At this point you will be overwhelmed by the difference in tone. However, if you don't mind spending another couple of hundred for absolute perfection, replace the power transformer so you can get a rectifier tube happening. The tube rectifier will give you natural tube compression and envelope. This would, of course, necessitate installing a tube socket as well. I would use either a Super Reverb power transformer or a Kendrick 2410P. Both of these are nearly identical. Use the GZ34 rectifier tube (like the original Vibroverbs) or for even more compression try a 5U4GB. Let your ears decide with an A/B comparison.

Please explain the "floating baffleboard" in the tweed Bassman? Do any other models use it?

The floating baffleboard used in all Fender tweed amps (except the Dual Professional) and in most other great sounding amps is very simply designed, yet very effective. A floating baffleboard takes advantage of the fact that the baffleboard itself can be a sound source, if it is allowed to vibrate. On a floating design, the baffleboard is attached on two opposite sides. This means that the other two sides are not attached to anything and are free to move. Therefore, as the speaker vibrates the baffleboard, the board itself vibrates (resonates) with the speaker. The wood is flexible enough to do this. Now for this to work properly and most effectively, the wood needs to be thin enough to resonate. Fender tweed amps use 5/16" plywood (which is no longer available today), 1/4" is too thin and will be ripped by the speaker mounting screws, 1/2" is too thick to resonate properly. So what's a fella to do? There is a type of plywood called 3/8" MDO that is used by sign painters and actually measures 5/16" thick. That's what we use at my shop.

If a baffleboard were attached on all four sides, it would be so rigid that the extra mounting screws would actually stop the board from resonating. The speaker would still resonate, but there is more surface area when you combine the surface area of the baffleboard with the surface area of the speaker.

In a Fender Reverb amp, is the reverb much of a drain on the signal? If so, is it possible to modify the circuit to reduce this, while retaining all of the reverb?

There is no drain on the signal of a Fender reverb amp. The 6G15 stand-alone reverb unit does have a one or two decibel signal loss because of the nature of the circuit. However, in all Fender amps, there is actually an extra stage of gain to mix the reverb and the dry signal together. This extra stage causes all reverb model amps to have essentially more gain than their non-reverb counterparts.

If you are interested in signal loss, all blackface vibrato amps that use an optocoupler (neon bulb shrink-wrapped with a light-dependent resistor) load down the signal—even when the vibrato is turned off. This problem can be overcome by removing the right lead going to the intensity pot. The difference in tone is dramatic.

I have a 1961 Pro Amp Model 6G5. It has a 15" speaker that reads 10 ohms with an ohmmeter. What ohm speaker should be used? After changing filter capacitors, you stated in your first book *A Desktop Reference of Hip Vintage Guitar Amps* to bring up the voltage slowly on a Variac. Is this done with all the tubes in and a speaker connected?

Something is off—in your meter, meter leads, tinsel wires (connecting the voice coil to the speaker terminals), or the speaker. That amp used an 8 ohm speaker which should read approximately 6 to 6.5 ohms on a D.C. resistance meter.

When bringing up an amp slowly on a Variac, do it with the amp in the play mode with all tubes and speakers connected. I like to set the Variac for about 40 volts and leave it on overnight. The next day, I bring it up at 10 volts per hour, so that after an 8 hour day, the amp is up to the full 120 volts operating voltage. If you have just done a cap job, that's all you need to do, but if it is an old amp and you have not done a cap job and would just like to see if you can reform the caps and "bring them back," do the same thing except use a current limiter between the amp and the Variac for the entire procedure. Sometimes, old caps can be reformed this way, but you have to use that current limiter to keep the current leakage to a minimum while the dielectric is reforming. Of course reforming old caps doesn't always work, especially if the caps are bleeding.

I have a silverface Princeton that I put in a Kendrick 2112A output transformer, and a Celestion 10". I changed the coupling caps to Sprague orange drops and installed new electrolytics and replaced the tubes with NOS tubes. I've done just about everything I could to juice it up, and it still doesn't have anywhere near the crisp transient response and sharp definition of my Kendrick 2112. Not even close. Am I missing something here? I tried a Celestion Vintage 30 and, although it sounds a little stronger, it needs more response on the bass end. What can I do?

Try moving to Texas before doing all those mods! No, really, there is more to an amp's sound than the transformer. The Kendrick 2112 doesn't use orange drop caps and it doesn't use a Celestion 10" speaker, and the circuit is very different. For one thing, the Princeton is fixed-biased while the Kendrick 2112 is cathode-biased. The Princeton has more loss in the tone section because of the circuit design. Does the Princeton have reverb? If it does, there is another stage giving the Princeton three stages as opposed to the Kendrick's two stages. What about the output tubes? The Kendrick is shipped with American NOS military spec 6V6s. Do you have these tubes in your Princeton? Just the output tubes alone can make a huge difference in the tone of an amp. The power supply voltage probably doesn't match. The Kendrick will run at about 395 volts while the Princeton will be more like 425 volts. The Kendrick has no feedback loop while the Princeton does. A feedback loop will cost some gain and smooth out transient response.

If you need bottom end, I would suggest keeping the orange drop caps because they will give above-average bottom end, but get a 12" speaker that has punchy bottom end. Celestion 30s are not known for their bottom end. You might want to experiment with removing the feedback loop and rewiring the tone controls to exactly like the Kendrick (same as 5E3 Deluxe). This should get you in the ball park. Also switching the output stage to a cathode-bias setup would get you a little closer.

How can I make my brown Fender Pro sound more like a Super Reverb?

The truth is: there's no easy quick-fix on this one. Here again, the Pro uses a completely different circuit than the Super Reverb and the speaker configuration is also completely different. You could get very close by rewiring the preamp circuit to be exactly like the Pro and then use a 4x10" speaker cabinet. When you wire the preamp to be like the Super Reverb, you must include the reverb circuit in order to have the gain structure be the same. You would also need to change a few power resistors to match the preamp tube plate voltages of the Super. Since the Pro has an 8 ohm output transformer and the Super has a 2 ohm, you could still use 4x10" speakers by rewiring the 4x10" speakers to parallel/series (8 ohm) in order to match the speakers to the Pro amp.

Depending on the particular brown Pro, it may have five or six preamp tube sockets. If it has only five sockets and you wish not to cut the chassis to mount the sixth preamp tube socket, simply leave off the vibrato circuit of the Super when you rewire it.

Here's a better idea—the brown Pro is probably worth several hundreds of dollars as is. Sell the Pro in the classifieds and buy a Super Reverb with the money. You save a vintage amp from massive modification, and there are a lot of people who would love the Pro just for its cool pitch vibrato. And this solution would cost you much less money.

Is there a single mod or two that will improve the sound of my new Fender Deluxe Reverb re-issue? I realize that it will never sound like an original due to the printed circuit board, but is there anything that can be done to get it closer?

A dramatic difference can be made to your amp with a couple of easy mods. First get rid of that speaker. For a British sound with lots of midrange, try a Celestion. If you are looking for that full-range Texas tone, you'll need a Kendrick. And second, get some good American NOS 6V6s. Those two mods alone will improve the sound of the amp beyond belief. Also, these mods are easily reversible if you ever want to sell the amp. You could put the old speaker and tubes back and you will still have your original tone investment of tubes and a speaker to use on some other amp in the future.

You suggested that someone replace the power transformer in a Vibroverb re-issue in order to use a rectifier tube. Couldn't you use a bridge rectifier on a Vibroverb, or any other amp with a solidstate rectifier?

Absolutely not!! There are two types of bridge rectifiers, tube and solidstate, both of which use a non-center-tapped secondary winding. The voltage on this secondary winding would have to be approximately half the voltage of a center-tapped full-wave rectifier winding (stock Vibroverb) in order to produce the same D.C. output voltage. This is because, in a bridge rectifier, the entire winding is used for each cycle; whereas, only half the winding is used on each cycle of the center-tapped full-wave rectifier. Even if the voltages were compatible, to use a tube rectifier, you would still need a rectifier heater supply winding, which the Vibroverb transformer simply doesn't have.

I have a blackface Fender Champ and Vibrochamp. What can I do to make them sound like a tweed Champ? Also, what can cause the power transformer in my Gibson Les Paul Jr and GA5 to drip wax? The GA5 still works, but gets so hot the wax is dripping.

Let's answer the last question first, and I think you already know the answer. The wax is dripping because the transformer is getting hotter than the melting point of wax. This leads us to another question—is this normal? Maybe and maybe not. There are a number of things that could cause those transformers to get hot—for instance: 1. The biasing on the output tubes is such that higher-than-originally-intended plate current is overheating the secondary winding. It could be that the amps were originally designed for 110 volts and you are running them at 120 volts. Some output tubes just idle higher than average as well.

2. Your wall voltage could be a little high.

3. The transformer could have rust which causes conductivity between adjacent laminates. This causes eddy currents where a current is induced in two or more adjacent laminates. (The transformer "sees" this as another secondary winding with a shorted output, hence the heat.)

4. A turn or two of any winding could actually be shorted. This produces basically the same result as rusty laminates. It may not be bad enough to blow a fuse, but can cause overheating.

5. You may simply get heat from the amp being used for a long period of time. Heat can build up in an amp in proportion to the amount of time it is on.

This may not present a problem. I know of amps that have been used for years while dripping wax.

Next answer about the Champs—in a nutshell—get rid of the tone controls and wire the preamp like a tweed Champ. This will get you much closer, but to really nail the tone, make a new baffleboard out of ¼" plywood and make sure you have a good sounding 4 ohm speaker. In all Champ-style circuits, the tubes will make more difference than in any other amp. This is because it uses the most basic circuitry and there is nothing to color the tone except the tubes. Use NOS American tubes in your amp for best results. They will probably cost a little more because the demand is up and the supply is down, but the tone will be better and they will last longer.

The following is a series of questions, sent in by one reader, that relate to blackface-era amps—mainly a Deluxe Reverb and a Vibrolux Reverb. Perhaps this will clear up a lot of questions concerning voicing, and give you some techniques for tweaking the tone to your liking. Since most blackface amps used the same basic circuit, what would apply to the Deluxe and the Vibrolux would also apply to most blackface amps.

I installed a "presence" control (a 5K linear pot in line with the 820 ohm feedback resistor) mounted in the extra speaker jack hole. It rolls off too much midrange. There is still brittle high end created by the loop. I'm afraid that disabling the loop en-

tirely will have the same effect. My goal was to emphasize mids and eliminate the brittle treble. The mod works as a presence control, but at the expense of the mids and punch I wanted to emphasize. What do you recommend?

For starters, let me explain some considerations when installing a "presence control" in a feedback loop. A feedback circuit must be designed on a particular ratio. The feedback resistor and the presence pot form a voltage divider whose ratio determines the amount of feedback. A stock Deluxe Reverb uses a 820 ohm dropping and a 47 ohm load resistor in its feedback (voltage divider) circuit. Therefore, the voltage will drop in that proportion. 820 divided by 47 is 17.4468 to one. That means, if 18.4468 volts were on the speaker jack being fedback, the 820 ohm resistor would "eat up" 17.4468 volts and 1 volt would be on the load resistor. The amount of voltage on the load resistor will determine how much feedback you will be using in the circuit.

When you used the 5K ohm pot, you should have changed the 820 ohm feedback resistor so that the ratio would still be the same. For instance an 87K ohm feedback resistor would have given you a 17.4 to one ratio (87K ohm dropping divided by 5K ohm load).

As the circuit is now, you are not dropping very much and almost all of the signal is being fed back. This will make an amp sound very sterile. Right now your ratio is .164 to 1. This means for every 1.164 volts, the 820 ohm resistor is "eating up" .164 volts. 1 volt is across the 5K ohm pot and of course winds up as negative feedback. If you compare the feedback circuit you are using now to the stock values mentioned in the earlier example, stock has 1 volt feedback for every 18.4468 volts output, whereas your modification would have 15.85 volts feedback for every 18.4468 volts output. You can see now that you are using 15 times more feedback than stock.

I changed the 6.8K resistor that is grounded on the bass pot to a 10K ohm value, which did help to increase mids, but the "presence" control remains a problem. How do I tame this thing?

I think you know what to do about the feedback resistor (change to 87K or more), but you could actually replace the 6.8K resistor with a 25K pot and "dial in" your best sound. After you have the best setting "dialed in," simply take the pot out without disturbing the settings and measure it with a meter. Now find a resistor that is this same value and substitute it for the pot.

Also, you might want to put a 250K ohm pot in series with a 87K ohm resistor in your feedback circuit, and "dial in" the feedback amount you like with that particular amp. You could remove the pot and resistor without disturbing the pot setting, measure the total resistance of the two and install a resistor of that particular value.

You've mentioned that replacing the phase inverter input cap with a .02 or .047 cap will allow more midrange frequencies at pass. There are three .1 caps in the phase inverter circuitry as well as a .001 cap in line with the two 220K ohm resistors. Which one is the input cap?

The input cap is the .001uf. Changing this will alter the voicing dramatically. Any value that gives you what you want is acceptable. Smaller values make the amp have less bottom (or more bottom-end roll-off, depending on how you want to say it). Larger values will allow less bottom-end roll-off. If you use too large a cap value here, you will get too much bottom-end. This could "muddy up" your sound. I generally lean towards the .02 value in most applications.

Which caps in the preamp sections would you recommend to eliminate brittleness? Obviously the 250pf cap, but what about the 10pf, 500pf, and the 47pf (on the Deluxe Reverb volume pot) caps?

The 10pf cap that is across the 3.3 meg ohm resistor (in the reverb coupling circuit) is to bring back some of the high end lost by running the signal through a resistor that large. I would not recommend changing that, but I would recommend checking the resistor's value. You will have to remove the resistor from the circuit to get an accurate reading. My experience has been that the 3.3 meg ohm resistor will sometimes measure really low and the result will be high-end brittleness. In fact, I prefer changing this to a resistor that reads a little bit on the high side.

The 500pf cap is used to couple the preamp to the reverb drive circuit. Changing values here will alter the actual tone of the reverb. Be careful here because too high a value will cause the reverb pan to distort.

The 47pf cap across the volume pot is a bright circuit that is always on! Using a larger value will increase mids. This cap bypasses signal around the pot. A smaller value bypasses higher frequencies, a larger value bypasses lower frequencies around the pot. If you desire a lot of midrange, you will probably like something between a 5000pf and a 500pf cap value. The plexi-Marshalls used a 5000pf.

My Vibrolux Reverb was obviously used in Great Britain as evidenced by the 220 volt selector and the receipt (written in English currency). The amp uses 6L6s, but there are no 470 ohm screen resistors on the power tube sockets. The 1500 ohm grid resistors are missing on the power tube grids as well. There's a 470 ohm 1 watt resistor and diode next to the pilot lamp. What concerns should I have here? Should I install 470 ohm screen resistors on the power tube sockets? Can I leave the 1500 ohm resistors off, since the amp shows no sign of parasitic oscillation?
I would advise you to use 470 ohm 1 or 2 watt resistors. Make sure to re-bias the amp afterwards because the resistor will drop the output which can be brought back up by reducing the negative bias voltage.

If you are not having a stability problem, leave off the 1500 ohm grid resistors. The diode and resistor are probably part of the bias circuit.

This same amp (Vibrolux Reverb) continues to burn up 12AT7s in the reverb Driver section every 7-8 hours. I've replaced the caps and resistors in the reverb circuitry, but the reverb remains oversaturated at low settings and sometimes bleeds through with the reverb pot on zero. Should I suspect the reverb transformer or the 100K ohm pot? There is a 33 ohm resistor across the red and blue wires on the transformer.
It sounds to me like your reverb transformer is bad. The 33 ohm resistor might indicate that the primary was open and a tech tried to fix it by using a 33 ohm load resistor with a capacitor going to the secondary side. Also, possibly the cathode resistor is bad, or the cathode bypass cap is shorted. All of these things could cause the 12AT7 to fail prematurely.

Although the pot would not affect the life of the 12AT7, you might want to check the ground of the pot. If the ground is not good, you will get bleed-over even when the pot is turned off.

Please tell me what I can use to clean and restore the grille cloth on my '66 Deluxe Reverb. I don't want to ruin the material. Secondly, what would cause the amp to squeal when turning the reverb past #4 setting?

I am taken aback! I've had other readers asking me how they could make their new grille cloth look old. One way to clean that grille cloth is to remove the speakers, logo and baffleboard from the amp and take the baffleboard to the car wash. Yes, I'm talking about high pressure wash. Don't let it stay wet very long. Perhaps you could shake the water out and let it dry in the sun. The grille cloth is not really a textile product and therefore is not affected by water, but I would be very careful not to let the wooden baffleboard stay wet very long.

If your reverb is squealing loudly when you turn it up, my guess is that you are experiencing acoustic feedback. This occurs when the sound from the speaker vibrates the reverb pan. Check to see if this is the problem by removing the reverb bag/pan assembly from the amp. Leaving the cable hooked up, see if the problem disappears. If the problem persists, you probably need a new reverb recovery tube. That is the 12AX7 next to the reverb driver tube (12AT7). If your testing shows the problem to be acoustic feedback, you may remove the pan from the bag and place a couple of long weather-strips to the top of the pan. While the pan is out, check the coil on each transducer inside the pan to see if it is loose. If it is loose, make a small wooden shim out of a chopstick and wedge it between the coil and the core of the offending transducer. Do you have the cardboard bottom to the pan installed? Do you have a bag for the reverb? Not having a bag, loose transducer coils, or a resonant pan can all cause this acoustic feedback squealing.

What is the difference between my 5F2-A Princeton and the 5F1 Champ, besides the tone control and the extra filter cap on the Princeton? Do these amps need the screen resistor that you mention on page 25 of your book *A Desktop Reference of Hip Vintage Guitar Amps*? What ohm speaker should be used? Would there be any benefit in adding a variable feedback pot mod as I've seen for the Bassman?

Beside the tone control and the extra cap, another difference is that the Champ probably sounds much better. The tone control exacts its losses and therefore the Princeton amp doesn't have much gain by comparison. I would recommending using a screen resistor as described on page 25 in my book. You could replace the feedback resistor with a variable pot and do some listening tests from there. The amp should be louder with less feedback.

I have a Fender Bassman and 2x15 cabinet (circa. '69). From what I understand, the amplifier is great for playing bass guitar or guitar through. The side labeled normal is for normal guitar playing, and the other side is for bass playing (I assume). The question I have is this: The side for the bass input has a high level of background noise. It sounds like AM noise on a radio. When you plug in a guitar or bass on this side, it seems to get louder and scratchy. It definitely gets louder when the volume or treble knobs are turned up. The normal side of the amplifier has the same noise but not nearly as loud. I can live with this. What can I do about it? Does a capacitor need replacing? Please let me know; this is driving me crazy.

I don't think you need to replace a capacitor to eliminate your noise. This sounds to me like you could have a noisy plate load resistor on the second gain stage of your bass channel. To find this resistor, find pin #6 on your first preamp tube. Follow the wire from pin #6 back to the board and it will be connected to a resistor whose value is 100K ohms. Replace this resistor and you will have probably eliminated the problem. My guess might be correct, but if it isn't, try a new 12AX7—one that you are certain is quiet. If changing to a quiet preamp tube still doesn't do it, try replacing the other 100K ohm resistor that is physically connected to the first one you changed. This one will probably have a .01 capacitor going across its leads. Leave the capacitor in the circuit, or listen to it both ways and decide which you like better. (You will get more high end with it out of the circuit altogether.)

A word of caution: Make sure the amp is bled of all voltage before you begin. An easy way to bleed the 1969 Bassman is to unplug the unit from the wall and put the standby switch in the "play" mode (up).

I've recently run into a problem while working on my Vibrasonic Reverb amp (master volume). I was changing some of the plate

load resistors to 110K and cathode resistors to 1.8K and the amp went dead. Almost! I can still get a very weak signal with the amp jacked up. Could you please give me some suggestions?
You need to do a little troubleshooting. First, I would check every plate of every preamp tube (pin #1 and pin #6) for high voltage. You are looking for 160 volts or better. This is a very dangerous procedure because it must be done with the amp "on" and running. It is possible that when you were changing plate load resistors, you somehow disconnected the high voltage supply from the plate circuit. In this case, you would have no voltage at all on pin #1 or pin #6.

Next, I would suggest you check every cathode of every preamp tube for low voltage. This procedure is also dangerous because you will have to check the voltages with the amp "on" and running. These are going to be pin #3 and pin #8. You are looking for voltage approximately 2 volts or less. If any of these have high voltage, your cathode resistor is either not grounded properly, or not hooked up to the cathode pin at all!

One final thought: If you have no voltage on the plate pins #1 or #6, either the plate wire or plate resistor isn't soldered properly, the high voltage supply isn't hooked up, or the resistor is open.

High plate voltage on 6V6 tubes in Deluxe Reverbs has become a concern with us. We had one silverface unit with 488 volts biased at 16 mA of plate current per tube, using a plug-in solid-state rectifier. Plugging in a 5U4 reduced the plate voltage by 30 volts, and rebiasing to about 32 mA plate current reduced the plate volts to 425, as per your book's recommendations. The 5 volt heater circuit was pulled down to 4.86 volts by the 5U4's increased current requirements. Do you feel that this is dangerous to the power transformer? What is a lower limit for the heater circuit volts? Conversely, how much plate current can we reasonably pull through a 6V6 before entering the "danger zone?"
The silverface Deluxe Reverb amps used a 5U4GB rectifier tube anyway and I don't think the voltage reading .14 volts low is going to make much difference—your meter could be off that much. There is yet another rectifier tube that will drop the B+ voltage more, and it only requires 2 amps of filament current (as opposed to the 5U4G's

3 amps of current). You may like to try this tube; it is called a 5R4-G. This tube should bring your B+ down a few volts and the filament current up a little. As far as rectifier filament voltage, I would think a ½ volt one way or the other would be a pretty good rule of thumb. If it is too low, it will affect your tone, because the tube will not get hot enough to sufficiently excite those electrons and your power transformer could overheat as well.

How much current can a 6V6 take? It will vary from brand to brand, but the rule of thumb is: If it has glowing cherry red plates, it is too hot!!

I'm restoring a friend's Quad Reverb, mid-70's vintage, and he'd like to keep the master volume. Overall, this thing is basically a CBS Twinkie in head-form; so, I figured I'd cop the blackface AB-763 schematic (yes, I did explain why it still won't sound like a '65 Twin). I'd like to know your thoughts on the stock circuit location of the master volume control (just ahead of the phase inverter?) Also, how well will this thing co-exist with the blackface changes in the circuit? Your opinions on this pagan necessity would be greatly appreciated!

So you couldn't talk him out of that master volume. If you leave it in as stock, and do a blackface mod, it will load the input impedance of the blackface-style phase inverter. Perhaps you should try some of the Ken Fischer master volume circuits that were originally in the 1989 Angela instruments catalog (pages 32, 41,42, 43 and 44) and currently reprinted in my book—*A Desktop Reference of Hip Vintage Guitar Amps*—(pages 184, 193, 194, 195 and 196). He has four different types and type #1 is my fave.

Concerning your Princeton phase inverter mod to eliminate the vibrato; would you consider publishing a similar mod for the Vibro Champ?

I can give you some cool ideas to try on your Vibro Champ, however; the whole point in eliminating the vibrato in the Princeton was to gain the use of an extra triode so we could make a phase inverter with more gain. This would not make sense for a Vibro Champ because there is no phase inverter in a Vibro Champ, or in any other single-ended amp for that matter. For some cool Vibro Champ and Champ mods, see chapter "Easy Mods for Your Blackface Champ or Vibro Champ."

Can you suggest any modifications for Fender amps that would make them better suited for acoustic instruments? I've found that 5Y3 rectifier tubes sound better and that the lower gain of 12AY7s and 12AU7s help also. Amps with midrange controls seem to sound better with no mids. Could this be reduced even further by changing cap values?

That's a lot for one question. You're on the right track on your research with the tubes. I suggest plugging a piezo tweeter into the extra speaker jack. This tweeter can be set on top of the amp and does not need a crossover. It will give that sharp, crisp high end definition that acoustic guitars need.

Blackface amps with no mid control usually have a resistor on the bass pot that goes to ground. Reducing the value of this resistor reduces mids and to replace the resistor with a straight wire is equivalent to the mids turned off. I would probably temporarily replace the resistor with a 10K pot and adjust it by ear for the best tone. After getting it set, I would then take it out of the circuit without disturbing the setting and measure the resistance. Replace it with a resistor of that same value and you're set. If it sounds best turned to zero resistance, simple replace it with a straight wire.

I would not necessarily recommend changing cap values in the tone circuit, but here's a cool trick to try with your bypass cap on the first and possibly second gain stage. Leave the 12AX7 tube and disconnect the bypass capacitor for that triode section. (This is the 25 mfd. at 25 volts cap that is connected to pin 3 and/or pin 8.) This will reduce the gain of the tube while giving more headroom and compression. I would only try this on the first and/or second gain stage. If you do it to both stages, you will lose considerable gain. Remember to disconnect only one end and be careful with the lead so you can put it back if you don't like it.

I recently put a new power transformer in my blackface Pro Reverb amp and discovered that it wasn't center-tapped. Are

the new transformers different from the old ones? I wired it up exactly like the layout says and it sounds great. The tech put a 50 ohm resistor from each side of the light socket to ground. Also I noticed a .002 cap across pin 1 and 8 of one power tube. What is this for?

First, yes—the new transformers are different from the old ones. I'm glad your amp sounds great! When you say it wasn't center-tapped, I assume you meant the filament winding wasn't center-tapped. If the high voltage winding wasn't center-tapped, the amp would not have worked when wired stock. Most blackface Pro Reverb amps had a center-tapped filament, but some did not. None of the silverface Pro Reverbs had a center-tap and silverface Pros had .002 caps across pins 1 and 8 of each power tube, as well. This would make me suspect that your amp might be a transitional model, perhaps with some silverface circuitry and yet a blackface! These amps used a .002 cap on each power tube. The cap was used to suppress a parasitic oscillation that occurred in the silverface. Actually, not all of these amps had parasitics. The cap was put on to facilitate quick amp testing and expedite shipping. If these caps are removed, the amp will sound much better—providing the parasitic oscillation doesn't occur.

Although your tech used 50 ohm resistors from each side of the filament winding to ground, a stock Fender winding would have used two 100 ohm resistors—one from each side of the filament to ground. This puts ground reference in the middle of the A.C. voltage resulting in preamp tube hum cancellation. You must remember the 12AX7s in almost all amps are wired humbucking. The tube itself has a filament heater that is 12.6 volts with a center-tap. When the ends are connected and used as one lead (pins 4 and 5) and the center-tap of the tube (pin 9) is used as the other lead, the 12.6 volt heater becomes a 6.3 volt heater that is humbucking. To get maximum hum cancellation, the ground reference must be in the center which is why your tech used the two resistors. There are two advantages using resistors instead of a center-tapped heater winding. First, two resistors cost less than center-tapping a transformer; second, if an output tube shorts plate to heater, the resistor will act like a fuse and blow, preventing the chance of blowing the transformer winding. Transformers certainly cost more than resistors!

I have a blackface Bassman (#AB165) with a 2x12 cabinet (Jensen C12Ns). I've seen plenty of mods and tricks to make them good guitar amps. I'm interested in getting a bass soon and was wondering if it would make a good bass amp. Will the speakers hold up? Would the cabinet sound better with the back off? Are there any tricks I could do to the head to make it a better bass amp? If it's not enough power, couldn't I wire a line out for a separate power amp?

I would suggest using some other speakers. If you play your C12Ns real loud with a bass, you will probably blow them, especially since they are very old and the voice coils and/or voice coil formers have probably deteriorated. Leave the back on the cabinet because it will stop phase cancellation of lower frequencies. Also, a line out for bass would probably not be the best way to go because it will contain all the distortion present in the output of the amp. A preamp output is a much cleaner way to run a bass when using an extra power amp.

I have a Fender 75 that has been my sacrificial learning platform. I converted it to EL34s, took the ultra linear taps out of the output transformer, installed a presence control, and it was working well until I screwed something up trying to install a negative feedback control. I never really liked the sound of the 75 anyway, so I stripped it down except for the power supply, cleaned the circuit board and am going to try and rebuild it by some of the old Fender schematics in your book—a driver circuit here, negative feedback loop there, vibrato circuit as a gain stage, and reverb from yet somewhere else, with a bias control instead of output tube matching, and a presence control. The power taps from the filter caps are as follows: (1) +516 volts (power tubes, from the power switch), (2) -75 volts (bias supply), (3) +474 volts (pre-amp & driver). Are there any special considerations when combining these different circuits from different amps?

Without knowing your specific design, I am lost for advice, but I will tell you this: High voltage increases gain and alters the frequency response of the tube. Higher voltages can make a tube sound very harsh and abrasive. The vintage circuits in my book do not use voltages as high as you have. Lower voltages will make a tube sound warmer and smoother with less gain. You may want to design your own power supply, especially for the preamp tubes.

I am very interested in the Champ mods and am installing them in one now. One of my favorite sounds is the Clapton/Bluesbreaker tone. What can be done to a silverface Princeton Reverb to approach that sound?

Many modifications can be performed on that amp; however, the exact tone that you hear will depend on: (1) that particular Princeton, because they don't all sound the same (2) the particular speaker in the amp—doesn't matter if it's stock, there is still a variable here, (3) the particular guitar you are using (hint—make sure it has humbuckers).

Tone is dependent on everything in the signal path including the power supply of the amp. Therefore, I can not give you a "magic formula fix-all mod" that will make your particular Princeton Reverb have the Clapton/Bluesbreaker tone. I can give you some suggestions to try, one at a time, and let your own ears decide. This advice is given in the same spirit as an optometrist who gives you two lenses and allows you to decide which looks better. In this case, however, you are trying different mods and choosing the tone that pleases your ears.

Here are some mods to try:

1. Refer to my "Save that Princeton" Chapter and do the phase inverter modification described. It is almost exactly the same circuit as the Bluesbreaker and will definitely improve sustain and gain.

2. Try a different speaker, maybe even switch to a 12" speaker. It will fit if you modify the hole in the baffleboard. The speaker is very important. Perhaps a 80 watt Celestion, a Kendrick Green Frame, or a Peavey Sheffield will give you a British voicing similar to Clapton's 12" speaker tone. Stay away from the Celestion 25 or 30 watt because, although they sound good in a 4x12 array, I doubt you will be satisfied because of the lack of bottom-end these speakers have in a 1x12 application. If you choose to stay with a 10" you might try a Celestion or a Kendrick. Make sure to use an 8 ohm speaker regardless of the brand.

3. Beef up the power supply. Replace the stock four-section can electrolytic capacitor with two UK made 50uf/50uf at 500 volt can electrolytic capacitors. These type of caps not only sound British, but they will give more aggression to the attack.

4. Try different tubes. I would probably stay with 6V6s; however, I would use a fresh pair of NOS American-made tubes. R.C.A., Tung-Sol, G.E., Sylvania, Westinghouse are all very good brands to try. Another tube you might try, even though it may be practically impossible to get, would be either the Mullard EL37 or the Genalex KT66. Both of these use .9 amps filament current. I would stay away from an EL34-type tube unless you plan on installing an auxiliary filament transformer because the EL34 draws about 1.5 amps of filament current. This could overheat and ruin your power transformer.

5. If you stay with 6V6s, leave the rectifier tube stock because the 6V6s probably can't take anymore voltage. If you change to the KT66 or Mullard EL37 or even to EL34s, you might try a solidstate rectifier replacer, which will add aggression to the attack, tighten the bottom-end and ultimately sound more British.

After looking over several blackface Fender schematics, I have noticed that some of the preamp tubes share cathode resistors, usually 820 ohms. I know from Ohm's Law that the resistance needs to be approximately halved when this is done so the current flow will stay the same through both sections as when a 1.5K resistor is used for one section, but why are they sharing resistors? Is there a coupling action, like in the standard Fender/Marshall long-tailed pair phase inverter? Or is it just to save on the parts count? Will it affect things at all if I give each section its own 1.5K resistor and 25uf/25v cap?

No desirable coupling action takes place here and you are right—this practice does keep the parts count down. I feel that the amps sound better with separate cathode resistors and separate bypass caps. However, when they are separated, not only is the resistor approximately doubled in value, but the cap should be approximately halved in order to keep the same frequency response. The bypass cap is supposed to have a capacitive reactance of one-tenth the cathode resistor value at the lowest anticipated frequency. A good design would be to engineer for the octave below the lowest anticipated frequency, (an open "E" string on a guitar is 160 hertz, the octave below is 80 hertz) allowing the A.C. signal to go through the cap while the D.C. goes through the resistor. If the capacitive reactance is too high (capacitor value too small), the A.C. will go through the resistor instead of the capacitor, causing degenerative feedback in the lower frequencies and ultimately defeating the purpose of a bypass cap. A 25uf cap has a capacitive reactance of 79.5 ohms at 80 hertz. If you are using an 820 ohm resistor, this capacitive reactance is approximately one-tenth the resistor value. But if you separate to a 1500 ohm resistor for each section and use a 25uf cap, the frequency response goes down to 43 hertz or you could use a 13.2uf to get 80 hertz. Since a 13.2uf is not a standard value, I would go to a 15uf and bring the frequency response down to 71 hertz. Anyone tune down to E flat out there?

I have a '67 Dual Showman I'd like to use for bass guitar amplification, playing blues in small clubs. What brand of tube and tube configuration would be the best to produce the warm "Motown" sound through this amp? It will be coupled with a '64 Jazz Bass and an Ampeg SVT 1510 box (15", 10" and tweeter).
I'm not familiar with that speaker cabinet. If it is 4 ohms, I would run a full quartet of 7581A output tubes (G.E. or Sylvania). If the cabinet is 8 ohms, I would run a pair of Tung-Sol 6550s and only put them in the inside two output tube sockets. The two outside output tube sockets would not have any tubes in them because we are only using a pair. Try the bass guitar running into the normal channel and change the first preamp tube to a 12AY7. This will not only increase headroom, but there will be slightly less gain. When biasing those 6550s, don't be afraid of current. I would idle them up around 60 mA per tube plate current when using a pair on the Dual Showman.

I have a Pro Reverb from about 1968. When I turn on the tremolo, the amp's response gets more mushy and distorted. Also, the tremolo isn't as dynamic (broad) as a Princeton tremolo. Is there anything I can do about these two problems?
The Princeton has a different type of tremolo than the Pro. The Princeton modulates the grid bias of the output stage to create the tremolo effect, whereas the Pro uses an optocoupler to ground out the signal to create the tremolo effect. The catch is this: A circuit that sounds good on a low-powered amp may not give the same results on a high-powered amp.

I would suggest replacing the optocoupler in your Pro to see if that solves the mushiness problem. Perhaps a new optocoupler will get rid of the mushiness but it will never sound like the Princeton. You could rewire the tremolo to Vibroverb 6G15 specs and this should solve both problems, but only if you are not playing real loud.

Incidentally, I would not recommend rewiring the tremolo circuit like the Princeton, because the Princeton circuit is modulating 6V6s and you are using 6L6s. The Vibroverb 6G15 tremolo circuit will work best with the 6L6 Pro circuit.

When using the Vibroverb-type tremolo, biasing is critical because the oscillator tube is modulating the bias of the output tubes. If the quiescent plate current is adjusted very high, the tube will distort on one side of the tremolo oscillator voltage swing and not the other. This doesn't sound very good. On the other hand, if the quiescent plate current is set too low, the amp will sound cold. This is where the listening test is a must. If you play very loud, you are almost certain to clip one side of the oscillator swing. Maybe this is why Leo went to the optocoupler design. The optocoupler doesn't cause any clipping, but just doesn't sound as nice overall.

I recently picked up a '66 Deluxe Reverb amp. When you flip the standby switch to the "on" position, I hear what sounds like the reverb springs rattling through the speaker. It does this if the reverb volume is either fully on "10" or fully off, "0." Also, the amp sometimes makes a soft but noticeable noise like small electronic fizzles. They come and go very quickly, but if I let the amp sit with the standby switch switched "on" for some time, it stops making the noise. Is any of this normal? What should I do?

None of what you describe is normal. I don't think your reverb pan is making the noise. When the reverb knob is turned to "0," reverb signal is being shorted to ground (lest the pot is not grounded properly). To double check that it is not a problem with the pan, disconnect both cables that go from the amp to the pan. If the problem persists, it is definitely not the pan. I have heard noises similar to that before. My experience says that it's probably one or both of two possible problems.

The first problem is dirty preamp tube sockets and/or dirty tube pins. Clean the sockets and pins and re-tension the sockets if necessary. This will probably solve your problems.

Secondly, sometimes there will be carbon deposits on the standby switch internal contacts. This can make that "whooshing" sound when switching to the "on" mode. This can be repaired either by cleaning or replacing the switch. Spray some tuner cleaner inside the switch and turn it "off" and "on" quickly several times. Repeat this procedure and see if it did any good. You may need to replace the switch if cleaning doesn't work.

As far as the "fizzles," there are a few possible causes. Sometimes there is internal arcing in the plate load resistors, in which case replacing all the 100K ohm resistors in the preamp will provide the cure. Although not likely, another possible cause is a capacitor going out or arcing. This is more difficult to isolate and troubleshoot, but then again it is highly unlikely that this is your problem.

You could have compound problems that are causing these symptoms. For instance, you may clean the sockets to find that the problem improves, and then you replace the standby switch to find the "whooshing" gone and the fizzle" better, later to find that replacing the plate load resistors gets rid of the rest of the "fizzling." When repairing a 30-year-old amp, compound problems are the rule rather than the exception.

I have a 1962 Pro Amp. Of course, it had a replaced speaker. The tone is not quite what I wanted; it sounds a little like a tin-can. I found a JBL 15" and that seemed to help. I changed the midrange resistor to back off the mids and that helped a little, but side by side with my '64 Concert, it just doesn't cut it. I love the vibrato. How can I open this thing up a little?

Also for your information, here's a tip. When I got the Pro, it had been painted. Everyone said just recover it. I bought a huge roll of duct tape and by sticking lengths of tape to the painted Tolex and ripping it off, I was able to pull the paint off. (Sounds a little like bikini waxing, doesn't it?) It took several sessions, but the brown Tolex looks great (maybe it was protected by the paint over the years).

Thanks for the duct tape trick. It seems we find another use for duct tape every day.

In regard to your tone problem, I have a few suggestions. For starters, I might suggest rewiring the speakers from the Concert to 8 ohm (parallel/series) and plug them into the Pro amp. This will present the question that needs to be answered: Should you be working on the speakers, the amp, or both?!

I'll bet the '62 Pro amp needs an overhaul. After all, it is 34 years old and there are probably many defective or worn out components. There's not much sense in trying to modify an amp that is sick. You will only be tone-tailoring the symptoms instead of the problem. I think that once you get the amp chassis overhauled and figure out which speaker/speakers you need, you will be 95% of the way there. The rest will just be a matter of using the right tubes and having the amp biased properly.

I have a 70's Twin that's a bit of a problem. When the tremolo is "off," the intensity knob acts as a second Master volume pot — both with or without a footswitch hooked up to the R.C.A. jack in the back. The tremolo works fine when it's up and running, but I'd like to be able to leave the intensity knob on something less than 10. But if I do, it cuts off signal (proportional to its position) to the master stage. I tried replacing the intensity pot itself, but it's the same deal. The opto-isolator seems to be working fine. Any suggestions?

The intensity pot is wired "cattywhompus." ("Cattywhompus" is a technical term used to describe something that just ain't right!) The intensity pot has three leads. One end is grounded. Leave it be. Of your remaining two leads, swap them and your problem is fixed.

Pertaining to silverface Fender amps which may oscillate when various "crutch" shorting caps and feedback loops are removed, you have told us to try and duplicate blackface wiring. To solve the problem, can you point out any critical areas in this regard? We commonly cut out a lot of extra length on the lead wire going to the tubes and pots. Is this part of the idea?

Yes, yes, yes. Ideally, grid wires should be short and plate wires longer. Ah, but the previous stage's plate wire becomes the next stages grid wire — only after the coupling cap. Another word of caution, mount the 1.5K grid resistors (that go to the power tube grids) with the body of the resistor directly up against pin #5. You don't want any lead at all between pin #5 and the body of the resistor. Ground all cathode resistors directly to the grounding buss. If you are doing modifications, keep the signal path in sequential order. For example, don't mount a post-preamp gain stage physically in front of the preamp section. It should be after the preamp section, as far as physical layout.

I'm a young guitarist who enjoys tinkering. A few months ago I bought a blackface Bassman head (AB-165) out of the local paper, and I am interested in modifying it to a more 'guitar friendly' state. I have heard of a "Bluesbreaker" mod that can be performed on a Bassman. I assume this involves EL34s? I would be greatly interested in any information regarding this procedure and in any steps to brighten up and add highs to the normal channel. I have a moderate level of technical knowledge (at least enough to keep me from harm) but any simple diagrams or sketches would ease my mind a great deal. One final question—On the cabinet label, the model number reads AA-165, but the second 'A' was manually crossed out and a 'B' written in pen above it. As I understand, there was never a model 'AA', so would this be a factory error? Is it one of the first following the AA-864? Would this help date the amplifier?

Whew! Was that one final question? First things first. The "Bluesbreaker" modification was invented by John McIntyre at Music Tech in Alberta, Canada. That mod appeared in *Guitar Player* magazine in February 1993. It does involve EL34s and also uses a unique channel blending-type preamp. John performs this mod; however, if you would like to do the work yourself, you will need a copy of the Feb 93 *Guitar Player* article—complete with schematics, etc.

If you would like to brighten up the normal channel, there is a capacitor that shunts the plate load resistor that is actually bypassing high-end to ground. Find V3, which is the third preamp tube. If you follow the wire going from pin #1 back to the board, you will find that it goes to a 100K plate load resistor. Across this resistor is a 500 pf. cap. Carefully remove the cap and you will dramatically improve high-end. Also there is a similar technique for brightening up the bass channel. Find V1, which is the first preamp tube. If you follow the wire going from pin #1 back to the board, you will see that it goes to a 100K plate load resistor. Across this resistor is a .01

mfd. cap. Remove the cap and you will improve high-end on the bass channel.

In regards to your final three questions, I doubt there was a factory error. I have never seen any info on a AA-165, but that doesn't necessarily mean Fender never built any. It was not uncommon for Fender to scratch out and pen different info on the tube configuration labels. Perhaps they had those labels already printed, modified the design and didn't want to throw them away. Waste not, want not! The fact that the label was manually changed does not help date the amp.

To my knowledge, the AB-165 was the amp following the AA-864. I do not know for sure, but have always suspected that the numbers relate to when the amp was designed. For instance, an AB-165 would have been designed in January, 1965. My guess is that the prototype would have been an AA and if the amp was modified later, but used the same design, the number became AB. Your particular amp had the funky feedback circuit with feedback in two places — from output transformer to phase inverter and from output tube plate to grid. This type of feedback is the least desirable circuit as far as tone is concerned. The "Bluesbreaker" mod would change that.

I recently acquired a 1964 Vibrolux Reverb. It has Oxford speakers. The tone is fantastic... fat, warm and very clear. My question is this: should I replace the speakers with re-issues to preserve the value o the amplifier, or are the Oxfords so important to the tone that I will be disappointed with the changed speaker tone? Also, I think the Vibrolux sounds warmer and fatter than my 1965 Super Reverb, louder too. How is that possible?

This is a very personal decision that only you can make. The question you should be asking yourself is whether you are a collector or a player. I personally do not like Oxford speakers—re-issue or original. But the re-issue speakers do sound close to the original. An original will have been "broken in" and will sound a little smoother with less edge, but you may want to replace with re-issue speakers and store the originals just in case you ever want to sell your amp. It will bring hundreds more dollars if you have good sounding original speakers. This extra value would probably more than pay for your re-issue speakers and you will have the re-issues left over after sale time. If you keep playing the originals you are constantly at risk of blowing one or both of them, especially if you crank the amp. If you are strictly a player, you could just stick with the originals and buy some more originals for big bucks when you blow the ones you have.

I think there are other modern 10" speakers that sound better than the Oxfords. In my opinion, the Mojo tone, Eminence and Kendrick speakers all have better note definition, sweeter break-up, and better efficiency. These more efficient speakers make the amp seem to be more touch sensitive.

Your 1965 Super Reverb needs service. If it were in good working order, the Vibrolux would not be louder. Also your comment about the Vibrolux being warmer and fatter doesn't make sense unless the Super Reverb is in less than optimum condition. I'll bet it needs a cap job, has a few burnt or drifted resistors, needs tubes, biasing and the speakers have probably lost their gap energy. You might even have one or more speakers wired out of phase—which would absolutely kill its tone!

I have a 1961 Brown Deluxe in mint condition stamped KJ (Oct. 61). Everything looks and dates out to be the original, but I have my doubts about the speaker, even though I was told it was original. By the looks of the amp, you would believe it to be, but on the rim of the speaker you have the normal numbers 220949 and P12Q etc., but the speaker dates out to be Dec. 59 and also there are some numbers on the speaker I haven't seen on any other blue Jensens. The code is S0130 on one side and S0109 on the other side of the rim. Do you know what these numbers mean?

Although it is possible that your speaker is original, I doubt that to be the case. The early Brown Deluxes were shipped with Oxford speakers with the brown plastic magnet cover. Later Brown Deluxes were shipped with Oxfords with the brown metal magnet cover. I say it is still possible that your speaker is original because Fender was very inconsistent about their products. If they needed to ship an amp that day and were out of speakers, it is possible that they found a speaker under a table and simply used it for that particular amp rather than wait for the standard part. Fender was not known for wasting time or parts. I have seen amps that were wired with similar but different value components and surmised that they were out of that particular part that day and used what they had to get the product out the door.

As far as the S0130 and S0109 number on your speaker frame. No one seems to know for sure what this number means. I do not know. I have researched this by calling all of my friends who would know and none of them know. I even called Jensen and they didn't know. I talked to some speaker design engineers with years of experience and they didn't know.

I did get some good guesses as to what it might mean, among them: the basket manufacturers code, the internal batch number indicating a particular batch, the order number, the shift number of the crew working on that order.

If anyone out there knows about this "SO" code number or can help solve this mystery, please let me know. Gerald Asks You!!

I recently purchased a 1964 Fender Deluxe Reverb. It is stamped "Export 117" on the tube sticker that is pasted to the inside panel. On the chassis there is a red voltage selector switch. This looks like Fender factory work. Did Fender Export much electrical equipment during this era? Does this extra "adjustment switch" affect the basic wiring or tone in any way? Also the grille cloth has a 2" tear and a 1" tear. Would the value be the same if I leave the torn grille cloth or replace it with old new cloth? What kind of speakers originally came with this amp?

Yes, Fender did do considerable exporting of its products during the 60s. Actually the power transformers on the export models were built for 50 hertz (as in Europe) instead of 60 hertz (as in U.S.). This means there is slightly more core stack on an export transformer because it is expected that the transformer will be operating with a lower frequency A.C. line. When you operate this type of transformer on a 60 hertz A.C. line, the transformer is actually a little more efficient. Everything else being equal, this transformer will probably run a little cooler, possibly last a little longer, and is more heavy duty.

Regarding grille cloth, do you expect to sell the amp soon? Do you like to look at the tear? If you answered "yes" to either of these questions, leave the grille cloth alone. I personally don't like looking at tears, and if I could replace the cloth with some that looked original (and of course the original was not torn), I'd go for it.

In answer to your final question, the 1964 Deluxe Reverb amplifier came stock with Jensen Special design speakers.

I have a Fender Bassman Amp. An adjustable bias circuit has been installed and all the capacitors have been replaced. I notice that if I overdrive the amp with a pedal, the amp shuts down. I can't find the tech who put in the bias mod. I think the model is AB165. It has three 7025s and a 12AT7 with two 6L6GC power tubes. Do you have any idea why the amp is shutting down? The amp is a head without speakers. Also is there any way I can bring up the wattage? It is supposed to be a 50 watt amp, but the tech said it tested at 42 watts.

The AB165 is my least favorite Bassman. Your "shut down" problem is probably output transformer saturation or a parasitic oscillation that is oscillating at a frequency higher than you can hear. In the case of the parasitic oscillation, you interpret this as the amp shutting down, even though it is working very hard to amplify those high frequencies that none can hear.

Substituting a known good output transformer will probably fix your problem, but if it is a parasitic oscillation you may need to alter layout, lead length, grounding, etc., in order to cure the problem. (See the chapter "Twenty Ways to Stabilize an Unstable Amp" in this book.) Parasitic oscillation is not an easy problem to troubleshoot and pinpoint. You will need the help of an expert.

If I owned an AB165, I would most definitely rewire the feedback circuit to AA864 specifications. This would, among other things, involve removing both 220K interstage feedback resistors that connect the plates of the output tubes with the plates of the phase inverter (12AT7).

There is also a 470K ohm interstage feedback resistor that should be removed. There are only two 470K ohm resistors (yellow, purple, yellow) in the whole amp. One is on the edge of the board near the bass channel treble control (leave this one alone) and the other is in the middle of the board near the first one (this is the one to remove). You will notice an improvement in gain, power, and tone if these modifications are performed.

I have a blackface "Bandmaster" head (not sure of the year as there was some minor foolishness by the previous owner) that makes one heck of a thud when the standby switch is moved to the "on" position. There is also a small flash of light when the standby switch is activated. I have tried leaving the amp to warm up for a good ten minutes prior to activating the switch, but it doesn't make any difference. I am afraid that the speakers may be harmed if I continue to use the amp in this condition. I would like to have an appreciation of the cause for when I take the amp to a technician for repair.

This is a very common problem with Fender amps. The reason is very simple. The standby switch is a 3 amp 250 volt switch. The standby circuit has somewhere close to 500 volts on it at the time of switching from "standby" mode to "on." The result is a tiny arc of electricity, like a miniature lightning bolt, just before the switch makes internal contact. Although this is usually not a problem, after years of this occurring, carbon will build up on the contacts and make the problem worse. There are two possible cures. You could spray a little tuner cleaner inside the switch and turn the switch off and on several times. This may clean the carbon from the internal contacts and make the problem better. I doubt if this remedy will be enough in your case, because it sounds to me like you have a severe version of this problem. The other cure is to replace the switch. Don't even think about using a 600 volt switch, because it won't fit in the chassis—600 volt switches are very large!

I would like to build a box to bypass the speaker of my Deluxe Amp and send the signal directly to headphones without hurting the amp, the headphones, or my eardrums. Any ideas you have for this circuit would be appreciated.

Your timing is right. There is a chapter in this book that is exactly what the doctor ordered. The name of the chapter: "Headphone Hookup For Your Tube Amp."

I just picked up a 1966 Bassman in excellent condition. The speaker enclosure was stuffed with fiberglass batt insulation and contained one Jensen C12NA gold label and an Oxford 12M69 (coded 465-242). The Oxford is a gold frame. It sounds good to me, but I'd like to know why the insulation and what or whom was the Oxford made? Was it made in '52, '62, or '72? Also, is there a reference book that deals with various amplifier speakers made in the U.S.A.? Thanks.

Insulation is used in a closed-back cabinet to break-up internal sound waves (damping). The result is that the speaker "thinks" it is in a bigger cabinet. A cabinet with batting performs as though it has 10 to 20% more air volume. This is due to complex acoustic conditions which are beyond the scope of this book. Speaker cone motion should faithfully reproduce the applied electrical signal. Any tendency of the cone to vibrate at frequencies which are not in the input or to continue to vibrate after the signal stops will color the sound and adversely affect its quality. Almost all closed back cabinets use some type of batting.

There is no way to know for sure about your question concerning the Oxford's age. The speaker code doesn't include the decade of manufacture. This is where educated guesswork comes into play. It was probably a 1962 model.

I am not aware of any book that deals strictly with American guitar amplifier speakers. I touched on the subject in my book, *A Desktop Reference of Hip Vintage Guitar Amps*.

When I really crank my Pro Reverb, I usually have to turn the Bass down because the tone gets muddy. I tried a solidstate rectifier and extra capacitance in the power supply, but my clean sound doesn't "swell" as nicely. I'm thinking of getting a better output transformer with better low end response. What do you think?

I'm really glad someone finally asked this question. Bass response is easier to get at high volumes than low volume levels. With almost all amps, the bass control will need to be turned down when the volume is turned up. Conversely, the bass should be turned up when turning the volume down. It is like a "loudness" control on a high-end home stereo. When the volume is down, you turn on the "loudness" switch to add more bottom. (This compensates for the fact that low-end is easier at louder volumes.) When the volume is turned up, the "loudness" switch should be defeated.

I really don't think you need an output transformer with more bottom-end. Such a transformer would be even muddier than the one you have (at high volume with the bass control turned up). Have you ever been in a recording studio and noticed how you could actually get clearer and punchier kick drum and bass guitar by reducing the low-end EQ? Yes, increasing the low-end can muddy-up the tone!

There is a possibility that you may need a speaker that produces clearer bottom-end. Try cutting the bass control as the volume is increased and see if the speakers you have will give you enough bass clarity to satisfy you. If it doesn't, try another speaker. Speakers that are known for their incredible bottom-end would include the Electrovoice, Kendrick Black or Brown frame speakers, or the Celestion G12 100H.

If this still doesn't give you enough bass to satisfy you, you may need to go to a closed-back extension cabinet. The Pro is open back and therefore subject to more low-end phase cancellation. A closed back extension cabinet may be just the ticket if you need even more bottom.

In your book, in the section on easy tricks to juice up your vintage amp, it says that in two channel blackface Fender amps, the preamp tube of the normal channel can be removed to increase gain and distortion in the vibrato channel. Also under "Mods for the Pro Reverb," it says you can switch the phase inverter tube from a 12AT7 to a 12AX7 to increase gain. I received a '65 Deluxe reissue as a gift and would like to try these changes; however, I am aware that Fender reissue amps are not exact replicas of the originals. So, the question is: will doing these mods hurt my amp in any way?
Both of these are very safe mods that can be tried on any amp without fear of damage. As always, try the mods one at a time and perform critical A/B listening tests before proceeding to the next mod. This way, you will be able to hear the difference of each mod individually and make your decision if it sounds better or worse.

GIBSON

I own a Gibson GA-40 Les Paul amp circa 1958. My problem is a constant hum and a roaring noise as I turn up either volume knob. I replaced the rectifier and power tubes to no avail. Will replacing the 5879s and the 6SQ7 tame this or is the transformer shot? People have told me to look for goo oozing out of various component parts but I don't see any goo. I need help and probably a lot of cash to get this baby sounding like it should. Can you be of help?

If the hum is about the same pitch as a B-flat, you probably need a cap job. In this procedure, all the electrolytics in the amp are changed. If you have not had this done for your amp, it is definitely time. Electrolytics use an electro-chemical reaction to store energy, just like the battery in your car. Thirty-five years without changing is just too long. A cap job on this amp should cost about $80. The roaring sound is probably either dirty volume pots, dirty depth pot (on vibrato circuit), or dirty input jack switch contacts. You might try cleaning the pots with a little tech spray and inspect the jacks to make sure the shorting switches (on the jack) are making good contact. My first guess is to check your depth control pot. In your amp, there is D.C. across that pot and D.C. is very hard on a pot. Cleaning probably won't help this component, but a replacement would only cost a few dollars. Your transformer does not appear to be a problem. The 6SQ7 is only an oscillator tube for the vibrato circuit and does not appear to be part of the problem either.

I've heard that the mid-60's Gibson Falcon is basically a cheap Deluxe Reverb—a virtual tube clone. Is this so?

There are some striking similarities between the Falcon and the Tweed Deluxe (5E3) but not the Deluxe Reverb. The Deluxe Reverb is fixed-bias with an optocoupler-type vibrato (tremolo) and treble and bass controls, two channels, and a grounded grid style phase inverter. It uses 12AX7 preamp tubes. The Falcon is cathode-biased with output tube grid bias modulation tremolo, only a single tone control, one channel, and a distributive load phase-splitter. It uses 6EU7 preamp tubes. The Tweed Deluxe has no tremolo or reverb, has two channels, and a distributive load phase-splitter and is cathode-biased. It uses 12AX7 preamp tubes. Although the Falcon is similar to the Tweed Deluxe in that it is cathode-biased, uses 6V6 output tubes, similar tone control, and uses a distributive load style phase-splitter; the gain structure, biasing of preamp tubes, type of preamp tubes, coupling cap tone shaping and power supply voltages are unique.

The only thing I could say to be wary of concerning Gibson vintage amps, is that they sometimes used odd-ball tubes that are difficult and expensive to obtain.

MARSHALL

I just bought a 1973 P.A.20 Marshall Head and a 4x10 Marshall cabinet loaded with G10 L35s at 8 ohms. The head is set for 16 ohms. Will changing the head to 8 ohms work well, and how much power increase can I expect going from 16 to 8 ohms? Is there a distributor of Marshall products a person can deal with directly to order speakers, etc.? Will these speakers work well? There are a several different ways that 4x10s can be wired. Are you sure that these speakers are wired for 8 ohms? When you measure the speaker cabinet on an ohmmeter, it should read about 6.25 ohms if it is indeed wired for 8 ohms. The power transfer will have less loss when the impedance of the head matches the impedance of the cabinet, but a 100% mismatch (16 ohm head setting with 8 ohms cabinet) will not be that much of a power loss (only two or three dB's). As always, I would suggest you go with what your ears tell you. On a Marshall circuit, or any other British amp, the primary impedance of the output transformer is much lower than what one would expect; therefore, you might like the 4 ohm setting better, this setting having a higher turns ratio and therefore a higher reflected impedance on the output tubes. When A/B testing, listen for harmonic content and overtones. Trust your ears.

Marshall products are distributed in the U.S.A. by Korg. They will not deal directly with individuals, and only sell to authorized repair stations and dealers. Look in the phone book in your area for a dealer or authorized repair station. Your speakers will work with your setup, but there is a difference between functioning well and performing well and that's a judgement that you will have to make.

I have two Marshall heads, a JCM 800 Lead series with 6550 output tubes, and a JCM 900 MKIII with EL 34s. When I checked the bias at pin #5 of each amp, they each read minus 50 volts. Does the factory set these at this reading? Can you tell me the proper bias voltage for each amp?

Every amplifier that Marshall builds must have the bias set before it leaves the factory. The proper bias voltage is dependent on the type of output tube of the amp, the brand of output tube, the circuit the tube is used in, and the plate voltages developed by the power supply.

I contacted Mitch Colby, vice-president of Marshall, and he tells me that Marshall biases each amp individually while observing a waveform on a scope. It is pure coincidence that each of your amps happens to be reading minus 50 on pin #5. Usually 6550s require higher bias voltage than the EL34s unless the EL34s are Chinese (in which case they would require more bias voltage than other EL34s). You probably have Chinese EL34s in your JCM900 MK III.

I have a Marshall JCM 800 50 watt 2x12 combo with 2 inputs. The sound seems overly brittle even with treble, presence and midrange all the way off. Are there any modifications I could make to increase bass and low mid-response while "mellowing out" the high-end?

Assuming that you are talking about the JCM 800 with master volume, and assuming you are using the "high gain" input rather than the "low gain" input, I would recommend changing the .68 uf bypass cap (that goes to pin #3 of the first preamp tube) to a larger value. Try a 25 uf at 25 volts for starters. This will increase bass and low mid-frequencies. You could go to a higher value to increase bass a little more (try a 50 uf or 100 uf). To mellow out the high-end, remove the 470 pf cap that is in parallel with one of the 470K voltage divider resistors going to pin #2 of the second preamp tube. If you still want more bottom end after this, try changing the both .02 uf coupling caps, that connect to pins #1 and pin #6 of the phase inverter tube, to .1 uf at 400 volts.

The original .02 caps squeeze off some of the bottom end and changing to .1 caps will let some of this bottom end through the amp.

I have a Marshall amp model 2205 which is a 50 watt with reverb and it never has seemed to be near as loud or punchy as other 50 watters. Is this model supposed to be tamed down because of its reverb circuit, or did I just get one with a weak transformer or something? I have replaced the tubes more than once and the filter caps recently; this changed nothing. I haven't swapped either transformer yet because I'm not sure if they interchange with other model 50 watters. Is there anything I can do to this amp to make it scare me?

These Marshalls are my least favorites because of the preamp tube distortion. In this particular Marshall, the preamp tubes are cascaded to produce preamp tube distortion. This kind of distortion is rather over-compressed sounding and somewhat buzzy. It is not the circuit that is found on any plexi or four-input metal-face Marshall. Of course the preamp section could be modified to be more like a plexi- or metal-face (four input) circuit. The idea would be to remove the master and overdrive the output stage instead of the preamp stage. This modification would probably scare you, especially if you are accustomed to the way the amp sounded when stock. I would not recommend trying this type of mod unless you are sure that you know what you are doing.

What is your opinion of the (Marshall) Vox AC-30 re-issues? I had a chance to demo one the other day and was impressed with the sound. What are your thoughts?

Although I have not personally heard this amp, I have heard nothing but good about it. I spoke with Mitch Colby the other day, (Mitch is vice-president of Korg/Marshall/Vox) and he informed me that the circuit was absolutely original but with the welcome addition of a standby switch. A dealer will be sending me one shortly so that I can write instructions on how to retro-fit the Kendrick Model 100 reverb add-on module. I'll let you know my opinion of the tone once I get a chance to hear it.

My question concerns early Marshall amps, late 60s-early 70s, 100 watt with high plate voltage of around 520 volts and the use of

EL34s. I prefer the sound of Siemens or Mullard brand tubes; however, they don't hold up long, not like GE 6CA7s which take a beating. I've heard of several mods to drop the B+ in the 460 range. Do you know of such mods and which is the best sounding? I can remember back to the 60s when Marshalls didn't even use screen resistors on their amps. For one thing, I would make sure that the amp in question has a 1000 ohm 5 watt screen grid resistor. This will limit the screen current and excessive screen current is a common cause of tube failure. I do not recommend any mods for lowering the plate voltage of a Marshall— other than replacing the power transformer. Even though the amp was originally shipped with Mullard EL34s, I would not recommend using the Mullard in this amp. For one thing, the cost of Mullards would be very expensive and the failure rate would be high—just like the originals. This wouldn't have been a problem in the 60s when Mullards could be bought for a few dollars, but at today's prices the failure rate would be something to consider. I prefer Phillips 6CA7s in this type of amp. They sound good and they can take a beating too. Other good choices would be the 7581A or the KT66. Both of these tubes can operate at high voltages without a problem and sound very good. The 7581A might be a little too clean for some taste, but it can sure handle the voltage. Any of these tubes mentioned can be fitted into the Marshall with only a minor bias adjustment.

MISCELLANEOUS

I have an Ampeg B15N bass head that I use for guitar. It has 6SL7 preamp tubes and 6L6 power tubes. Are there any modifications I can do that will make the amp more responsive, and also to add a touch more midrange? Would you recommend converting the power tubes to EL34? I use the amp with two 12" guitar speakers in a closed back cabinet and I would like to know the proper way of hooking up these speakers, since the original Ampeg speaker plug has four wires on it in what appears to be an unusual configuration. Thanks for your time.

For starters, the #2 input of the first channel has more mids than the #1 input. I wouldn't recommend changing the tubes to EL34s, but would suggest modifying the preamp instead. There is a tone shaping circuit between the first gain stage and the volume control. I would rewire to remove the tone-shaping circuit (disconnect or remove C4 in the B15N schematic and the circuit will be disabled). Next, short out C5 with a straight wire so that it will be bypassed (shorted) and change C9 from a 500 pf to a larger value. (Any value from 2000 pf to 5000 pf selected for tone by actual listening tests.)

Of course, channel two could also be modified in a similar fashion. These modifications should give you more gain, more midrange honk, and more response as you requested.

The 4-pin XLR connector that Ampeg used for a speaker connection looks like an unusual configuration but it is actually a safety device to keep from running the amp without a speaker load. Pin #1 is the positive speaker lead and pin #4 is the negative speaker lead. Pin #2 and pin #3 should be shorted together on the speaker cabinet jack. Here's how it works. Inside the amp, pin #3 goes to ground and pin #2 goes to the cathode resistor of the phase inverter. If you try to run the amp with no speaker plugged up, then pin #2 and #3 will not be connected. With the phase inverter cathode not grounded, signal cannot pass to the power tubes. When the speaker is plugged in, the cathode resistor becomes grounded which enables the phase inverter and of course pins #1 and #4 are hooked to the speaker. This is kind of like an idiot-proof standby switch built into the speaker plug.

I have an Ampeg Gemini II amp with 7591 output tubes. Since these aren't readily available, what would be necessary to convert it to 6L6s. I know the later model Gemini II's used 6L6s. Do both of these amps use the same power transformer?

I'll answer the last question first and say that the later Gemini II's used a slightly different power transformer and the power supply voltages were slightly different. This would not make much difference concerning your ability to change from 7591 output tubes to 6L6s.

To perform the conversion, simply:

1. Take everything off of pin #8 (if anything is connected to pin #8)

and move it to pin #4. If anything else is already wired to pin #4, leave it connected.

2. Take everything from pin #5 and move it to pin #8. Some amps will only have a ground wire and others will have a cathode resistor, but whatever is on pin #5 must be moved to pin #8.

3. Move everything from pin #6 to pin #5.

4. Place some 6L6GC tubes in the sockets.

5. If the amp is fixed bias, you **must reset bias** to avoid damaging the 6L6GC tubes. If the amp is cathode-biased, it will probably have a 140 ohm cathode resistor and an electrolytic cap. These will be on pin #5 before they are rewired to pin #8. **You will have to increase the resistor value to 250 ohms at 10 watts** (leave the capacitor value as is) or you will blow something up!

Do not attempt this conversion unless you are certain that you know what you are doing! Bring this information to your favorite tech if you doubt your skill in performing this mod.

I own an Ampeg VT40 (top-mounted controls), that needs re-tubing. First of all, can the 7027As currently employed in the power section be changed over to Sovtek 5881s? If so, what will that do to the character of the sound, which has always been great until recently when it finally became apparent that new tubes were needed? Secondly, a couple of preamp tubes are proving hard to find, specifically the 12DW7 and the 6K11. Please tell me where I might find these or what can I substitute for them?

Do not attempt to change to Sovtek 5881 tubes. The 7027A is a tube that can handle 600 volts on the plate, 500 volts on the screen and 35 watts of plate watt dissipation. The Ampeg VT40 is running both the screen and the plate just shy of 600 volts. The Sovteks would turn to ashes as they fried and roasted. If you just have to convert the output tubes, you might try 6550s. Although the 7027A has unnecessary additional pin outs for both the screen (pins #4 and #1) and grid (pins #5 and #6), the 6550 does not have pin #6 or #1 connected to anything and uses pin #4 for the screen and pin #5 for the grid. The 6550s other pins for heater filament and plate are identical and they would probably work in the 7027A socket without any rewiring. I've never actually tried it and I don't have a VT40 in my shop now to inspect to check out the actual wiring of the amp, but going by the info on the schematic, the 6550s should drop right in with only a minor bias adjustment (adjust the value of R38 to set bias).

I would not recommend changing output tube types—stay with the 7027As. They do sound great in the Ampeg and have mucho power.

You can get all the tubes you described, including the 7027As at any full-service vintage restoration shop. You won't find them at a music store or an electronics retailer because they are not a high volume item and are only used in a few older amps.

Are 7027 tubes for Ampegs as rare as I think they are? I found some old R.C.A.'s, but they have poor tone! What are the best replacements? Ampeg VT-22s and V-4s are super loud and super clean-sounding amps (with little, if any, distortion).

The 7027A tubes come in two varieties, namely the 7027 and the 7027A. Both are fairly rare. The Ampegs you named must have the 7027A, which can take much more voltage and has 10 extra watts plate dissipation. In fact, the regular 7027 would blow up in the Ampegs you mentioned. I've always thought the R.C.A. 7027A was an excellent sounding tube, when biased correctly. In the V-22 and V-4, a quartet will deliver 125 watts of pure, undistorted power. Although this may be too much headroom for some people, it is excellent for keyboard, bass guitar, and clean guitar. I would not recommend replacing the tubes with a different type. A 7581A would substitute and probably work just fine with only a minor bias adjustment. The pin out on the 7581A and the 7027A are very similar and, the way Ampeg wired the sockets, either would fit without having to rewire the socket.

I have heard of people using 6550s in these type amps, but changing the screen resistors to 1000 ohm 5 watt and bias adjustments are necessary. Since the 6550s pull twice the filament current, I would not feel comfortable using these.

I own a late 70's Ampeg V-4B bass amp and I want to convert the power tubes from the stock 7027 type to the 6550 type. Please keep in mind that I want to retain the original transformers, if possible. I would consider doing reversible modifications and/or using smaller transformers (B-25 type?) to reduce

the weight. I have no objections to a lower power output.

Compared to the stock configuration of 4 x 7027, push-pull Class AB layout, what are the pros and cons of the following:

1) 4 x 6550 Class AB
2) 2 x 6550 parallel single-ended Class A
3) 2 x 6550 push-pull Class AB
4) 2 x 7027 parallel single-ended Class A
5) 1 x 7027 (or 6550) single-ended Class A ?

What should I consider when dealing with negative feedback, power supplies, biasing, and output impedance? What change-over would be the least destructive? Also, can a Bassman 10 be easily converted to 6550s?

Whew, that's a lot of questions! Let's get down! You can save a lot of headaches by forgetting any single-ended design. You would need a special transformer that has a gap in the laminates in order to avoid the core becoming magnetized and we're talking big bucks here. Also, the parallel single-ended would require some power supply mods (probably a huge choke) to avoid arcing across the power tube sockets.

The most practical conversion would be to switch to 6550s push-pull Class AB. This would require only 6 resistors. There would be no modifications to the feedback loop. Simply change all four screen resistors from the stock 470 ohms at 1 watt to 1000 ohms at 5 watts. These resistors are marked R41, R42, R47, R48 in the V-4 schematic.

The other two resistors to change are in the bias circuit. Change R50 to 82K and change R49 to what ever value that will adjust the bias correctly with your 6550 tubes installed.

If you are using 4 x 6550s, the output impedances marked on the impedance selector are correct. However, if you are only using 2 x 6550s, the impedance of the speakers should be **twice** what the impedance selector indicates. That is to say, set the impedance selector to 4 ohms if you are connected to an 8 ohm speaker. This will correct the impedance matching of the output stage.

Bassman 10 amplifiers are basically converted to 6550s the same way. Change the screen resistors to 1000 volts at 5 watts and reset the bias. You will not need to mod the bias circuit to bring the negative voltage supply into range, but you must reset the bias.

Recently I purchased a Spearsonic #EC-511 Echo Delay. It looks like a mini-head and uses a tape cartridge that looks like a small 8 track tape. Do you know where I can get instructions and/or replacement tapes? I can't figure out what the playback knob really does.

I can't tell you who sells that tape, but I can offer a couple of suggestions. Take the tape to a radio station and show it to the engineer there. Radio stations use cartridges that look like 8 track cartridges and may have a supplier that sells the exact tape you seek. If that fails, you could always roll your own. Buy some high quality tape that is the same width. Cut the tape in the cartridge and splice the end of some new tape on it. You will have to figure which way the tape rolls so the the splice is on the take-up side. Carefully pull on the other end of the other cut end until it loads the new tape on the cartridge internal spool. You will know when you are through when you see the splice come back out the other side. At that time you will just need to splice the tape and spin it until the slack is taken up.

If this doesn't work, you haven't lost anything except a little time and a little tape, assuming it's not working now anyway.

I recently purchased an old Silvertone "Twin Twelve" amp head and have the following questions:
1. The amp head has a very small built-in spring reverb which seems to distort due to too much reverberation depth. All the tubes test good. Any suggestions?
2. Which tube is the reverb tube—one of the four 12AX7s or one of the two 6FQ7s (6CG7)?
3. Is the 6FQ7 an unusual tube?
4. One of the 6L6s has white powder on one side of the glass. Is this defective? (The tube lights up, and the amps works, but the tube tests dead.)

I'll answer these in order.
1. The reverb springs in this particular unit probably never sounded all that good, the reverb tank being the least desirable of all reverbs. However, you could increase the value of the cathode resistor that goes to pin #8 and pin #3 to ground. This is an 820 ohm ½ watt resistor. Try changing to a 1500 or 2700 ohm ½ watt and let your ears decide which sounds better. This will reduce the drive considerably.

If you really wanted to customize, change the 1 meg grid resistor that goes to pin #7 with a 1 meg pot. One end goes to the circuit, the wiper goes to the grid pin #7 and the other end to ground. Turn the pot until your ears say it's as good as it is going to get. Remove the pot without turning the shaft and measure from wiper to end and wiper to other end. Replace with two fixed resistors of the appropriate value wired equivalently to the way the pot was wired.

2. The 6CG7 on the end is the reverb tube.

3. The 6CG7 is not all that unusual. It is pretty similar in performance to a 12AU7. It has an amplification factor of 20.

4. The 6L6 is bad. The white powder indicates that the tube has lost vacuum.

I recently found a home-made speaker cabinet that appears to be late 50s or early 60s. Inside I found a 15" speaker that I would like to identify. The magnet plate reads: Trademark True Sonic Model 206, SN# 955, AX 16 ohm, Stephens Mfg., Culver City Ca. U. S. A. The speaker also had some type of transformer type box riveted to the housing and another screwed to the inside of the cabinet wall. This box looked like it went to a tweeter that had been removed. Does it have any vintage value?
The transformer riveted to the housing is probably a 70 volt line transformer (for PA use to convert 70 volt line to 8 or 16 ohms). I say this because you are describing an extension cabinet and not a combo amp, therefore it could not be an output transformer. The other box, that looked like it went to a tweeter was probably a crossover device. I believe you have already identified the speaker. It was made in Culver City, California, by Stephens Manufacturing. It is a 16 ohm 15" and was marketed under the name "True Sonic" as Model #26. I don't care to speculate on vintage value, but remove the transformers and listen to the speaker, it might be a real find!

What kind of amp are you working on (in the picture)? I understand that your speakers have been upgraded. Can you tell me your opinion on your speakers and which Jensens and which Oxfords do you recommend for a 1965 Fender Twin?
The amp in the picture is a chassis of a Kendrick 4000 Special. Less

than 100 of these were made. 120 watts of power was just too loud for most people; however, they were also exceptional for bass and PA.

Yes, the Kendrick 12" Blackframe speakers were sonically upgraded about a year ago. The original design, like old Jensens, had weak mushy bottom-end, and we felt that the low-end could be improved to become bolder, clearer, tighter and punchier. Also, we were not satisfied with the efficiency of either the original Jensen or the Kendrick. After 18 months of trial and error research, we came up with the exclusive Kendrick Bell-cone. This cone was shaped like the end of a trombone bell—like a big paper horn. Not only did this design achieve everything we had hoped for, but the speaker became much louder and richer. I knew I was on the right track when I put a prototype speaker in a blackface Deluxe Reverb, played a note and had the sensation of water in my ears. You know the feeling, when you go swimming and the water gets in your ears and just fills them up; that was what the Bell-cone was doing to me. The Kendrick 12" Blackframe Bell-cone speaker is what I like the best in a '65 Twin. To quote the *Guitar Player* magazine review (Jan 93), "Switching to the Blackframes was like experiencing the Twin through a 4x12 cabinet. The sound instantly became bigger and fuller..."

Which model Jensens sound the best? The least desirable are the Field coil models. Made in the late 40s to early 50s these speakers used an electromagnet instead of a permanent magnet. Almost as bad are the 60s models. These models usually start with a "C" (which stands for "ceramic" magnet) or "EM." The problems with these are the weak magnets and the thin paper cones; they make all the problems that Jensens have, even worse.

My favorite Jensens are the Alnico Jensens made in the 50s. These speakers came in different models, the top of the line being the P12N. Other good sounding ones, in ascending order, were the P12R, P12Q, and P12P. (Stay away from the P12S and P12T, they are unusable.) I would prefer the P12N in a mid-60s Twin.

I personally do not like Oxford speakers and therefore cannot recommend any of them. Because they lack efficiency and definition, they never were very popular.

Why are the splitter/driver tube plate load resistors for some pre-CBS Fenders and Marshalls unbalanced (100K and 82K)? I

have asked a couple of techs this one and they all wax poetic or become catatonic. Is this a quick-fix shotgun remedy for getting more balanced operation of each section of a 12AT7 or, conversely, to purposely unbalance push-pull and contribute to asymmetrical clipping and tonal response?

I get asked that question a lot. Let me preface the answer by explaining that each section of the tube is not being used in the same configuration. The first section is used as a grounded or common cathode configuration. This configuration has good voltage and power gain and high input and output impedances.

The second section of the tube is used as a common or grounded grid configuration, which has substantially different characteristics than the common cathode configuration used in the first section. For instance, it has a low input impedance, a somewhat lower power gain but a high voltage gain. Unlike the common cathode configuration, it does not produce a phase reversal from input to output. Use the two sections together and you have a phase inverter. Here's how it works.

The first section, which is modulated by the grid, is a common cathode configuration; it produces phase reversal. The second section (which is modulated by the cathode resistors used in the first section) is a common grid configuration; it does not produce phase reversal. Since the gain characteristics of these two circuits are different, the plate load resistors are chosen so that both outputs are balanced.

What's your opinion about the Sovtek 6V6GTA and 5AR4? Everyone agrees the Sovtek 5881 is a great 6L6-style tube, but I've heard lots of people say that the Sovtek 5AR4 and 6V6GTA "aren't quite there" yet. You see more of these devices in a week than I see in a year—so are they worth it? The price is right.

The Sovtek 5AR4/GZ34 is a good rectifier tube. It has its limitations, especially when you compare it to a Mullard GZ34. For instance, it cannot take the current that a Mullard can take so don't use one in a tweed Twin (80 watt). But they work just fine in a 50 watt amp. Also the warm-up time is faster, but that's no problem if you have a standby switch.

I have done extensive testing on the Sovtek 6V6GTA and I can say that they sound like they were made in China. Also they do not hold up in higher voltage amps (blackface Deluxe). One thing for sure, the price is right on both of these tubes.

I have rust on my output transformer. Couldn't I:
1. Apply rust dissolving solution with a soft brush;
2. Bath the laminations in Blue Shower (electronic cleaner) to clean off the rust dissolving solution;
3. Apply anti-corona black varnish? Do you think this might work?
One must consider the real problem with rust on a transformer. The real reason that rust is undesirable is because it can cause conductivity between adjacent laminates. Conductivity means wasted power because of induced eddy currents. (Eddy currents occur when the shorted laminates look like another secondary winding to the magnetic field of the transformer. Instead of all the secondary signal being received in the real secondary, some goes to the laminates that are shorted and travels round and round the laminates. This causes heat in the laminates and loss of power developing in the real secondary.)

The best thing to do is assess the severity of the situation with the rusted output transformer. If the transformer is rusted to the point of conductivity between adjacent laminates, it is best to have the transformer re-laminated. This could be done for $25 to $50. If the core material is identified properly, the transformer will sound authentic.

If the rust is not really bad, I would spray it with some Rustoleum flat black paint, taking care not to get any on the windings or insulating material (because of the carbon that is present in black paint.)

You could knock off some of the rust first by using non-conductive sandpaper (the kind that does not have metal in it). Be careful not to sand with the grain of the laminations or even across the grain, but **lightly and in a circular motion.** You do not want any parallel burrs or perpendicular burrs. After the surface rust is removed, a light coat of Rustoleum would be appropriate to protect from future rust.

I do not recommend Blue Shower, because it helps conductivity and we certainly don't want any conductivity between adjacent laminates.

Of course, you could always de-laminate, de-rust, paint each laminate individually and then re-laminate.

What are the elements that make up a good tube tester for amps? Is a Mil tester better than a B&K?

Please see the chapter, "Cool Tools for Amps." Good tube testers for guitar amplifiers would have to test tubes at high voltages and test for noise and microphonics. I do not recommend any commercial testers because most commercial testers test at relatively low voltages and do not really work very well for guitar amp tubes. However, they will let you know if a tube is shorted or gaseous—big deal.

There are no tests for tone or microphonics on commercial testers. Preamp tubes can be tested for transconductance (gain), but tone and microphonics tests can be performed simply by plugging the tube into the intended circuit and listening.

This really happened, I am not making this up: A client of mine sent me some output tubes to match and preamp tubes to check gain. The client had listened to these tubes and matched them *by ear*. The tubes were taped up in what he thought were matched pairs and the preamp tubes were ones that he had listened to and thought to be balanced (both sides had equal gain). Upon checking these tubes, they were matched, and the preamp tubes were balanced. This is the ear method and it worked for him!

If you have the matcher from the "Cool Tools" chapter, you should be able to match pairs easily. I test my preamp tubes by ear and I have a homemade tester to check actual gain characteristics.

When installing a 12" speaker in a 10" cabinet, is it necessary to enlarge the opening in the baffleboard?

If you want the low-end that the 12 has to offer, yes. Also, depending on the speaker, extra baffleboard may cause annoying buzz.

Could you please help me regarding the make and year of manufacture of speakers in my 1965 blackface Fender Super Reverb combo. There are very few old amps here in New Zealand and there are no reference points to consult. However, I have identified one replacement speaker which is an Oxford 109942-465-917 10L6-2. This is either a '59 or '69, I presume from the 917 numbers, but the other three speakers are a mystery. I have looked at Aspen Pittman's Tube Amp Book III which defines

Jensens and Oxfords, but I can't determine these. They are: 064121 137 6910N and 064121 137 7438. If you can help me, I would be most grateful.

You are correct about that Oxford speaker. The 465 is actually Oxford's E.I.A. source code and the 917 indicates the seventeenth week of either '59 or '69. All of the other speakers are C.T.S. speakers and are indicated by the 137 (which is the E.I.A. source code for C.T.S.). The first speaker code would indicate a manufacture date of the 10th week of 1969 (what a great year that was, eh?) and the other is the 38th week of 1974.

I recently added a 1966 (pot-dated) Premier 90 Reverb unit to my collection. It utilizes 6EU7 and 6AQ5 tubes. The only transformer that I can find any identification on is ink stamped 630004, 55A16 and 8316319. When I play through the unit, it adds a bit of distortion to the signal. Are my pickups too hot for it? Also, when playing through the unit, you can hear a very high-pitched sound that sounds like a "bomb" dropping. This only happens every 10 or 20 seconds or so. Should I replace the caps and resistors? Should I rewire the 6EU7 socket for a 12AX7 or leave it?

Assuming that the songs your band plays are all longer than 10 seconds, it sounds to me like it's time for a cap job. The "bomb dropping" sounds are most likely bad electrolytics making noise. All the electrolytics should be replaced. When electrolytics are in need of replacement, the supply voltages can drop dramatically. This would kill your headroom and introduce the distortion of which you speak. It can also cause extra current to flow through the power transformer and cause it to overheat or blow.

Also, the power supply resistors should be checked for accuracy of value. They can drift way out of tolerance and cause insufficient power voltages. In fact, in a worse case scenario, I once measured a 10K resistor that read 820K.

I doubt that your pickups are too hot for it, and also doubt the need to rewire the 6EU7 to accept a 12AX7. At this point, changing the design is not the answer, but replacing bad components would be the way to go.

I built your tube matcher described in a previous chapter and noticed voltages reading anywhere from .32 to .8 volts. The ones that read higher voltages seem to break up more. Is this normal?

You are correct. The tubes that read the higher (cathode) voltages are actually drawing more current; therefore, they will usually saturate easier because at idle they are already closer to saturation.

I own a Mesa Boogie amp that does not produce any sound. It went down at a gig right in the middle of my lead part. No sound comes out, zilch, nada. I replaced the tubes, checked the fuse, and checked the speaker wires, but still no sound. A friend had this happen to his Boogie and found that a printed circuit board trace had broken due to heat. What am I missing here? Any help will be appreciated.

Although it is very difficult to diagnose any amplifier without looking at it, my guess would be that you have one or more traces burnt off of the printed circuit board. This is a very common problem with printed circuit board amps, because the trace is a thin copper foil that will at times act like a fuse and simply burn up. Amps wired with real wire do not have this problem. Simply inspect the printed circuit board and look for burnt markings. You might need to get a schematic and trace out the circuit using the schematic, because sometimes printed circuit boards will have certain traces burned completely off of the board without any clue of what was there. Replace the damaged/burnt trace by soldering a wire in place, where the trace use to be. Now that you have replaced the trace with a real wire, be aware that there was something in the circuit that caused too much current to go through the trace in the first place and whatever that was (probably a shorted output tube or bad filter capacitor) must be replaced. I have repaired dozens of Boogies with burnt traces and would guess there is a 95% chance that this is your problem.

I recently found a Magnatone 431 guitar amp with one 12" and a small (3"or 4"?) speaker at a yard sale. The output transformer is missing. What steps could I take to locate a transformer that might work?

The schematic that I have for a Magnatone 431 shows only a 10" speaker and a 6UH8 output tube. If your amp has that 6UH8 output tube, it wants to see 11K load (primary impedance) going into 4 ohms. This translates to a 52.44 to 1 turns ratio. The closest thing I know that is readily available is a 2 ohm tweed Bassman or a Super Reverb output transformer. Since the Bassman is higher wattage rating and has more core material, you will get less copper loss and less coupling loss. This should more than make up for the slight turns mismatch. If you want to try the budget approach, get a Fender re-issue Bassman transformer. If you want to get a real paper bobbin transformer, get a used original or a Kendrick 2410A transformer. If you want to spend big bucks, have one custom wound to exact specs.

Lots of amps use capacitors with a design called "extended foil," usually marked by a bar or line on the outside of the cap. Will the orientation of the caps affect tone?

Although the orientation of the cap is important, it will not directly affect tone; misorientation will, however, result in more noise/hum in the circuit. To correctly orient the cap, mount it with the outside foil side pointing towards the most grounded side of the circuit. This will cause the outside foil to act as shielding. If you don't have a marking on the cap, usually they are made so that when you are reading the writing on the cap, the right lead will be connected to the outermost layer of foil.

Here's a cool test you can make to be sure. Wrap a cap with a layer of aluminum foil and make sure the foil does not touch either lead. Get a capacitance meter and connect one end to the foil and the other end to one of the cap leads and get a capacitance reading. Now measure from the foil to the other lead. Whichever lead reads the most capacitance will be the outermost layer and therefore should be connected to the most grounded side of the circuit (not the high voltage side).

What are the advantages vs. disadvantages of "active" tone controls in amps? For that matter, what is an "active" tone control? Were any vintage amps ever made using active electronics?

Good questions. If you asked several different people, you probably won't get consistent answers, but here are my definitions.

"Active" tone circuits are the opposite of "passive" tone circuits.

In a "passive" circuit (also called "subtractive" tone circuits), certain frequencies are grounded out, leaving certain other frequencies alone. By grounding out highs, you leave more bass. This is the type of circuit used on a Stratocaster guitar, for instance.

In an "active" tone circuit, certain frequencies are boosted more than others. Therefore, we can produce more treble by adding more high-end to the circuit or more bass by adding low-end, etc. You might think of "active" as amplifying or adding frequencies where "passive" is subtracting frequencies.

The advantage of "active" circuitry is that it is possible to have a greater change in dB per octave. This may not necessarily be all that desirable in a guitar amp.

Although I cannot think of any vintage amps that use purely "active" tone controls, the Ampeg B15N and certain other amps use a tone circuit that both "cuts" and "boost" various frequencies simultaneously and therefore would be both "active" and "passive" by my definition.

I've found that putting new components in old amps changes tone, sometimes not for the better, in my opinion (e.g. an "aged" carbon composition 820 ohm feedback resistor measures 1K ohm and gives softer clipping.) What effect do you think aging components has?

Let's start by making a distinction between "aged" components, components whose value has drifted, and malfunctioning components. If you like the sound of a 1K ohm feedback resistor, why not use a new 1K carbon composition in place of the 820 ohm? Circuits don't know color codes, therefore, you don't have to go by the schematic, but use instead an ohmmeter to measure and find a replacement that "reads" the same value on a meter as the component you like. This will work fine in some cases, however there are other cases where drifted values actually hurt the tone. For instance, I remember an extreme case where a 10k ohm power supply resistor measured 820K ohm. This reduced all preamp voltages to the point of very low gain with no headroom and no punch. "Constipation" is the one word that comes to mind.

Coupling capacitors are another story. Different brands and styles all sound different and sometimes not for the better. Electrolytics capacitors are equally important as well and will affect tone, especially if they are not formed properly. I would never recommend keeping a malfunctioning or questionable part in an amp for the sake of keeping it original unless I was never going to play the amp or if I planned to put it in a museum. Leaky filter caps can cause power supply failure and/or power transformer failure/burnout. I have dealt with a few collectors who want their amps repaired but don't want to change any parts, in the interest of keeping them original. If the part is malfunctioning, it is not original anyway because no amp originally had a malfunctioning part in it when it was new. Here's an interesting question: Would you prefer your '59 T-Bird with a rotten radiator hose, bald 1959 rotten tires, and a malfunctioning battery when taking it out on the freeway? To quote a famous Texas ex-presidential candidate, "I rest my case."

One of the most introspective tube techs I know said, "Fast slew rate is not necessarily desirable in a guitar amp." What do you think he means?

Slew rate is a reference to how fast the output follows the input. If an amp had an input of 1 volt and an output of 10 volts, it would take a small amount of time for the 10 volts to appear at the output, after the 1 volt appeared at the input. The longer the time, the slower the slew rate. Tubes amps inherently have a slow slew rate because the output transformers are an inductive load which resist changes in current because they are inductors. Transistor amps do not have output transformers and therefore are inherently much faster. The power supply has a lot to do with slew rate. For instance, a tube rectifier is much slower than a solidstate rectifier and more filtering in the power supply is faster than less filtering. High-end tube audio gear generally uses massive filtering in the power supply to increase the slew rate. On the other hand, guitar amps do not use such massive filtering.

Amps that have a spongy feel to the attack have slow slew rates. This is heard as tube compression. Faster slew rates give more punch, slower gives more compression. You said you wanted both???

What makes a bass tube amp different from a guitar tube amp? If a bass speaker cabinet is used, can a Fender Bandmaster or a

Sovtek Mig 60 be used to play bass guitar through, without damaging the system?

With a bass speaker cabinet, you can use any tube amp without damaging the system. You may not however, get the tone you require for bass.

Bass tube amplifiers are generally designed with beefier power supplies to tighten up the low-end and keep them from farting out. Generally this would include a solidstate rectifier with higher value filter caps and a power transformer with bigger diameter wire used on the high voltage winding. All of these things reduce the impedance of the power supply, thus giving more instant power available for reproducing those hard-to-reproduce lower notes.

In addition to the beefier power supply, slightly larger cathode bypass capacitors are used in the preamp to bring up the bottom, and different value tone capacitors are used to accommodate the somewhat lower center frequencies of the tone circuits. Sometimes the preamp cathode resistors are left unbypassed to increase headroom.

Sometimes higher voltages are used on the power tubes, here again to increase headroom, and a better output transformer with more inductance is necessary in order to fatten up the bottom.

The Bandmaster will work. However, they do have a rather cheesey output transformer and I would recommend replacing that to get a better tone for bass. You need something with a lot of iron in it. I am not that familiar with the Mig 60, but I know it has very high plate voltages and would probably work well.

One last thing, bass amps are generally biased with higher bias voltage and so they idle a little colder. This is to give more headroom. 6550 output tubes are usually preferred for bass, due to having quite a bit of headroom and excellent bass response.

I am ready to give it a shot at replacing the filter caps in my amp, but have heard from several sources that I should charge the caps slowly with a Variac. I don't have a Variac so would it be possible to make a similar device using a high-powered light dimmer? If so, how would this be wired? Couldn't I use my digital multi-meter to monitor how much power is making it to the amp?

No, no, no. The Variac, unlike a light dimmer, is actually a variable transformer. Before my company started pre-forming 100 caps at a time on a cap charging/forming table that I designed, I would put an amp on a Variac overnight at about 40 volts A.C. The next day, I would increase the Variac voltage by 10 volts each hour. By the end of the day, it would be fully formed and ready to go. There are many people that do cap jobs without forming them. You have other options, such as buying pre-formed caps or going ahead and investing in a Variac. A really good one with a built-in ammeter and voltage dial, can be gotten for under $85. If you're going to do cap jobs in the future, the $85 is a worthwhile investment. On the other hand, if you are only doing caps jobs once in a blue moon, the pre-formed caps are the way to go.

Why do some Fender amps use a transformer choke in the power supply and others a 1K ohm resistor? Can any amp be converted to cathode-bias by disconnecting the negative voltage supply and adding a 250 ohm resistor from cathodes to ground? Last question first, there's more to cathode-bias than a 250 ohm cathode resistor. See the chapter describing in detail how to convert an amp to cathode-bias. If you disconnect the negative voltage supply, you must ground the grid return resistors or the grid will have no relationship to the cathode and you will have the tubes drawing mega-current. This could reduce usable tube life to seconds or minutes. The grid return resistor value will probably need to be different than the fixed-bias values, depending on the amplifier being converted. Also, although many cathode-biased amps used 250 ohm cathode resistors, that value is related to the particular circuit and you may need a much larger value depending on the rest of the circuit. Also, you will need to select a value for a cathode bypass capacitor.

Some amps use a resistor instead of a choke because chokes are expensive compared to a resistor. The smaller, less expensive amps generally used a resistor and the more expensive amps used chokes. They do sound very different. There is more clarity and punch with a choke, more distortion and compression with a resistor.

A lot has been said about two 6L6 tubes putting out 60 watts. I have a Music Man RP-100 that has two Tesla EL34s that measure 31.5 volts at the output before clipping (into an 8 ohm load).

This seems to be a lot more than 60 watts. Please explain.

Good question. Let me explain. If you were dealing with pure D.C. and a purely resistive load, then by Ohm's law 31.5 volts across an 8 ohm load would give you exactly 3.9375 amperes of current (voltage divided by resistance). To get power you would multiply amperes by volts and would come up with 124.03 watts.

However, an amplifier is not dealing with a D.C. output but an A.C. output. This means the voltage current and power are constantly changing, so where do we begin measuring?

When an a.c. voltage is applied to a resistor, the resistor will dissipate energy in the form of heat just as if the voltage were D.C. The D.C. voltage that would cause the identical heating in an a.c. excited resistor is the root-mean-square (RMS) value of the a.c. voltage. This value of any waveform can be determined by integral calculus, but for a pure sine wave, the peak voltage divided by the square root of 2 is the RMS voltage. Or said conversely, the peak voltage times .707 equals RMS voltage. The peak to peak voltage is twice the peak voltage. Thus your 117 volt wall outlet provides 117 volts RMS, 165.5 volts peak and 331 volts peak to peak. Also for a sine wave the average voltage is equal to the peak to peak voltage divided by pi. This means your 117 volt outlet has an average voltage of 105.2 volts.

Everything that we just said about a.c. voltage applies to a.c. current as well, provided the load is purely resistive. To get average power, you will have to multiply peak voltage times peak current and divide by 2. RMS voltage (peak voltage divided by 1.414) times RMS current (peak current divided by 1.414) equals average power (same as peak voltage times peak current divided by 2). This example would show your Music Man as delivering 62.015 watts average power. This is provided that you are using a sine wave as source signal and a purely resistive load. Did I explain more than you wanted?

Preamp performance varies enormously. The problem is existing rating systems for preamp tubes only target microphonics, not musical tone; therefore, the only way to find the best tube is to go on a fishing expedition. This procedure can be very expensive and time-consuming. Is there any way to measure musical performance of preamp tubes similar to the

method used by certain wholesalers for power tubes? (Certain wholesalers measure transconductance and plate inductance.) If not, what guidelines can you give for selecting the best sounding preamp tube for a tweed Champ?

Your assumption that transconductance (gain) and plate inductance measure musical performance is a myth—that is if you include tone in your definition of musical performance. A tube can have lots of gain and still have poor sound quality. Just like a guitar string can have good tensile strength and still sound dead.

In my opinion, tone is in the ear of the beholder and the best method of testing a tube for good musical quality is to listen to it.

I have an old Pignose practice amp that I like to use as a preamp. The speaker has become very worn-out and buzzy sounding. I would like to try disconnecting it and installing an 8 ohm " load" in its place. Where can I get a resistor that will fit within the amp and do the job correctly?

If the speaker is trashed anyway, why not cut the spider and cone to remove the voice coil? The voice coil would be a perfect dummy load. If you don't want to sacrifice the speaker, almost any place that sells resistors would have a 8 ohm 10 watt resistor.

I have a '59 Fender Bassman re-issue with a Kendrick output transformer and Kendrick speakers. I'm planning to install a 5U4GB rectifier and adjustable bias pot. I would like to use a Marshall Power Brake with the amp. Four R.C.A. connectors are used between the speakers and amp; the power brake uses standard plugs. Would you recommend using an adapter or installing a phone jack connection into the amp?

I would recommend removing one of the R.C.A. jacks and installing a ¼" phone jack in its place. Re-wire your speakers with a wiring harness that hooks the speakers in parallel and goes to a 2' long speaker cord with a ¼" plug at one end. When you are not using the Power brake, simply plug the speakers in the jack.

I noticed that the vibrato control on my 70s Vibrolux amp was seen by the guitar signal as a 50K path to ground. Since it looked somewhat like a master volume, I thought that I would eliminate it to see if liveliness or playability improved. I think it did. **My question is, can I use a 50K pot with a switch to completely disconnect the vibrato when not needed, then switch it on to get normal control function? I'm worried about a huge pop signal getting into the driver circuit. Any considerations or techniques?**

Very brilliant observation! The vibrato control loads down the signal and therefore sounds much better out of the circuit. While you are at it, you might as well change each of the two 330K resistors in the driver grid circuit to a 1 meg ohm. This will help it even more. The cool way to hook up the switch you asked about is to take the right lead from the vibrato control (looking at the back of the control—this lead is opposite the lead that is grounded) and disconnect it. Put a 10 meg resistor in series with the pot lead and the wire that was removed. You don't have to use a 10 meg, a 5 or 6 meg will do nicely. Now put your switch across the resistor so that when the switch is closed it shorts the 10 meg resistor and the amp is dead stock. Of course, when the switch is open you would have the 10 meg plus 50K as a path to ground but it is so high an impedance that you can't hear the difference.

I have two questions regarding bias. Using the transformer shunt method on my '66 Bandmaster and '65 Deluxe Reverb, the measurements were 19 mA and 16 mA respectively, for both power tubes using a Fluke 8060A meter. On my '61 Concert and '65 Super Reverb, the bias on the two power tubes are quite different. For example, one reads 22 mA and the other reads 35.3 mA. Are the 19 mA on a 6L6 and the 16 mA on a 6V6 too low? What is the significant difference in bias between two power tubes in one amp? Does this indicate mismatched power tubes?

Let's get a couple of things straight. First of all, there is no right or wrong plate current for a 6L6 or a 6V6. It really depends on the amp's plate voltage, class of operation, and what kind of tone you want. Amps with higher voltages will require less plate current generally speaking and conversely amps with lower plate voltages will generally require higher current. For instance, a vintage Fender tweed amp with, let's say, 325 plate volts may sound killer idling at 60 mA and cold as a witch's #*! idling at 10 mA. On the other hand, an amp that is running 500 volts on the plate may sound great idling at 15 mA and could easily blow up if you ran it up to 60 mA.

In addition to all of this, an amp that is set to idle at higher current is already closer to saturation and will break up easier. Just as an example, when I designed the Kendrick harp amp, I idled the output tubes at 60 mA and used 460 volts on the plate. This makes the amp break up very easily which is what harp players wanted. On the other hand, when I set up Jimmy Vaughan's Kendrick amps for his tour, I adjusted the idle current to 20 mA because he wanted a lots of headroom and did not want the amp to break unless he was really pushing it hard.

I have seen some Class A single-ended 6V6 amps that typically use lower plate voltages and idle at 70 mA.

If your tubes are drawing different plate currents, they are mismatched. If the difference is only a few mA, don't worry about it because they will drift around a little, up and down one mA or two, as they get hotter or cooler.

I came across a little no-name 10" combo with three tubes no one can identify. Does 18G6A, 60FX5, 36AM3 ring any bells? That amp sounded great for about two minutes, then the rectifier screamed and smoked.

Sounds to me like you have got a series filament winding and no power transformer. In an American tube name there is a code and the first digit or digits will represent how much filament voltage is needed to operate the filament. Since power transformers cost mucho deneiros, some designers would simply use a rectifier, a power tube and a preamp tube that all draw the same current and whose filament voltages equal the sum of the a.c. wall voltage. Notice that 18 plus 60 plus 36 equals 114 volts. If you string the filaments in series and then rectify the wall a.c., you could get by without using a power transformer. My guess is that the filter caps were/are leaking massive current which caused the rectifier to overheat and blow. By the way, don't stand on concrete barefooted while playing a guitar with a string ground. If you replaced the filter caps and replaced the rectifier tube, my guess is that the amp would work fine. The 36AM3 is the rectifier, the 60FX5 is the output tube, and the 18GD6A is the preamp tube. They all use 100 mA of filament current. The 60FX5 puts out about 5 watts.

I have a Danelectro DM-10 amp with a 8" Kendrick speaker. Can I use the 5 watt single-ended power transformer to change this amp to Class A operation? As far as I can tell this amp is a Champ circuit.

Surprise, surprise, surprise. Champ circuits are already Class A. Any amp with only one output tube is Class A, but not all amps with two output tubes are necessarily Class AB. Also, I think you meant to say, "5 watt single-ended output transformer," not power transformer. Output transformers are single-ended or push-pull, but I have never heard anyone refer to a power transformer as single-ended.

What would make a '59 Bassman sound like the pots are dirty? The amp has all new electrolytic caps and new plate resistors. I've tried new pots but the sound is the same.

Although you changed your pots, there is a slight possibility that when soldering in new ones, some rosin from the solder splattered into the pot, thus causing it to sound dirty. Pot cleaner won't help because the rosin won't budge. If you are sure this is not the case, the other thing that makes a pot "sound" dirty, when it is not dirty, is D.C. voltage. There are only two ways that D.C. voltage could appear on those pots (assuming the amp is wired stock). Either one or both of the .02 coupling caps are leaking current or you have a bad second gain stage preamp tube which is leaking current back from the grid. Here's the fast way to fix it. Making sure both pots are turned all the way up and with the amp on and in the play mode, measure the voltage across each pot. Now pull the second preamp tube (counting from the side closest to the input jacks). **Important note:** When you pull that tube with the amp on, it is very likely to start a semi-violent oscillation, so it might be a good idea to turn all tone controls and the presence pot to the lowest setting first! Now measure the voltage on each pot again. If the voltage problem goes away, then the tube should be replaced with another good 12AX7 and voltage on the pot re-checked to confirm that the problem is resolved.

If the D.C. voltage is still on the pot after removing the tube, then one or both of the coupling caps are bad. These are the two small .02 caps that connect the plates of the first preamp tube to the volume pots. To check these, simply unsolder the .02 cap lead going to the pot and measure it for voltage. Now do the same thing with the other .02 cap. This will tell you which one is leaking.

There is a much simpler way to fix this—try the Texas Shotgun Approach. Simply replace both caps and put in a known good preamp tube. If it works, it's fixed.

I am a missionary in Uruguay, South America. For many years I have had good success with using a step-down transformer to convert guitar amps designed for 110 volts/60Hz to work on 220 volts/50Hz. In a few months I will be returning to South America with a few mid-60s Fenders that I have recently acquired. Are there any precautions that should be taken when using a tube amp with a step-down transformer?

Also, I will be storing several amps for several years here in the States in a high humidity environment. How should these be packed to insure that no damage occurs?

First things first, do not store amplifiers in a high humidity environment. High humidity can cause transformers to rust, circuit boards to become conductive and you will have strange parasitic oscillation problems; cloth insulation can become conductive, carbon resistors can soak up moisture, and the speaker paper will become mushy. This is in addition to corrosion of sockets, switches, jacks, etc.

If you just didn't have a choice, I guess you could put each amp in a plastic garbage bag, seal it, and put it in a huge box filled with dried rice or dried beans and then seal the box with a larger plastic bag. (I've never actually tried this.)

The step-down transformer is a good idea, but there are two things to remember. First, step-down transformers are rated in watts and I would be very careful to use one that is approximately twice what the amp consumes. An amp with a 2 amp fuse will not consume more than 250 watts average power; therefore, a 400 or 500 watt transformer would do nicely.

When an amplifier that is designed for 60 hertz is used in a 50 hertz power supply (such as in Europe or South America), the efficiency of the transformer will not be quite as good. Also 50 hertz power will mean the filter caps will not filter out the hum quite as well. This is only very slight and will probably not present a problem. I would be more concerned about the storage problem.

What are the drawbacks of not forming filter caps after a cap job? I've got two Marshall Super Leads and both have unbalanced phase-inverters. One side is being driven harder than the other resulting in premature wear of the power tubes on that side. What can be done?

Let me answer the cap question first. An electrolytic cap is very similar to the battery in your car. Both use conductive plates and an electro-chemical reaction to store energy. Consider this: Take two brand new car batteries that have never been charged and put one battery on a slow "trickle charge" and charge it for a day. Take the other one and charge it up by running 16 volts at 60 amps through it for 5 minutes. Which one is gonna last longer? Which one will perform better?

The truth is that all capacitors are made with two conductors separated by a dielectric (non-conductor). An electrolytic capacitor, when new and never used, has no dielectric. The dielectric must be "formed" when D.C. voltage is place across the leads! If this is formed slowly you will end up with a fully formed capacitor and ultimately better performance.

If a cap has not been used in a long time, it is best to "reform" it. This is achieved by putting the capacitor in a circuit with full D.C. working voltage, but with the current limited to 10 milliamps. This could take several days, but performance can be greatly improved.

If you are certain that your phase inverter is off balance, the best way to balance it is to use a dual trace scope hooked to the grids of the power tubes (pin #5). Put a signal generator into the amp and increase volume while monitoring signal of both traces. (It is best to use a new phase inverter tube selected for balance.) Choose an appropriate value plate load resistor to even things up by selecting a value where both traces clip at exactly the same time. This is going to mean reducing the value of the plate resistor on the higher gain side.

If you don't have a scope and want to go the low-tech approach, changing the "long-tail" resistor in the cathode circuit of the phase inverter (usually a 10K) to a larger value will make the outputs tend to self-balance more. Try a 22K. This will reduce the gain somewhat. You could also just try using a smaller plate resistor on the side that is wearing out the fastest. (Drop it by 15K or 20K from stock).

I own a 1967 Fender blackface Bassman #AB-165. The bass player in

my band used it until it began losing power after it had been on for about 30 minutes or longer. I peeled it open and found that it had torched a screen resistor. Both screen resistors were replaced with 470 ohm 3 watt resistors. I also took the opportunity to replace all of the plate resistors on the preamp tubes with 100K 1 watt metal film type resistors and I replaced all of the power supply caps downstream of the standby switch with 20 uf 500 volt units. Power supply resistors were changed to 2 watt flame-proof types. The first stage caps were replaced about 3 years ago with 68 uf 350 volt units and left alone. It was also discovered that a previous owner had replaced the silicon rectifiers and had forgotten to solder one side, giving me the effective use of only half the power transformer and half wave rectification. This problem was "rectified" as well.

Upon powering up the amp for the first time, it gave off (and still does) a sizable hum for about two seconds after the standby switch is flipped on. The inboard power tube (a Sovtek 6L6) also has a hot spot on its plate. The bias seemed a little unstable and weak, so I changed the cap from the stock 50 uf 50 volts to a Sprague 100 uf 63 volts. The bias now meters at -39.5 on both grids very solidly, but whichever tube is in the inboard socket still glows with a hot spot near the seam. For good measure, I replaced the .022 uf 400 volts coupling caps with matched CDE .02 uf 600 volt caps, but with no change. I still have the hum at turn on and the hot spot. I have measured output tube plate voltages as high as 472V, and, for what it's worth, the amp is now really loud, not noisy (besides the hum at turn on), cleaner, and a bit more brittle on the top end when cranked. Any ideas?

I was also wondering how necessary it is to power up new caps slowly with a Variac. The techs that I've talked to around here say no, but they don't know what's wrong with my amp either. Also, what is the purpose of the 220K resistors that connect the plates of the power tubes to the plates of the driver tube, and should they be removed? None of the other Bassman tops in my schematic file seem to have them, but I can't see Fender wanting to make the resistor companies richer.

The good news is that your amp can be made to perform and sound better. For starters, get rid of those 100K metal film resistors and replace them with carbon composition 100K 1 watt. This will improve tone dramatically. Second, get rid of the Sovtek 6L6s, they are not a good tube for your amp; use the Sovtek 5881, Kendrick 5881, or NOS American tubes. Your bias voltage is way too low for the amount of plate voltage. Just guessing, I would say you need more in the neighborhood of minus 50 volts. The actual voltage can be determined with the shunt method described in my first book, *A Desktop Reference of Hip Vintage Guitar Amps*. I would also recommend changing the 68 uf 350 volt main filter caps to 100 uf 350 volt caps. Make sure and use a 1 watt 220K resistor across each of these caps to equalize the voltages across these caps.

Those other 220K resistors that go from the plates of the output tubes to the plates of the phase inverter are a negative feedback circuit that works in conjunction with the feedback circuit that goes from speaker to phase inverter grid. I would recommend getting rid of them and changing the wiring of the phase inverter to match the schematic of the AA864 Bassman. This would only require removing the two 220K resistors, a 47K resistor, a .1 cap and adding an 820 feedback resistor and a 100 ohm load resistor. I would also change the phase inverter grid input cap to a .02mfd. I recommend rewiring the bias circuit to match the AA864 as well. These mods would put the feedback in the cathode circuit instead of the grid circuit and would also give you the opportunity to actually bias your tubes. You cannot really bias your output tubes with your stock circuit now, but only balance them—no good!

If oscillation occurs when you change the phase inverter setup to be like the AA864, simply swap the transformer primary leads that go to pins #3 on each power tube with each other. This may or may not be necessary. Also, charging caps slowly will make better sounding caps that will last longer. Some companies, Kendrick for example, sell preformed caps to save time for techs.

Do you have any advice for cleaning existing wiring for soldering? I'm working on a Silvertone 1481 and a Sound City 120-R Mark IV. No matter how far back I strip the power and output leads, there's still that black gunk all over the copper. I'm assuming it's part of the jacket that's dissolved into the copper over the years. I'm weary of petroleum products and alcohol

because of possible damage to what's left of the jackets, but there's got to be a way to make them take solder!

It's gonna be a hassle, but my advice is to replace the wire with good wire. If the leads are from a transformer, the endbells of the transformer can be removed and new leads can be soldered directly to the magnet wire inside. Just take your time and document where everything goes and you will have no problem.

What is a good way to date my silverface Deluxe Reverb? What is a good way to clean and condition old Tolex?

The best way to date a silverface Deluxe is to check it out and look for date codes on the inside chassis. The tube label, speaker codes, pot codes, model number, and even the circuitry can be a tip off. Also the cabinet design and/or chassis design can be a tip off as well.

Cool tip for cleaning old Tolex. Remove the chassis and speakers and bring the cabinet to a high pressure car wash. Get some of that spray-on tire cleaner and spray the Tolex down. Shoot it with that high pressure water until it comes clean. You may have to repeat this a second time. Spray out the grille cloth as well, but don't get any tire cleaner on it. Dry the cabinet and douse it with plenty of Armorall. In a few hours, the Armorall will soak in. Douse it good again and it will soak in. Keep spraying the Armorall, letting it soak in and re-spraying until it won't soak in anymore. *Voilà*, it will look brand new.

I am experiencing odd harmonic overtones in all my guitar amps since I moved to my new home. They are non-musical and sound kind of like when filter caps go bad. The house was rewired before I moved in. The electrician said the power company had the option to replace the transformer or put some kind of booster at the pole to upgrade the new service to the proper amperage. I have tried everything I can think of to cure the problem including servicing all the amps, tubes, caps, resistors, etc. I even tried isolation transformers, power conditioners, and disconnected all other circuits, cordless phones and other sources of possible interference and still I have this problem. Is it possible something in the wiring is causing this problem?

If you have tried a power conditioner and it was of ample wattage, then you must be getting some kind of electromagnetic interference. Is there a military base or a radio transmitter nearby? Are there some high tension lines in the immediate area? Perhaps you could drive a few blocks away and ask someone (that uses a different utility pole transformer) if they would let you plug in your amp to see if the problem still exists. This would at least tell you if the problem is in the electricity or in electromagnetic interference. You could have your power company come and check your electricity to see if it is 120V/60 Hz. If it is not 60 Hz it could be that your amp's filter caps cannot filter out the lower hertz.

How do you test a pot to tell if it is audio or linear taper and which is best for replacement in amplifiers?

A linear potentiometer, when turned halfway up, will be exactly half of its resistance value. If this were plotted on a graph (percentage shaft rotation vs. percentage resistance), it would end up being a straight line, hence the name "linear." So to check a pot to see if it is linear, you will need an ohmmeter to measure resistance. Turn the shaft until it is exactly halfway up, measure the resistance from the wiper (center lead) to one end and compare this to the actual resistance across the pot (one outside lead to the other outside lead). If the resistance from the wiper to end is half the resistance of end to end, you have a linear pot.

Humans don't hear linearly, we hear logarithmically; therefore, an audio pot will plot out on a graph as a curve (percentage shaft rotation vs. percentage resistance). Depending on the brand of pot (and consequently the actual curve) the wiper to ground resistance will only be about a quarter of the end to end resistance when the shaft is rotated to half-way up position. This is tested in the same manner as the linear pot and the measured resistance (when the shaft is rotated to the half-way position) will give you the clue needed to determine whether the pot's taper is audio or linear.

Either taper pot will substitute for the other; however, the audio pot is generally used for volume and tone controls (hence the name "audio") and the linear is usually used for trim adjustments (bias adjustment trim pots, hum balance controls, etc.) I did say usually, but not always. If you use a linear pot for a volume control, the volume will seem to come up all at once. In fact, a trick that many manufacturers have used is to put a linear pot on a volume control circuit (Bassman re-issue, many older Marshalls). When the volume is turned up to, let's say, 4, the pots resistance is somewhere around what an audio pot would have been on 7 or 8. This gives the illusion of having much more power than what's really there. It seems to make the amp more impressive; however, the amp will lack the smooth range of volume that it could have had with an audio pot.

I was looking at the schematics for the blond Vibroverb and the black Vibrolux Reverb in your first book. What's the difference?
If Fender ever made a blond Vibroverb, I have never seen one. Perhaps you are speaking of the brown Vibroverb 6G16 in which case there are major circuit differences. The five major differences are:
1. negative feedback loop on the 6G16 uses more negative feedback.
2. phase inverter stage of the 6G16 has much more gain. Although the plate voltage looks close to the same (within 20 volts), the cathode resistors are different and the 6G16 "sees" an extra 61 volts across the tube. This increases headroom in that stage and gain.
3. tone circuits are completely different. The 6G16 uses tapped treble pots for more of a hi-fi sound.
4. reverb circuit is different and coupled into the circuit differently as well.
5. coupling cap on the phase inverter input of the 6G16 is twenty times smaller than the Vibrolux Reverb. This affects the overall EQ of the amp.

I purchased a Kendrick demo tape. Terry Oubre is a smokin' player! Does he have an album out and how could I get a copy? I was listening to the tape for pure enjoyment, but a friend borrowed the tape and won't give it back!
Yes, Terry Oubre is a smokin' player. He has been in the studio for the last few weeks working on a brand new demo tape that will demonstrate all Kendrick amplifiers made, including the new Texas Crude line. He has played on many albums as a studio musician, and recently released a new CD entitled "Future Blues." That new demo tape will be available by the time this gets to press. It's about an hour of 28 smokin' examples.

What tone cap is similar to the old "Astrons?" I prefer not to use 600 volt orange drops.

I like the Mallory 150 series caps as "Astron" replacements. These are polyester and tubular foil.

Here is a somewhat wacky question for you. I have a Musicman Seventy-Five head which I think sounds amazing. I run it through a Kustom 3x12 cabinet, the tall padded model. The cabinet is 5.3 ohms. When I run the head at the 4 ohm setting, it sounds great but I get a ton of squealing feedback when I crank it up. I generally use a guitar with 2 single-coil pickups. Maybe you will just tell me my setup is crazy, but if you could explain my feedback problem and give me a suggestion for eliminating some of it, I'd appreciate it.

This problem could be related to the inductance of the speakers changing the phase angle of the feedback loop or maybe the amp has a reversed output transformer lead. Does the amp do this with other speakers? If not, I would suspect that your amp just doesn't like those speakers. If the amp still squeals with other speakers, the problem is probably related to the output transformer primary being wired backwards. Disconnect the feedback loop. This is the wire that connects from the 8 ohm transformer lead back to the circuit. Did the problem go away? If it did, I would try reversing the primary leads of the output transformer. These are the leads that terminate on pin 3 of the output tubes. Did the problem get better or worse? If it got worse, put it back. If it got better, your amp is fixed.

Are screen voltages usually lower or the same as plate voltages? If not, in what cases are higher screen voltages used? I ask because I have a few Gibson amps and looking at some of the schematics, they have "higher than plate" screen voltages. Shouldn't the screen resistor drop the voltage somewhat?

I prefer amp circuits where the screen voltage is less than the plate voltage. While the plate inside the tube has more mass than the screen, I don't think the screen should have higher voltage because this could encourage abnormally high screen current. There will be some voltage drop across the screen resistors, but only if there is screen current passing through resistors at the time it is being measured. Obviously, the more current going through that screen resistor, the more voltage drop.

I recently obtained a Silvertone amp-in-case, model 1457. This was a flashback to my youth considering that the guitar was built by and is virtually identical to a Danelectro. It has two lipstick tube pickups, stacked pots, Masonite body and nice playable neck! I am delighted with the purchase. It makes a perfect "bedroom guitar." However, the amp, which is rated at 50 watts and drives a single 8" speaker, does not breakup even when turned to 10. I would like your suggestions as how to increase gain and install a master volume control. The amp has vibrato which I never use, so I could disconnect the circuit to increase gain. Since there is no circuit board layout, due to the small size of the cabinet (approximately 8"x12"), it presents a confusing picture to the eye. Parts are crammed together in close proximity with several small circuit boards attached at various places. The area under the tubes (12AX7, 6V6GT, 12AX7, 6X4—all new) is packed with resistors and wires. Ideally, I would like to disconnect the second input, put in its place a master volume control pot and retain the vibrato circuit while increasing the gain. What do you suggest? If you cannot locate a schematic, I will make you a drawing of the layout. Thanks for your help.

That amp was made in two versions. One version was like yours and has a power transformer. The other version used different tubes and did not have a power transformer. (One side of the A.C. line actually went to the chassis ground which ultimately was connected to your guitar.)

Let's get a couple of things straight. The amp of which you speak is not a 50 watt amp. It may consume 50 watts of power in order to operate, but it will only put out around 4 or 5 watts!

Second, if I were a gambling man, I would give you two to one that the filter caps need replacing. I see about 1000 vintage tube amps a year that come into my restoration shop, and I would say that more than 900 of these have original filter caps that should have been changed a decade or two ago. There's probably a burnt or drifted resistor or two and possibly a leaking coupling capacitor as well (this alone could kill your gain as it will throw the bias of the following stage off.)

Third, do not attempt to do service work on an amplifier unless you are absolutely certain that you know what you are doing.

Instead, bring it to a competent tech and have him check it out. To a competent technician, that mass of confused wires makes perfect sense and is quite simple to troubleshoot.

After you have gotten all of the defective components replaced, you will be able to hear the amp for the first time and see what it really sounds like. I have heard those amps and they have a nice singing tone with a spongy envelope and would be perfect for playing some Ry Cooder licks on slide, and my guess is that you will be happy with the gain when the amp is in proper repair.

My questions relate to output impedance. You have told us that vintage Fenders are very impedance tolerant, whereas Marshalls are not. We "mess" with a lot of oddball amps and would like to be able to determine in a non-destructive way if we might mismatch output impedances. Can we determine this from the schematics or is this a complex function of voltages, transformers and other components? What are the possible effects of moderately higher impedances other than lower power? How about moderately lower impedances? I have been told that tube amps can be fairly tolerant of even a shorted output, although I'm not about to experiment with any of my amps.

Impedance can become somewhat complicated, but here are some basics to help you understand impedance better.

Your output tubes are fairly low current and high voltage devices. This makes them high impedance. Your speakers, on the other hand, are high current/low voltage devices which make them low impedance. The purpose of the output transformer is to match the high impedance of the tubes to the relatively low impedance of the speakers. (Of course, the transformer also keeps D.C. current off the speakers as well.) It accomplishes impedance matching by having two coils of wire wound on an iron core to a specific ratio of turns. This ratio is determined by a mathematical formula (turns ratio equals the square root of the primary divided by the secondary). If the transformer is wound for a 4000 ohm primary and an 8 ohm speaker load and you connect a 4 ohm speaker to it, the tubes will now "think" they are running into a 2000 ohm primary. This is because the ratio of the transformer does not change. When you consider that 2000 ohms is exactly halfway between a short circuit and 4000 ohms, you may wonder if this affects the tubes? My experience has been that tube life is shortened significantly. Turning the amp up loud makes it worse, and if the amp uses very high plate voltages, it's even worse.

In the case of a Marshall, as with most British amps, the primary impedance of the transformer is wound for less impedance than you would expect to find in an American design. This further complicates the problem of mismatching to a lower impedance speaker load, especially when you consider that Marshalls have fairly high plate voltages and people usually play them loud.

Conversely, if you connect a 16 ohm speaker load to the transformer in the previous example, the tubes would "think" they were operating into an 8000 ohm impedance. This would have the effect of limiting the current from the tubes and would effectively increase tube life. The envelope would sound a little spongier and the tone would possibly be more compressed.

I sometimes buy used tubes and test them myself. Many times these tubes have no numbers on them. Is there any way to identify tubes that do not have any numbers on them?

It is not easy to identify a tube that is unmarked. If you are familiar with tubes, you might make an educated guess. To help with the guessing, look at the inside of the tube and determine how it is wired in terms of pinout. You can also use an ohmmeter to check continuity of pins. If you are reasonably sure of what tube you have, you can run gain tests to confirm or disprove your guesswork.

I have a Ward's Airline Model #62-9012A. I have no schematic, but the tubes are 12AX7, 6V6GT, and 6X4 as marked on the chassis. I would like to modify it for a tweed Champ sound. What parts would I need? Should I change the rectifier and socket from the 6X4 to a 5Y3 or 5V4? What type of resistors and caps should I use and where can I get them? What are some good 3.2 ohm speakers to try?

I would recommend that you leave the 6X4 rectifier tube stock because the transformer does not have a 5 volt winding and you will not be able to change rectifiers without a 5 volt winding. The

6X4 uses a 6.3 volt winding. The rest of the circuit is very similar to the Champ but not exactly the same.

Perhaps you could draw a schematic of the amp and compare it to the Champ schematic. (Use the 5F1 schematic) See if you can duplicate the Champ circuit in the preamp and output stage. You may even want to change the filter caps, but leave the 6x4 alone.

Use only carbon composition resistors and tubular foil Mallory 150 series coupling caps. You will not need very many.

Also that 3.2 ohm speaker that you are looking for is actually a 4 ohm impedance but the D.C. resistance is 3.2 ohms. Since there have been no 4 ohm 8" speakers made in years, I developed one myself and have been using them on the Kendrick 118 amp. The speaker is available through my company as well as any Kendrick speaker dealer. So are the other components.

Where could I obtain an Electronics Industry source book?

There are a few good source books, but they can get rather expensive to purchase. You might try your local library. If it is a large municipal library it will most certainly have the information at no charge. If you would like to purchase books with this information, Harris Publishing at (800) 888-5900 publishes a "EITD Buyer's Addition" (Electronic Industry Telephone Directory) for a price of $55. A four volume set called the "Electronic Engineer's Master Catalog" is available from Hearst Business Communications. The cost is $99 and their telephone number is (516) 227-1300.

What causes reverb or vibrato to bleed through even when switched off with a pedal? How can the problem be corrected as non-invasively as possible?

When a reverb is switched "off," generally, the switch shorts out the return signal coming from the pan. This would make the load resistance zero ohms; therefore, no signal could be developed across it. If there is stray resistance in the circuit, that resistance will act as a load resistance and signal can be developed across it. Some examples of stray resistance would include the resistance between the jack and plug of the footswitch, the footswitch jack to ground resistance, any resistances on the solder joints, and the resistance of the footswitch cable itself. This problem can be corrected by locating where the resistance is occurring and making sure the connections are good (zero ohms). Such is generally not the case with vibrato/tremolo circuits. In almost every case, a vibrato footswitch kills the oscillator in order to stop the effect. When you stop the oscillator, there can be no bleed-over because the effect is stopped.

There are a few amps that are wired differently and leave the oscillator on at all times. The footswitch connects the vib/tremolo circuit to the signal path. This is a good idea, in a way, because it allows the vibrato circuit to be disconnected from the signal path when not in use. This type of circuit could bleed-over when switched "off." Depending on the particular amp, the bleed-over could come from two places. One place would be inadequate filtering of the power supply such that the oscillations drag down the power supply and thus alter the gain of the overall circuit. The other place could be too much capacitance in the footswitch cable and/or switch. This stray capacitance can cause A.C. signal to pass through the circuit even when it is switched off! To determine what is causing this, unplug the footswitch and plug a dummy plug into the jack. Notice whether the problem suddenly gets better or stays the same. If it stays, you may want to bring the amp to your technician for further evaluation. Perhaps the amp has inadequate filtering (bad caps) or needs more decoupling (install an extra "L" section on the power circuit feeding the oscillator tube plate.) Or perhaps you would like the footswitch circuit rewired to kill the oscillator.

How do you choose the appropriate value resistor when converting an Ext. speaker jack to a "line out?" I am talking about the hot to ground resistor. Please help.

When installing a "line out" modification, you won't know what value load resistor to use unless you take into consideration the speaker impedance, the "line out" signal voltage desired and what size dropping resistor you are using. If you use a dropping resistor that is too high, you will lose definition and highs. For this reason, let's assume that you are using a 2.2K dropping resistor. This is the resistor going from your "hot" speaker lead to your "hot" Ext. speaker jack (soon to be your "line out" jack). This leaves two variables, namely: What is your speaker impedance and what is the "line out" signal voltage desired? Why not skip the math and go the practical approach? Knowing that you will need a value of 500 ohms or less, why not use a 500 ohm pot wired as a rheostat. (To wire as a rheostat, connect the wiper and one end to ground and the other end to the "hot" lead of the Ext. speaker jack.) Now hook the newly created "line out" jack to whatever device you desire and set the pot to the place where it sounds the best. This is the correct place, you heard it for yourself. Now, wasn't that a lot easier and more practical than getting out the engineering formulas? You could measure the pot value and replace it with a fixed resistor or leave the pot wired as an adjustable volume control for the "line out."

I've had my Kendrick 2212 for a couple of months and I'd like to tell you I like it better every day. You should do a 30-40 watt class "A" amp à la Vox.

What makes a good tube tester? Where can I buy one? How much does it cost?

I'm very glad you like your amp. I already make a 35 watt class "A" amp called a Texas Crude 35 watt combo à la Kendrick!

What makes a good tube tester is one that tests at guitar amp voltages. Unfortunately, you will not find one on the market. I wrote a chapter called "Cool Tools for Amps," on how to build one that interfaces with a guitar amp by plugging into the amp's power tube socket. This device can be built for under $20.

My question involves the loss of highs when most amps' volume controls are turned down. I would guess that the volume pot, as a resistance voltage divider, would be fairly linear with frequency. If so, the fall off in highs must be due to interaction with the following stage. Would it be possible to design a "bright" capacitor circuit with a ganged volume pot to keep the response uniform at different volumes? Does this loss of highs apply to voltage dividers used as a "line out" across the speaker? Can this application be capacitor compensated? Lastly, what resistance values might be used to create a high impedance "line level out" on a Champ amp, wired across the speaker?

The volume pot for all practical purposes is a resistor with a fairly high value (usually 1 megohm). When you run a small signal through a large resistor, the first thing you lose is high frequencies. Many designers put a compensating cap across the two ungrounded leads of a volume pot to bypass highs around the pot so that you will not lose highs when the amp is turned down. This is not perfect and will actually have a little more highs when you turn the pot down. The lead channel on a Kendrick, Marshall, Fender and many other amps use this circuit. The difference is in what capacitor value is used. Also this circuit is used on guitar volume pots for the same reason.

Loss of high-end is not a problem on a "line level out" because whereas a volume pot is dealing with a very small signal running into a 1 megohm pot (usually 1 meg but sometimes 500 K), a "line level out" is generally dealing with high-powered signal off the output transformer and using a dropping resistor a fortieth or fiftieth of the size of a volume pot (usually 2.2 K). Therefore, you will rarely have a high-end problem with the "line level out."

Why would you ever need a high impedance "line level out?" A high impedance output would easily be "loaded down" when it was connected to another device, even if the other device was high impedance. Outputs should be low impedance, inputs should be high impedance. This is obvious when you consider what is actually happening when an output of any device is hooked to an input of another device.

If your output impedance is high, let's say 1 meg, and you hook it up to another amp's input (usually 1 meg) the source impedance is now 500K because the output impedance is in parallel with the input impedance of the other amp. This is halfway between the original output impedance of 1 meg and a direct short (zero ohms). Now let's say your output impedance is 600 ohms and you hook it up to a 1 meg input of an amp. Your source impedance is now 599.64 ohms—almost no change. This is why test equipment is very high impedance, so it won't load down the source impedance. I use a Fluke 8060A multimeter and I believe the input impedance is about 20 megohms. When I take a voltage reading, the meter does not have enough conductance to affect the source. And this is why you would never have use for a high impedance output.

There is a guy in one of the local electronics stores who works on old juke boxes and at one time he bought a bunch (40 or so) of

American made 6V6GTs. He offered to sell some to me, so I asked him if they were matched. He didn't know. What determines matched sets and how could I match them? Would current draw at the same voltage match them up?

To answer your question, there are several ways that one could match tubes. You could match by transconductance, plate current, and cathode current. I have experimented with all of these and found that cathode current is the easiest and most reliable. For all practical purposes, cathode current equals plate current. What goes in must come back out and grid current is negligible. If you were asking me about high end audio gear I would say match for transconductance also, so that the waveform would be equal as the tubes were driven harder and would clip at the same time. However, my actual listening tests have shown that tubes matched for transconductance, when used in a guitar amp, do not sound as fat. In fact, the tubes sound warmer and richer if they are matched for cathode current or plate current but transconductance is not matched perfectly. This became very obvious when I designed a circuit that deliberately mismatched the drive voltages feeding the tubes. Because the drive voltages on a push-pull amp were deliberately made non-symmetrical, the tone fattened up. (This is an actual control knob on the Kendrick Harmonica amp. It is called a waveform symmetry control!!)

I recently purchased my third Traynor amp, a model YGM-1 Guitar Mate Reverb Combo. It has two 6BQ5s and four 12AX7A tubes. I think it is a 1968 model. It appears the metal canister caps are leaking. The labels on the outside are no longer readable so I can't tell what value they are. Any ideas? Where do I get replacements? Is it possible to get this amp to a tube-type rectifier?

Some Traynor amps have the schematic inside the "lid" of the amp cabinet but some do not (like the YGM-1, for instance). How does someone get Traynor schematics?

Any hot-rodding tricks for Traynors to make them "sing" a little better?

I don't have any Traynor schematics; however, I have worked on many Traynors and my experience has been that every Traynor I have ever seen used simple, classic, Western Electric circuits. That's why I would guess that your filter caps are 16 uf at 475 volts. Any reputable restoration shop should stock these caps.

To add a tube rectifier, an auxiliary transformer is needed. Either a 5 or 6.3 volt at around 4 to 6 amps will work nicely depending on what type of type rectifier tube you decide to use. Instructions on how to hook it up are beyond the scope of this chapter.

If you need a particular schematic, you can advertise for it in the classified section of *Vintage Guitar* magazine. Ads are free for subscribers, and the subscription rate is $23.95 per year (U.S.). Their telephone number is 701-255-1197.

As far as hot-rodding goes, I can tell you this in general: Any amp that uses 6BQ5s or EL84s can be retrofitted with a 7189 or a 7189A without any modifications to the wiring. These are a direct plug in. The 7189 is more rugged, can take more voltage without failure and sounds very sweet.

Other hot rod mods would depend on the actual circuit and what it is you want that the amp isn't giving. Your tube rectifier idea would help the "singing" qualities of the amp by adding more tube compression, but you may lose a couple of watts in the process.

I have a 1973 Traynor Bassmate combo with 1x15" speaker, two EL84s, two 12AX7s one very large transformer and another about half the size. Only three knobs: volume, treble and bass. Point-to-point wiring using what appears to be good quality components and neat and tidy work. It has a good solid, round sound, but is a bit too dark and lacks gain and bite. I've already replaced the tubes and the amp seems to be working up to its capacity. I feel this amp has a lot of potential due to the size of the transformer, and the fact that I've had a lot of other period Traynors that kick butt! I would appreciate any suggestions on what might be done to re-configure the tone shaping circuit and hints on boosting gain.

You are very astute! Traynor did make some unsung heroes and generally had very good quality transformers and excellent workmanship. Since it is a bass amp, it is probably lacking in gain to keep the tone clearer and it probably has coupling and bypass caps that are chosen more for bass frequencies. Without a schematic of this amp,

I can only give you some places to look that may or may not give you what you want. The EL84s are obviously the output tubes and one or more sections of one 12AX7 is obviously the phase inverter. The other 12AX7 is then your preamp sections, with perhaps the volume control and tone controls in between the sections, or the tone circuit could be after the second section.

Re-biasing the cathode of the first preamp tube (pin #3) might be a good way to increase gain and adding a bypass cap if there isn't one already. Try using a 1K cathode resistor with a 25 uf at 25 volt cap for your first gain stage cathode-bias resistor. Remember that the minus side of the cap goes to ground. The plate voltage on this tube may need to be raised a little by changing the power supply resistor (that feeds the plate resistor) to a smaller value.

If you think we're going in the right direction, try the same thing for the second stage. This will be the same tube, but the cathode is pin #8.

Now let's voice out some of the bottom and get a little mids and highs in there by changing the coupling caps to about half of what's stock. If the stock coupling caps are .1 uf, change them to a .047 uf or .022 uf. This should liven things up a bit.

Amps that use EL34s generally run at lower voltages, so some power supply resistors may need to be reduced in value in order to get the preamp plates with high enough voltage to get some gain from them.

Be on the lookout for interstage voltage dividers that intentionally reduce gain. By removing the voltage dividers or changing the ratio of division, you could pick up some useful gain in the process.

Don't try everything all at once! Instead, it is better to try one or two mods and listen to what you have. If you're on the right track, try some more, but if you are straying from the intended results, put it back before you try another mod.

Experiment with some higher or lower value capacitors in the tone circuit, depending on if it's passive or active, you may need larger or smaller values. This is where a $10 capacitor substitution box comes in real handy.

I have a couple of old Traynor tube amps and I have questions about how the EQ works. I am familiar with the Marshall/Fender-style tone

stack from your excellent book, *A Desktop Reference of Hip Vintage Guitar Amps*. On the YBA-1 Bass Master, there's a 100K feedback resistor to the phase inverter from the speaker output—a "high range expander" control from that resistor to ground consists of a 25K pot in parallel with a 4.7K resistor, the wiper of the pot is grounded thru a .1 mfd. cap, the end of the 25K pot is not connected to anything. I assume this is a variation on Fender's "presence" control. Am I right? What's the cap for? Bypassing highs to ground since this is a bass amp? What's the "low range expander?" Is it just a mid control? How does this work? Also, how do I add a bias pot to this amp?

You are correct; the "high range expander" is simply a presence circuit. There were some Fender Bassmans and Concerts wired exactly the same way. You are probably more familiar with the presence circuit that used a 5K pot without the 4.7K shunt resistor, which I think is a better circuit, sonically. The reason Fender and Traynor used the 25K pot with the 4.7K resistor in parallel was to keep D.C. voltage off the pot. If D.C. is on the pot, the pot will be "scratchy." With the cap as the only ground in the pot, only A.C. is moving through the pot. The D.C. goes through the 4.7K resistor. (Remember the 4.7K resistor is actually part of the cathode circuit and quiescent current is developing a D.C. voltage on the top of that resistor any time the amp is on.) This type circuit might keep the pot from being scratchy, but it makes the "high range expander" about half as effective. The reason is simple: the A.C. has two paths to ground (thru the resistor or thru the pot/cap) instead of one. I prefer the 5K pot without a 4.7K shunt resistor. It may be a little scratchy (because of the D.C.), but it has superior range. Also, you should know how that circuit works: When signal is fedback through the 100K feedback resistor, it is 180 degrees out of phase. This causes phase cancellation in the cathode circuit of the phase inverter, which will cancel some of the gain. Here's the catch. If certain frequencies are louder than others, the feedback will be more on the louder notes, which will phase cancel a little more than normal of the louder frequency. This helps "even out" the frequency response. The cap shorts some of the high end to ground. Just how much gets shorted to ground is determined by the setting of the "high range expander/presence" control setting. If high end is grounded out, then those particular frequencies do not get phase cancelled and are therefore louder. In other words, as more high end from the feedback circuit goes through the cap, less high end gets phase cancelled in the cathode circuit of the phase inverter and therefore more high end ultimately shows up on the output. Pretty cool, eh? A "presence" control is actually like a treble control whose frequency range is higher than the normal treble control.

You are correct, the "low range expander" is simply an ordinary mid control and it works the same as the mid control on early Marshalls, Bassmans, Tweed Twins, etc.

Bias pots are always added in the negative voltage bias supply circuit. The pot is simply a voltage divider so that the voltage can be adjusted. I highly recommend a Cermet element pot for this application for two reasons. First, all adjustable bias pots have D.C. voltage on them and ordinary pots cannot hold up with that much D.C. (note: the D.C. in a bias supply is many times greater than the D.C. in a "presence circuit"). Secondly, bias pots need to be stable and the Cermet element pot will not change. Those pots are used to calibrate digital gas pumps because once adjusted, they will not move unless you readjust them manually.

Specifically, in your particular amp, replace R30 with a 50K ohm cermet element pot (one end goes to ground and one end goes to the top of C15. The wiper of the pot should go to both R22 and R25. With the amp in the "standby" mode, adjust the pot until you see minus 40 volts on pin #5 of each output tube socket. Proceed with biasing and you are done. Do not go to the "play" mode until you are certain you have bias voltage on pin #5 of each power tube socket.

We have installed "spike suppressors" on the power tube sockets of a number of our amps. This is described in Aspen Pittman's book and we have seen it in Ampeg amps from the 60s. I understand that this came from Ken Fischer. This involves three 1N4407 diodes in series from the plate to ground, reverse biased (cathodes facing the tube and anodes to ground.) Do you feel it is worthwhile in terms of tube life and is it inaudible? Where are the voltage spikes on the plate coming from? Is the same design applicable for a wide range of plate volts, or should different diodes be used? Our amps have from 375 to 500 volts on the plates.

I am familiar with the mod you describe. It will change the tone somewhat, but not much, and it can be used regardless of the plate voltage in your particular amp. This mod is useful for high gain/high wattage amplifiers. The "spikes" are induced in the primary winding of the transformer. When current abruptly ceases to flow in an inductor (and an audio transformer primary has lots of inductance), the energy stored in the magnetic field returns instantly to the circuit. The rapid disappearance of the field causes a very large voltage to be induced in the primary. Usually the induced voltage is many times larger than the applied voltage, because induced voltage is proportional to the rate that the field changes. When the inductance is large and the current in the circuit is high, large amounts of energy are released in a very short time. The diode suppression circuit shorts this high voltage to ground. Actually, the diode does not have to go to ground. It could just shunt each side of the primary winding, with the cathode facing the tube socket (transformer end winding) and the anode connected to the center tap. You may have seen this same type circuit in the coil of a relay (diode reverse biased across the coil) so when the relay's switch was turned off, the induced voltage would have some place to go, other than arcing the coil.

While working on an Ampeg Jet recently, we checked the cathode-biased D.C. current on the 7591 tubes by measuring the voltage across the cathode resistor and dividing by the cathode resistance to get the current. We did this with the amp warmed up for 5 minutes and found the current to be in a range safe for the Deluxe Reverb power transformer which we had bolted in (70 mA or so). Just for laughs, we measured the current after one hour, using the same method, and found that the current was approaching the 98 mA level. Is this a common problem? How long can it take for an amp to reach "steady state" temperature? Does turning the non-standby amps off during music breaks help with the temperature at all?

There is no set time for a tube to reach "steady state." Certainly 5 minutes is not very long to heat up an amp unless it is really being cranked. You could probably leave one on for several days and it would still have a little bit of drift. Although a slight drift is normal, this sounds like excessive drift to me. The first thing I'd check would be the 1K resistor in the power supply (screen supply). This amp has no screen resistors and the 1K resistor in the power supply could be heating up and changing value. This, in turn, could change the screen voltage which would change the idle current of the tube. Some Jets use a 1K -2 watt resistor and others used a 1K -10 watt. I think I'd make sure and have a 10 watt because it is more stable in relationship to temperature.

One other thing to check might be the cathode resistor and the cathode bypass capacitor on the cathode of the 7591s. Correct value should be 140 ohm at 10 watt resistor and a 25 uf at 25 volt capacitor.

Yes, generally it is a good idea to turn non-standby amps off during a break.

You guitar players have all the cool amplifiers! In the 60s, we keyboard players used bass amps. Today, we're stuck plugging into the PAs. I'm told guitar amps are voiced for guitar. Their pre-amps tend to distort the hotter keyboard signal. Plugging into an effects return works out better, avoiding the preamp completely. Does the power section reproduce the fuller range of the keyboard? Are there any tube heads, modified or stock, that can be used for keyboard amplification?

Yes, guitar amps are designed (voiced) to color the signal. This is because the amplifier is actually part of the instrument. A keyboard amp generally does not want coloration but a true reproduction— more like a high-end audio type amplifier. For keyboard, you need headroom. The more watts, the better. The idea is not to distort the signal, but to gain a true reproduction of the signal. Any high-powered amp would be just fine, if you are plugging into the F/X return. The output stage of almost any amp will reproduce the full range frequencies of a keyboard, but stay away from guitar speakers. Virtually all guitar speakers intentionally roll-off high-end around 6K and you will probably want more range for the keyboard.

An Ampeg SVT or V4 would be a good choice and you will have to get your tech to install an F/X return (or power-amp in) for these models. Anything with lots of wattage and 6550 tubes would be good for your purpose.

Why do some well respected bass amps (e.g., Mesa 400+Bass, Sunn model "T" & 2000s, Rickenbacker B35) have chokes in the

power supplies, while other classics (e.g., Ampeg SVT, Fender Studio Bass, Bassman 135) do not? Would it benefit me to incorporate a power supply choke in my SVT or V4-B? If so, how should I select the correct rating?

This is a question with many answers, depending on the exact amplifier under discussion. The Bassman 135 has an ultralinear output stage and therefore the screen supply is actually tapped off the output transformer. This design requires no choke. The SVT does not really need a choke because it has a separate power supply for the screen supply. That is not the case with the Ampeg V4-B. However, the V4-B uses 7027A power tubes which do not draw very much screen current. Also, the 470 ohm dropping resistor in the screen supply circuit adds a little compression to the sound that the designers at Ampeg apparently liked. Generally, the choke is used to resist changes in current so that the screen supply is rock-solid. This may make the amp true in its reproduction, but truth may not be the designer's intention. My guess is to try a choke for yourself and A/B compare your particular amp to see if you like the choke or the resistor better. I would recommend that you use a choke that is at least double what your anticipated total screen current will be. Generally speaking, a 100 mA choke is adequate. Typical inductance values could be anywhere from 3 to 20 Henries—the more Henries the less fluctuation in screen current.

I want to replace the tubes in my Peavey 212 Heritage amp with Groove Tubes' 6L6GBs. The original 6L6s are many years old and I am getting a loss of power. How do I make the bias adjustment on this amp? I realize this is not a vintage amp now, but probably it will be soon. This amp sounded really good for several years. I would like for it to "do it" again.

It sounds to me like you have played that amp a long time without servicing it. If that assumption is right, you probably need something more than output tubes. Parts can go bad that are critical to an amp's performance. Your filter capacitors are more than likely in need of replacement and could be starving the tubes of both voltage and current. Resistors can drift way out of tolerance and ruin the self-biasing of preamp tubes (or in your case transistors), possibly choking them off. The end results are less gain and constricted tone.

Coupling caps can be leaking a volt or two of D.C., thus adversely affecting the bias of the next stage. Sockets can be dirty and/or otherwise not making proper contact with tube pins! There's a big difference between your amp merely working (not blowing fuses and some sound coming out) and really performing. I urge you to get your amplifier overhauled to restore it electronically, if you really want to restore the original tone.

To answer your question on bias, let's go back to the basics. In all fixed-biased amps, there will be a negative voltage fed to pin 5 of the 6L6 sockets. In a push-pull style amp such as yours, the two sides of the output stage must be isolated; therefore, a pair of 220K ohm or 100K ohm resistors isolate each side of the push-pull circuit (pin 5 on the output tube socket) from the negative bias voltage supply. (Please note: These isolation resistors are not to be confused with the typical 1.5K ohm to 28K grid resistors that sometimes appear on pin 5.) If you follow the circuit past the isolation resistors back to the actual negative voltage supply, there will always be either an adjustment pot voltage divider or a pair of resistors used as a voltage divider to adjust the actual bias voltage that is fed through the isolation resistors to pin 5 of each output socket. If you make one of the resistors larger, the voltage will go up, and making the other resistor larger will make the bias voltage go down. The converse is also true. (If you make the first resistor smaller, the voltage will go down and if the second resistor is smaller, the voltage will go up.) It is a matter of adjusting the voltage divider resistors to achieve proper bias voltage. There is also a filter cap (sometimes two) in the negative voltage supply that will be mounted with the positive lead grounded and the negative lead going to the negative bias voltage supply. It is important that this cap is good. If it is bad, your bias voltage will be unstable as the cap heats. In fact, if the cap shorts, your output tubes will be history. This cap in the bias supply probably needs replacing if the amp is over ten years old.

Now that you know how to change the bias voltage, you can alter the amount of quiescent current flowing through the tubes to achieve proper plate current. This can be monitored in a number of ways, including: the "transformer shunt method," "one ohm resistor on the cathode method," oscilloscope (not recommended), or if you know what you're doing, by ear.

I have been thinking of a few obscure questions that your readers may find interesting:

1. With the exception of the heater current, what stops the use of a 6550 in place of the 7027A? What benefit does the 6550 give you?

2. How could an amp be modified to comfortably accept the increased heater current draw?

3. I would like to see a step-by-step instructional series on the installation of a master volume in a V4/VT22.

4. How can you check an output transformer?

5. How about the installation of a bias pot for fixed biased amp? (eg. later silverface Fender.)

I would be very careful when attempting to substitute a 6550 in a 7027A socket. The reason is that the pin-outs are not exactly the same. The 7027A has both pins #1 and #4 internally connected to the screen and pins #5 and #6 internally connected to the grid. If the socket is wired using pin #6 as the grid connection (different from the 6550) and pin #4 as the screen connection (same as 6550), then your 6550 substitution will become a diode 'cuz it ain't got no bias! In certain 7027A amps, the screen voltage supply is connected to pin #4 and the grid is wired to pin #5 of the socket. In this case, a 6550 will work with only a bias adjustment and a screen grid resistor upgrade. You should never try placing a 7027A into an amp built for 6550s without extensive inspection of the socket. 6550s do not have anything connected to pins #1 or #6. In most cases, the socket is wired using either of those pins as a mounting post. Use a 7027A in an amp that is using pin #6 as a mounting post for the screen grid resistor and you will be putting the entire screen voltage supply on the grid of the 7027A. This will make it a diode and toast—in that order. The screen voltage design maximum for the 7027A is 500 volts whereas the 6550 is only 440 volts. The 6550 has the advantage of availability and it can handle greater plate watt dissipation. It might also be a better choice for bass guitar.

If you are planning to use 6550s in a V4/V22, I would recommend changing the screen grid resistors to 1K at 5 watts. The tube sockets are wired so that you will not need to rewire them. You will need to alter the bias voltage so that the tubes have more negative bias voltage. To do this, change R50 on the V4 to an 82K ½ watt resistor and change R49 to whatever resistor value gives you appropriate bias.

Amplifiers can be made to accept increased heater current, if you need it. The truth is that most amp designers will double the current rating of a filament transformer as a safety factor when designing a transformer. If that is the case with the amp you are modifying, you may already have enough latitude to use a tube that draws more heater current. I'm a practical kind of guy so here's the practical test. Measure the filament voltage with the substitute tubes temporarily in place. (You could even have the amp in the standby mode since you haven't had the chance to bias the output tubes yet.) If the tubes are drawing excessive filament current, the voltage will drop below 6.3 volts. If you make this test and the voltage is anywhere from 6.2 to 6.7 volts, you don't need a auxiliary heater supply. If you do need an auxiliary heater transformer, I would use one with twice the rating of the tubes and wire it directly to the output tube filament connections. Of course you will disconnect the original transformer from the output tube filament connections and use it only to operate the preamp tubes. Sometimes you may want to reverse this and use the auxiliary heater transformer for the preamp tubes, thus freeing up additional current for the output stage. For instance, when I convert a 1964 Super Reverb to EL34s, I typically disconnect the preamp tubes from the original heater winding and use a 4 amp 6.3 volt auxiliary transformer to operate the six preamp tubes. Since the preamp tubes take about .3 amps each, this frees up an additional 1.8 amps of filament current (from the original heater winding) available for the output tubes. Since an EL34 is about .6 amps of heater current per tube more than a 6L6GC, I have provided plenty of extra current with a safety margin to boot.

I don't think there is a step-by-step instructional video on how to put a master volume control in a V4 or a V22. I am currently working on a script for a video about amp restoration, modification and repair. Maybe I'll include your idea in the modification script.

There are many ways to check an output transformer. It really depends on what you are checking for. You could use a high-pot test to determine if there is any arcing between any of the windings and/or ground. Of course, you could also use an ohmmeter to check for continuity between ends of the same windings to make sure there are no open windings. You could also check to make sure there are no shorts between windings—again this could be done with an ohmmeter. You could check turns ratio with a signal generator (or Variac) and an A.C. voltmeter. But the easiest way to check, especially if you suspect that a problem exists with the output transformer, is to substitute a known good output transformer. If the problem goes away with the new transformer, it is a pretty safe bet that the output transformer was bad.

As far as installing an adjustable fixed-bias pot in a silverface Fender, I have already covered that in *A Desktop Reference of Hip Vintage Guitar Amps* in a chapter titled "Silver to Blackface Conversion for Twin Reverb." Although it was written for the Twin Reverb amp, the same conversion is used for any fixed-bias silverface Fender with the exception of the Princeton and Princeton Reverb.

I had a Mesa .50 Caliber head that had a miniature Accutronics reverb. The springs were no longer than 6" in length. The depth and quality of the effect was surprisingly good for such a dainty little inboard reverb. I moved on to a Mark III Simul-Class which has, as I'm sure you know, the full-size reverb pan. I was expecting a really great sounding reverb beyond what the miniature Accutronics had been capable of delivering—thinking bigger is better.

The Mark III reverb seems weak, in general, and really doesn't kick-in until a setting of 5 or 6. At half power, the reverb is relatively poor, and on full power, the reverb is fairly acceptable. Mesa explains that in order to achieve the gain of the Mark III, the reverb circuit had to be compromised. They suggested I run a .003 capacitor across the output from the reverb to increase the reverb's gain. The result did not offer any noticeable improvement that I can discern. Your comments and suggestions will be greatly appreciated.

When an amp has jillions of controls and settings, it is impossible to have everything work great. The more controls you have, the more compromises that must be made. The reason is very simple. What works best at half setting won't work well at full setting. What sounds good with overdrive may not sound good with a clean setting, etc. This is the reason why many amp designers stay away from the "Swiss Army knife" approach and simply do one thing with excellence. My advice to you is to decide if you want full power or half power and once the decision is made, stay with it. If you want full power, leave the reverb circuit alone. If you want half power, have your tech modify the reverb circuit to get more reverb return volume.

I have a few technical questions that I would like cleared up. I see output impedances of 4K to 6.6K recommended for two 6L6GCs in push-pull. How do I determine what's the best value? How will the amp's sound be affected?

Also, can I use a higher value plate resistor to lower plate voltage? Or, should I change the power transformer to a lower voltage. I do not want to use the Zener Diode technique.

In your opinion, at what point should the high frequency of a guitar amp roll off? I've heard some techs say to cut off high frequency as low as 6KHz.

How can primary impedance of an unknown output transformer be measured?

Let's start with the first two questions. You asked how to determine the best value output impedance? And how will that affect the sound? Your question sequence is backwards. In a guitar amp design, the output impedances are chosen to get the best sound, which answers both questions simultaneously. Generally speaking, lower values are generally used in higher current situations. The converse is also true. The difference in sound: Higher impedances can tend to get less high end and perhaps a little more bottom. Also the "kick back" voltages are higher, resulting in a spongier attack envelope. The converse is also true. Lower impedances can usually get a little more crunch and harmonics.

Tone is a personal preference. I remember a time when harmonica/vocalist Kim Wilson and I were doing some impedance research for the Kendrick harmonica amp. In blind tests, he consistently liked amplifier combinations where the speaker load was twice what the transformer required. This is exactly the same as doubling the transformer primary impedance, because transformers can't count or read spec sheets—the only language they understand is turns ratio. The turns ratio of a 4.2K (primary impedance) transformer with a speaker load that is twice the recommended transformer output impedance (say, for instance, an 8 ohm speaker hooked to a 4 ohm output transformer) is exactly the same as a 8.4K transformer with the speaker load matching the transformer output. So in effect, changing the speaker load changes the primary impedance of the transformer!

In response to your second question, NEVER increase the value of the plate resistor to alter voltage. Plate resistors are found in the preamp circuit. They are "fed" voltage by a power resistor. It is the power resistor that must be increased in order to decrease voltage.

Here's how to determine the exact value of the new resistor. Measure the voltage drop across the power resistor that will be changed. Remove the resistor from the circuit and measure the resistance. Divide the voltage drop measured by the resistance measured. This will give you the amount of current, in amps, that was flowing through that part of the circuit. Now decide how much voltage you would like the resistor to drop. Divide this voltage amount by the current that was flowing through the old resistor. Your quotient will be the resistance value, in ohms, of the new power resistor.

When you talk about amplifier roll-off, I am somewhat amused. Why? Because almost all guitar speakers roll-off at 6K. Vintage speakers sometimes roll-off at 5K. Since the speaker is what actually makes the sound vibrations, it really doesn't make any difference what frequency the amp rolls off, if the speakers can't reproduce those frequencies anyway.

When I first designed the Kendrick Blackframe speaker, I had the speaker roll-off at 11K. This range was and still is completely unheard of for guitar speakers. Although these speakers sounded wonderful in Kendrick and most other amps, there were some amps that sounded extremely harsh with the Blackframe speaker. The speaker actually reproduced the sounds that the amplifier was making—perhaps for the first time.

In response to your primary impedance of an unknown transformer question—see Part IV of this book, the chapter titled, "Everything You Always Wanted to Know about Transformers (but were afraid to ask)." Everything is explained in vivid detail.

Can you share your knowledge of currently available parts without getting yourself in a wringer with the manufacturers? You have done a lot more listening test than the average tech. Ken Fischer's recent note about horrible electrolytics made me wonder how many I have put into people's amps (and my own).

Some parts I wonder about: Illinois cap, Mallory, Sprague orange drops, Xicon, and of course Sovtek tubes, Czech tubes, and Chinese tubes. Anything from your end?

Some of the parts manufacturers you named are not manufacturers at all. For instance, Xicon is a private label that buys capacitors from other manufacturers and has their name printed on the products. Sovtek is the name of a sales operation in New York City that distributes parts made by other manufacturers. Illinois capacitors are made overseas and not in Illinois. Although many parts are available, there are inconsistencies in parts bought through private label sales outfits. The reason is very simple: Since these type of companies do no manufacturing of their own, quality control is virtually impossible.

Case in point: I used Sovtek Brand preamp tubes for many years and was always happy with them. One day, all of the amplifiers on our "burn in" table were making funny noises. There was a mysterious sizzling sound when the amps were put in the "play" mode and every now and then, there was an intermittent popping noise. Those ten amplifiers puzzled several technicians and myself for several days. We changed to other new Sovtek tubes, not even thinking that something could be wrong with the entire batch. Could it be a bad output or power transformer? All sorts of tests were performed by my technicians for several days, wasting hours of productivity. After many days, one of the technicians noticed that the plate material inside the preamp tubes looked slightly different from the plate material of the exact same tube type, but from an earlier shipment. We tried the older version and the problem went away. When Sovtek was notified, they were completely unaware that their supplier had changed the plate material without telling them.

Recently I had a bad experience with capacitors. In fact, because of this experience, I devised a capacitor tester that checks D.C. leakage.

Whether an electrolytic capacitor is a Sprague, Mallory, etc., is not any kind of guarantee to its quality. I spent some time talking with an actual capacitor manufacturer and I was told that the inside of an electrolytic begins to crystallize if it is not used within a year of manufacture. It can be reformed by baking the capacitor to re-liquefy the electrolyte inside. There are companies that track their inventory and return the unsold caps to the manufacturer if they are not sold within a year. The manufacturer reforms the cap by baking and returns it to the distributor. This is done every year until the cap is sold. The problem occurs when caps remain in someone's inventory for years and get bought and sold on the wholesale market without ever being installed in an amp. After a few years, the electrolyte crystallizes so much that it cannot be brought back.

As far as my opinion of current production parts, let me say this. The Russians just don't know how to make a cathode and for this reason their 6550s, 6L6s, and 6V6s will not hold up very long. The 12AX7s can be good or bad—depending on which batch you get. The 5881s are terrific and are probably the best sounding current production tube being made.

The only tube I ever liked from China was the 5AR4 that looks (and performs) like a Mullard. The Czech (Tesla) EL34s are probably the only good sounding, well made EL34s being made today. The Tesla preamp tubes sound good if you can find one that is not noisy. (I heard the 12AX7s are out of production now.) I have several thousand of them that I cannot use because they are noisy!

As far as caps, do not buy surplus or close-out electrolytics. It is an exercise in futility. There are many coupling caps available that perform well, but they all have their own sound. The orange drops remind me of a JBL speaker kind of voice. Big bottom, clear top, with transparent mids. The Mallory 150 series sounds more like an Astron original than any other current production cap.

I spoke with Ken Fischer, and he informed me that the capacitors he was referring to were the Illinois capacitors. These were in an amp he was reviewing and he substituted some good filters and was astounded at the difference.

I've got this 200 watt Hiwatt that has a patch cord between two of the four output jacks. It has been there since I bought it and when it is taken out, the amp works but at very, very low volume. It is good and loud when this patch cord is left connected. Nothing seemed to have been modified inside the amp, but even when it's working LOUD, it doesn't have the volume that my Ampeg V4 has through the same speakers (4x12 Celestions). Can you tell me what's going on here?

You have just described two or more output jacks that are wired with internal switches that short the "hot" side to ground. Sometimes an amplifier has one output jack wired this way so that in the event the amp is turned on without an adequate speaker load, the jack is shorted to ground through a switch that is a part of the jack (so the transformer will not be operated without a load.) This is a safety precaution to avoid output transformer damage. Evidently your particular amp has more than one jack wired this way and plugging the patch cord into these jacks "opens the switch" and ungrounds the "hot" of the output transformer. Almost all Fender amplifiers have a jack wired this way on one output and not on the other; therefore, you must plug the speaker plug into the one jack and use the other jack only if you are attaching a second plug for an additional speaker. I don't recall Hiwatt using two switching jacks on the output, but maybe they intended for an amp of that power to be played with two speaker cabinets only.

Though I've never A/B sound tested an Ampeg V4 with a 200 watt Hiwatt, I would think the 200 watt Hiwatt would be louder than the Ampeg V4. If it is not, I would suspect that perhaps the Hiwatt is improperly set up, has weak tubes, or needs servicing. I did notice in the picture you sent, that the Hiwatt had original filter capacitors. These are evidently about 30 years old and probably need replacing. I would also check the speaker impedance to make sure the head is adequately matched to its speaker load. Biasing may also affect the overall power.

You recently mentioned leaving an amp on for a day or so to form the dielectric on new electrolytic caps. Does this have any adverse effect on the amp — leaving it on for this period (heat, etc.)? Also, the concept of "burning in" power tubes, that even matched sets will drift if not "burned in" properly; please explain what this means and how to do it properly. Also, concerning the "light bulb in series with the amp" trick mentioned in your first book, *A Desktop Reference of Hip Vintage Guitar Amps*, to charge up filter caps; how long should you charge them this way and how can one tell if they are charged properly? Also, can you repeat the process if you feel it hasn't been done properly?

My suggestion to leave an amp on for a day to form the dielectric on new electrolytic caps was intended to be used with a variac in which the amp would not be getting full voltages and therefore does not heat up very much. This does not have any adverse effect on the amplifier, if done properly. I suggest leaving the variac on the 40 volt setting overnight and then bumping up the voltage 10 volts per hour the next day. After 8 hours of bumping up the voltage 10 volts per hour, you will be up to the full 120 volts.

The concept of "burning in" power tubes is very simple. When a tube is brand new, it is not used to the high heat of which it will be operating. If you monitor a brand new tube that has never "cooked," you will see the plate current drifting around. It may go up and then down and then up, etc. After a while, it will stabilize and quit moving around. This would be the right time to take a measurement for matching purposes. If you match the tube while it is still drifting, you are very unlikely to end up with a matched set later. Why? I don't know for sure, but there is no denying that the tubes will drift when they are brand new and not used to "cooking" and not fully warmed up. I would only guess that certain impurities that are present during the manufacturing process, are stabilized by sustained high heat.

Leaving an amp "on" for a while is the best way I know of "burning in" new tubes. This gets the tube used to operating at the temperatures it will be asked to operate.

The "light bulb in series" trick is simply a current limiting device that starves the amp for current. This is useful because it keeps the cap from leaking much current while still keeping D.C. across it. By the way, D.C. voltage (not current) is what causes the dielectric to form. I have heard of vintage radio collectors actually using the current limiter to re-charge very old caps by keeping voltage on these caps for extended periods but without the possibility of drawing too much current. If electricity was water, current would be gallons and voltage would be pressure. If a lot of current leaks from a cap, it will make the cap worse and it will be a drain on the power supply.

You can repeat this process if you feel it has not been done properly. And how do you know if it has been done properly? For one thing, your amp will sound good and have very low hum levels if it has been done properly. Also, the "B" note on the "G" string (16th fret) will play in tune. Of course, the exception to this is the Vox AC30 which will always have a funky, out of tune harmonic on this note—good caps or not.

REVIEWS
AND
CATALOGS

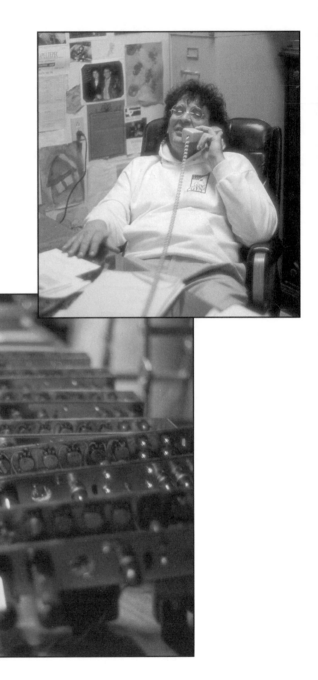

KENDRICK
REVISITED

VINTAGE GUITAR, NOVEMBER, 1996: BY WARD MEEKER

Since last we spoke in-depth with Gerald Weber (*VG*, October '92), Kendrick Amplifier has experienced significant growth and evolution. The company recently revealed its new amp lineup, the Texas Crude Series, and Weber is set to publish a new book, *Tube Amp Talk for the Guitarist and Tech*, (500-plus pages), in late October. We caught up with him at a recent guitar show and discussed these events.

Tell us about your new book, what you want to accomplish with it, and how it will differ from your first effort, A Desktop Reference of Hip Vintage Guitar Amps.

When I became the first tech writer for *VG*, in February '91, most guitarists had a limited knowledge of the workings of their tube amplifiers. Other knowledgeable tech writers like Dave Funk, Ken Fischer, and Dan Torres later came on board. Eventually, guitarists became much more knowledgeable about their amps. Shortly thereafter, many hobbyist tube amp companies started springing up in garages all over the country. So my first book assumed that the guitarist knew nothing. From that vantage point, it gradually explained how each component of an amplifier affected tone. I believe this ultimately led to guitarists servicing, modifying and fine-tuning the tone of their amps.

The new book is all new material that deals with fine points of servicing and tweaking your amp, performing modifications, a condensed version of the new Trainwreck pages as they appeared in *VG* magazine the last several years, and there is a section for the Texas Tone Del Maximo, which uses the 3-amp setup.

There's another section called Lagniappe, which is Creole for "...that little bit extra." It deals with interactive volume controls, tube rectifiers, filament circuits, reverb circuits, transformers with conversion charts, and simple homemade tools for servicing amps. Like the first book, there is also an extensive question-and-answer section, which was condensed from my "Ask Gerald" column in *VG*. And like I said, *Tube Amp Talk for the Guitarist and Tech* has all new material. Nothing is repeated from the first book. The purpose of both books is to elevate the guitarist's understanding of his tube amplifier. This information is needed whether the guitarist does his own amplifier servicing, or uses a tech to personalize his tone. Whereas the first book assumed the reader knew nothing, the second book assumes the reader is more knowledgeable than he or she was four years ago.

Talk about the decision to get into accessories, like the Kendrick Pipeline Reverberator, the ABC Box, the Six-Shooter tube mic preamp, the pure nickel strings, etc.

You refer to these as accessories, but we feel these are necessities. At Kendrick, almost every product we develop has been an unintentional, but happy, accident.

For instance, I developed the ABC Box with no intention of ever selling one. I needed a way to use three amplifiers without hum or impedance loading. I designed and built the first ABC Box strictly for myself. Eventually, people requested that I build them one, and it became so popular we added it to our product line. As far as I know, there is no other way to use three amplifiers together without tone alteration or impedance loading. If there had been, I would have probably never developed the ABC Box. It came from necessity.

The same is true of the 6-channel tube mic preamp we call the Six-Shooter. Many of the employees at Kendrick, including myself, have multitrack digital studios, and we needed a way to make a digital recorder have an "analog tube" sound. Again, I designed and built the first 6-channel tube mic preamp only for myself, never intending to sell one. Then, many of our employees decided they needed them for their studios, and this led to us offering limited production units of the Six-Shooter.

When using a tube mic preamp, the output of the preamp goes directly to the input of the digital recorder, so the microphone signal is warmed up and "analogged out" before it hits the digital input. When it is played back, the sound is multi-dimensional and smoother, similar to analog tape. If you don't use a tube mic preamp, the way most recordings are made, the microphone signal goes to a mixer and sees a transistor first; then it is recorded digitally. This ultimately sounds cold and lifeless. I strongly believe that when you're going for a million dollar guitar tone, the tube mic preamp is very much a necessity.

As for the Pipeline Reverberator, a descendent of the Model 1000 Reverb first introduced in 1990, the story is almost exactly the same. At that time, no one was making a stand-alone reverb unit, and because stand-alone units are far superior to onboard reverb, we felt it also was a necessity. Stand-alone reverbs, like the Model 1000, and now the Pipeline, have controls for the actual reverb drive, tone, and mix; so the user can fine-tune the actual reverb sound. This makes a significant difference from an onboard reverb, in which tone parameters are fixed, non-adjustable, and only a mix control is employed.

The Kendrick Pure Nickel Strings, we feel, were needed because virtually all the strings on the market were not pure nickel and therefore, had an objectionable high-end brightness that just didn't record properly. When you go to all the trouble to get your guitar and amp to sound right, you don't want to degrade you tone with strings that sound harsh.

Explain the evolution of the old line of Kendrick amplifiers and the new Texas Crude series. Why are you moving away from reissues?
I have never been satisfied with status quo. When I first started building '50s Leo-style amps, I was the first and only one building a point-to-point handwired Bassman-style unit. This was significant because it hadn't been done in 30 years. I shared my knowledge and research with anyone who was interested. This eventually led to a couple of things.

First, Fender introduced some handwired products. Then many small companies followed my lead. Actually, the first Kendricks were not just "copycat" amps, but had some extra features not found on the original tweed amps. For instance, there was an extra tube that could be switched in or out of the circuit to increase post preamp gain, thus increasing output tube distortion. There was also a bias adjustment pot, effects loop, and preamp output features not found on any '50s amps.

Those first six years were a major learning experience for me. I saw the importance and significance of each and every aspect of amp construction. Almost every component in my amps were custom manufactured–from the switches to the hardware to the speakers to the transformers–even the cloth-covered hookup wire was custom made to my specifications. I was the first to reverse-engineer original transformers, in order to duplicate original Triad designs.

After making and listening to hundreds of my own amplifiers and thousands of vintage amplifiers from our restoration shop, I had to ask myself a few questions such as, "Is this really the best that can be done?" "How good can an amplifier sound?" and "Can it get any better than this?"

This was about the time that I was designing a special amplifier for a well known Texas hero, the Reverend Billy "F" Gibbons. Billy was actually declared a Texas hero by our State legislature. They recognized his efforts by elevating him to the same stature as Davy Crockett, Sam Houston, Jim Bowie and other Texas heroes.

Anyway, this particular amplifier was one in which every aspect was scrutinized for tone—the cabinet, the power supply, the circuit, the speakers, the baffleboard. Every component, including the transformers–were all selected via time consuming trial-and-error listening tests. And I'll never forget the day Gibbons looked me in the eye and said, "Gerald, my Texas Crude amp has more tone than any amp I've ever played."

You cannot imagine the impact his words had on me! If I had dropped dead right then, my life would have been complete. Since then, I have designed and produced the entire Texas Crude series. And I truly believe they are the most responsive, toneful amplifiers ever made. I would put them up against anything in the world.

After having achieved all this, it seemed pointless to continue making '50s reproduction-type amps—even though I was selling 70 amps per month when I stopped taking orders for them. Besides, there are many companies now that are doing a fine job with "copycat" amps. The reputation of Kendrick earned us the right to have our own identity.

TEXAS CRUDE SERIES

KENDRICK AMPLIFIERS

PATTILLO HIGGINS
1863 - 1955

ANTHONY LUCAS
1855 - 1921

The 20th century officially began on January 1, 1901. Which means what happened just south of Beaumont, Texas nine days later could well be considered the first great event in 20th-century America.

That was the day an Austrian-born engineer named Anthony Lucas proved a Texan named Pattillo Higgins right, by hitting oil at Spindletop.

No one anywhere had ever seen a gusher like Spindletop. Its daily production of 75,000 barrels equaled half of the country's daily consumption. With a last laugh that shot 200 feet in the air, Higgins and Lucas had proved the so-called experts wrong. And from Spindletop flowed a whole new industry and image for Texas.

Flag, Star of the Republic Museum
Oil well and portraits courtesy of
Spindletop/Gladys City Boomtown
Museum, Beaumont, Texas

SPINDLETOP & GUSHER

BEAUMONT

AUSTIN

The Spindletop (100 watt) and Gusher (50 watt) amplifiers both feature four inputs, treble, middle, bass, presence, and bright and normal volume controls. Footswitchable post gain stage for more output stage overdrive. FX loop and cathode/fixed bias selector switch are standard. With a Kendrick ABC box (sold separately), you can switch between the bright and normal inputs for channel switching flexibility. Both the Spindletop and Gusher are available as either a piggyback or combo amp format. You choose either "Austin" (rectangular) or "Beaumont" (trapezoidal) cabinet styling. Two Kendrick 12" speakers are standard although the Gusher is also available as a four 10" "Austin" style cabinet. Both amps are perfect for stage performances or recording.

THE RIG & WILDCAT

THE RIG

THE WILDCAT

The Rig (35 watt combo) and Wildcat (25 watt combo) amplifiers each feature four inputs, bright and normal interactive volume controls and a single tone control. Cathode/fixed bias selector switch and external speaker output jack are standard. Line level output jack can be used for direct injection to power amp, mixing board or another guitar amp. Both amps available in "Austin" (rectangular) or "Beaumont" (trapezoidal) style cabinet. One Kendrick 12" speaker is standard, although either amp is available as a two 10" "Austin" style cabinet. Both amps are ideal for studio or live performance.

THE ROUGHNECK

The Roughneck (6 watt combo) practice amplifier features a bright and normal input, single volume control and line level output. Class A single-ended 6550 output stage with one 8" speaker. Ideally suited for lower volume applications such as professional and home recording, practicing and rehearsals. Available in "Beaumont" or "Austin" style cabinet.

THE LOWRIDER

The "Low Rider" Studio Bass Amp (100 watt combo) features two Kendrick 10" bass speakers in a "closed back" combo configuration. Designed for studio or low volume stage use. For both bright and both normal inputs, the #1 input (of each) is for passive pickups and the #2 input (of each) is set for -10db for active pickups. Footswitch operated FX loop with tube powered unity gain control, tube or solidstate rectifier are standard features.

THE "SONNY BOY" HARP AMP

The "Sonny Boy" Harp Amp (35 watt combo) amplifier features four inputs, tone control, grind volume and extra grind volume are standard. Unique waveform symmetry control fattens tone and reduces acoustic feedback. Unique continuous floating baffleboard. Designed especially for harmonica, not for use with guitar or any other instrument other than harmonica.

THE "PIPELINE" REVERBERATOR

"Pipeline" Reverberator features Reverb Drive, EQ, and Mix controls. Available in "Austin" or "Beaumont" cabinet style. Adds the ultimate reverb to your sound. Output tube driven "Hammond style" reverb tank.

THE "SIX-SHOOTER"

"Six Shooter" Tube Mike Preamp features six channels with balanced (low Z) and unbalanced (high Z) inputs and an unbalanced output for each. Don't cheapen your million dollar guitar tone by recording through a mixer. Using the tube mike preamp directly into your tape machine brings out the third dimension of your tone and eliminates adverse coloration, loss of high end, and unwanted filtering found in every mixer and recording console. Two rack space.

CONTINENTAL & TOWNHOUSE GUITARS

The Continental Guitar (left) features three single-coil pickups, maple neck, rosewood fingerboard, swamp ash or alder wood body, sunburst finish (custom colors also available). Vintage style tuners and vintage six screw / 5 spring vibrato, "state of Texas" inlay fretmarkers and custom wound pickups.

The Town House Guitar (right) features two humbucking pickups, mahogany neck and body, maple top, rosewood fretboard, sunburst finish (custom colors also available). Vintage style tuners and "stop" tailpiece, "state of Texas" inlay fretmarkers and custom wound pickups.

KENDRICK STRINGS

Pure nickel wound on a hex core giving longest life and richest tone. Core diameter engineered to produce more even tension between all six strings providing a more consistent feel when bending strings as well as facilitating the personal nuance in each players style. Available in Regular .011 thru .050, Light .010 through .046 and Extra Light .009 thru .042.

KENDRICK SOFT AND HEAVY DUTY CASES

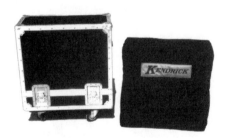

Kendrick Soft cases and Heavy Duty Flight Cases made with wear-resistant, military grade materials. Soft case is padded with special Velcro fasteners for bottom panel so that amplifier is completely encased in padding and outside protection. Heavy Duty Flight Cases designed for the rigors of touring. Extruded Aluminum with surgical steel rivets, tough outer shell, foam interior, recessed butterfly latches, heavy duty casters.

KENDRICK ABC BOX

Designed as a single input / three output (switchable) active audio distribution preamp, the Kendrick ABC box can be used for multi-amp setups, amp switching, or channel switching. Any one of the footswitchable outputs can be a "send" for a tuner or for studio use. The input and all outputs are isolated and buffered. Heavy duty switches are arranged in a unique triangular configuration allowing all three, any two or any one output to be switched with a single foot-step. Zero insertion loss and unity gain, all three output signals are identical to the input signal. LED status indicators. 120VAC (no batteries).

KENDRICK TUBES

GZ34 5881 EL34 12AX7

GZ34 5881   EL34   12AX7
Kendrick Tubes selected for tone quality and durability. Choose GZ34/5AR4, 5881/6L6GC, EL34/6CA7, and 12AX7/7025. All output tubes come in matched pairs or matched quartets.

KENDRICK SPEAKERS FOR GUITAR AND BASS

BLACKFRAME 8"

8" / 4 ohm Blackframe Guitar Speaker — Rich and full-bodied clean tone with creamy high-end break-up when pushed hard.

BLACKFRAME 10"

10" / 16 ohm Bass Loudspeaker — Full range, solid bass sound, extreme fidelity. Stunning three dimensional bass guitar reproduction.

BLACKFRAME 10"

10" / 8 ohm Blackframe Guitar Speaker — Classic sound of the 50s and 60s.

BLACKFRAME 12"

12" / 8 ohm Blackframe Guitar Speaker — Probably the most efficient guitar loudspeaker ever made. Delivers piano-string bottom-end clarity, transparent mids, and crystal chime top-end.

GREENFRAME 12"

12" / 8 ohm Greenframe Guitar Speaker — Classic low wattage speaker with a smooth top-end response, solid mids without the honk, smooth highs without the fizz, round bottom-end.

BROWNFRAME 12"

12" / 8 ohm Brownframe Guitar Speaker — Big, full-bodied clean tones when driven lightly, deep cello-like resonance when pushed into overdrive.

EXPERT REPAIRS & RESTORATIONS

Your tube amp may function, yet not perform! Electrolytic capacitors lose much of their effectiveness after 10 years, causing weak and mushy response and often times annoying hum. Resistor values drift, connections get corroded and components wear out. At Kendrick we are known for bringing life back to dead tube amps. We can make your tube amp sound better than it ever sounded. Also our recovering department can recover your amp in aged tweed, tolex or virtually anything you want. Speakers, grill cloth, handles, knobs, cabinet, hardware, logos — no problem. You tune your guitar every time you play, but when was the last time your amp was tuned? If you are like most people, it has been too long.

Here's how it works. Box up your tube amp with appropriate padding (remove the tubes and pack separately) and send it to us UPS. (Shipping is inexpensive - probably $10 - $20.00). Be sure to include your daytime phone number, shipping address and a letter stating what you want to accomplish with your amp. Upon receiving your amp, we will give your amp a thorough examination - checking all caps, resistors, transformers, voltages, tubes, etc. We will then call you with an itemized list of everything that is needed to bring your amp to optimum performance. There is no bench fee and our diagnosis is free! After you authorize the work to be completed, we will perform the work and ship your amp to you usually within 48 hours. Recovering or cabinet building will sometimes take a few more days depending on the job.

CUSTOM CIRCUIT DESIGN & MODS

I am frequently asked, "What kind of mods do you do?" My reply is always, "What do you want to accomplish?" There are many ways to alter tube amp circuitry, and when circuitry is altered, so is the sound. Do you want more sustain? Low end? Mid-boost? Dirtier? Cleaner? Hum reduction? Less noise? Browner tone? Dynamic response? Reverb? Gain? Punch? These are only a few of the possibilities!

We do not recommend any mod without knowing, "What result are you wanting to achieve?" Sure we do F/X loops, line-out, switch power tube types, etc. and we can do just about anything. However, when we know what you are wanting the amp to do, only then can we make recommendations to that end. Call us for a free consultation.

Ship to our factory address:

Kendrick Amplifiers

531 County Road 3300
Kempner, Texas 76539-5755
gerald@kendrick-amplifiers.com

Kendrick Brand Tubes

5881	GZ34	7025A Grade 1	7025A Grade 2
Milky smooth distortion with creamy top end.	Punchy with character'	Toney, fat, with high gain, can be used in all 12AX7/7025 sockets	Toney, fat, with high gain for use in all 12AX7/7025 sockets except the 1st gain stage of a high gain amp
Item # KEN 5881	Item # KEN GZ34	Item # KEN 7025A1	Item # KEN 7025A2

OTHER POPULAR BRANDS

Mullard	EL34*	Item # MUL EL34
	EL84*	Item # MUL EL84
	GZ34	Item # MUL GZ34
Phillips	6L6GC*	Item # PHL 6L6GC
	6CA7*	Item # PHL 6CA7
	12AT7WC	Item # PHL 12AT7WC
	12AX7WA	Item # PHL 12AX7WA
RCA	6L6GC*	Item # RCA 6L6GC
	6V6GT*	Item # RCA 6V6GT
	6550*	Item # RCA 6550
	7027A*	Item # RCA 7027A
GE	7581A (20% upgraded 6L6GC)	Item # GE 7581A
Tung Sol	6550*	Item # TS 6550
	5881*	Item # TS 5881

AMERICAN MADE PREAMP TUBES & RECTIFIERS

12AX7	Item # AM 12AX7
12AT7	Item # AM 12AT7
12AU7	Item # AM 12AU7
6SC7	Item # AM 6SC7
6SJ7	Item # AM 6SJ7
6SL7	Item # AM 6SL7
6SN7	Item # AM 6SN7
5AR4/GZ34	Item # AM GZ34
5Y3GT	Item # AM 5Y3GT
5V4GT	Item # AM 5V4GT
5U4GA	Item # AM 5U4GA

* DUETS, QUARTETS AVAILABLE. PRICE PER TUBE IS THE SAME.

OTHER KENDRICK GOODIES

KENDRICK "T" SHIRTS

0% cotton, black "T" shirts with the Kendrick logo on the back and our slogan "She will slap you, She
I kiss you, You will love her!", on the front. Specify size S, M, L, XL, XXL, XXXL. **Item # Shirt**

KENDRICK SPEAKER STICKERS

old foil with the Kendrick Blackframe Speaker logo.

Specify size: for 8" speaker — **Item # Stick 8**
for 10" speaker — **Item # Stick 10**
for 12" speaker— **Item # Stick 12**

KENDRICK DEMO TAPE

assette tape demonstrating the Kendrick combo amps and reverb. **Item # Tape**

REPLACEMENT PARTS & TRANSFORMERS FOR VINTAGE AMPS

omplete line of replacement parts, components, transformers for the do-it-yourself repairman. Call
r parts catalog and/or pricing. We can rewind any transformer. We stock most early tweed through
'is transformers.

A DESKTOP REFERENCE OF HIP VINTAGE GUITAR AMPS

500+ page book written by Gerald Weber. This book covers the hip circuits of the 50's and 60's.
chematics of dozens of amps. Tricks, mods, tips, restoration ideas. **Item # Book**

KENDRICK AMPLIFIERS

Aᴅᴅʀᴇss ғᴏʀ ᴏʀᴅᴇʀs ᴀɴᴅ ᴍᴀɪʟ:

531 County Road 3300, Kempner, Texas 76539-5755 U.S.A

PHONE 512-932-3130
FAX 512-932-3135
EMAIL gerald@kendrick-amplifiers.com

©1997 Kendrick Amplifiers (a trademark of MOJO TECHNOLOGIES INCORPORATED)

SOUNDCHECK
KENDRICK 2210 COMBO

GUITAR WORLD, MAY, 1994: By Chris Butler

Not only is the Kendrick 2210 the best-sounding, best-looking and best-built Fender-style amp I've ever run across, it also encapsulates two prevailing trends in musical equipment: a craving for the lost quality attributed to anything "vintage," and an appreciation for small production-run, handmade gear.

Kendrick is a "boutique" company based in Pflugerville, Texas, that's building dream amps from primo materials...and if the right parts aren't available (like super-efficient output transformers and black-frame Jensen-style speakers), they just tool-up and make 'em themselves. One glance tells you this beautiful amp is a labor of love—the antique tweed covering gives off a golden glow, the light brown grille cloth is pure Fifties, the steel-reinforced leather handle is hand-stitched and the triple-plated chrome on the control panel looks an inch thick. Inside its solid pine, dovetailed-jointed cabinet, you'll find hand-soldered, cloth-covered wire, a 16-gauge steel chassis and sweet-sounding Sovtek tubes nestled in ceramic sockets. The workmanship, of course, is superb.

Rather than trying to meet every guitarist's needs all at once, the 2210 concentrates on getting just one thing right: a classic, Tweed-era sound. There are contemporary embellishments—the caps and resistors have lower tolerance values than those originally used by Fender and thus will last longer, there's spike protection and transient power suppression to prolong tube life, a hum-reducing circuit and a few other modern mods—but nothing violates this Prime Directive.

The electronics are based on the famous 40-watt, Western Electric/Fender '59 Bassman circuit and are housed in a Fender

"Super"-sized, 2x10 open-back cabinet. The tube configuration represents a precise balance of power and tone: twin 5881 power tubes (military version 6L6s), a trio of 7025s, a tube rectifier (GZ34) and a single 12AY7, which has less gain but draws more current than 12AX7s, producing a richer, "tuber-y-er" sound.

Sharp-eyed technoids will have spotted something very un-Bassman-like in the tube list: an extra 7025. Discretely parked underneath the main amp chassis is a mini-chassis with an effects loop send and return, loop return gain/third gain stage control and slide selector switch, plus footswitch and preamp out jacks. Toggling the switch (or the supplied footswitch) assigns this extra tube to either the loop's return circuit or adds a third gain stage for mega-distortion.

Since speakers and output transformers determine half of any amp's sound, Kendrick has paid special attention to these components, too. As noted above, recreating classic sounds often requires classic parts, so Kendrick manufactures improved replicas of Alnico 5 Jensen model P-10R speakers. As expected, these black beauties sound incredible, reproducing singing highs and unmuddy lows effortlessly.

But even great speakers and choice amp circuitry will produce a sucky sound if linked together with an anemic output transformer. Once again, Kendrick found existing components unacceptable, so they hand-wind theirs one at a time. The 2210's transformer is the key to its awesome sound. The 2210 is fat, fat, fat...a total surprise since amps this small are expected to sound thin.

Another transformer-related benefit comes from only having two 8 ohm speakers rather than the traditional, 2 ohm, 4x10 Bassman configuration—thus the amp can work the speakers twice as hard. Credit goes to New Wave guitar god Tom Verlaine (of Television) who first asked Kendrick to make him a 2x10, since he was doing a lot of shuttling between recording sessions in New York City and needed a taxicab-friendly amp.

Road-testing the 2210 was cake...nearly everyone in my circle wanted to take it for a drive. The Gefkens, a band I share rehearsal space with and who are very sonically savvy, spotted it in my amp pile and were immediately curious. Guitarist Matt Azzarto, who normally uses a 50 watt JCM800 Marshall with a 4x12, liked how the 2210 thickened his single-coil-loaded Rickenbacker 330 and Telecaster, but his humbucker-equipped SG was too bottom-heavy. Matt also wondered if the 2210 might be over-powered for its size. "It's solidly built, but at high gain levels the cabinet buzzed like crazy...like [the speakers] wanted to jump outta there!" Alan Edwards of Chocolate USA also used the 2210 with an SG...only this time with positive results. In a marathon weekend recording session, Edwards cut both clean and skrawnky over-dubs on tracks destined for release by Bar None Records (They Might Be Giants, Freedy Johnston). "It's great!" he enthused. "I got a fuzzed-out, indie rock sound...then shifted to a clean setting for melodic parts. It's very versatile." Engineer Ray Ketcham praised its "thick and ballsy dirge-y sound.... And it didn't sound like a pedal." But he was most grateful for the amp's fast set-up time...a godsend on low-budget, no-time sessions.

Me? I've probably logged 50 hours playing-time with the Kendrick 2210, tried jumping its channels with a Y-adaptor, ran all kinds of cheesy effects through the loop...even laughed about it having the longest power chord I've ever seen...and still can't find anything to complain about. Maybe I'm getting old...er..."vintage," but once in a blue moon something comes along that's really the "real thing."

KENDRICK
"THE RIG"

GUITAR SHOP, AUGUST, 1996: By Lisa Sharken

Kendrick is a small operation out of Pflugerville, Texas, and the heart and soul of founder and chief engineer, Gerald Weber. Kendrick amps have been around for several years, known mainly for their vintage Fender-style tweed sound and look. Weber recently re-designed his entire Kendrick line, discontinuing the Vintage '50s series amps and replacing them with the new Texas Crude line. His new amps still have a vintage motif, right down to their pointer knobs and leather handle, but as the name implies, they're meaner and cruder-sounding. The Texas Crude line is offered with new trapezoidal "Beaumont," or square-shaped, "Austin"-style cabinets. Both cabinets are open-back and have a TV-shaped front with Kendrick Kreme grille cloth and Kendrick Nico-tweed covering. However, custom coverings are also available by special order.

I checked out an "Austin" version of The Rig, a 35-watt all-tube combo with one 12" Kendrick Black Frame speaker. It's hand-wired and has point-to-point circuitry that's designed around a pair of 5881 power tubes, a GZ-34 rectifier tube, and two 7025 preamp tubes. Kendrick Amps are shipped with the tubes packed separately, so the user has to install them and be certain they're in the right places before attempting to operate the amp. Inserting the tubes isn't too difficult, but it is necessary to stick your head inside to see which way the tubes fit. When you look "underneath the hood," you'll find a line-out, an extension speaker output, and a cathode/fixed bias switch that's hidden between the two preamp tubes. This switch changes the way the amp reacts and sounds by changing the bias. Cathode bias produces a tighter, more compressed tone that works well for more quiet situations like recording or performing in a small room. Fixed bias is usually preferred for playing louder and when you want more punch.

The Rig's controls are extremely simple: there are two channels—Mic and Instrument—and each channel has two inputs for high and low. The Mic channel is bassier and padded for using hotter instruments, while the Instrument is hotter and has more treble. There's a separate volume control for each channel and one tone control. And all the controls go to 12! Does that mean it goes louder? I think my neighbors can answer that one. Because The Rig doesn't have a Master Volume, it does have to be cranked for more gain.

I tested The Rig with a '65 Strat and a new Historic '59 Reissue Les Paul. Plugging into the Instrument channel first, The Rig delivered a delectable shimmering clean tone with a clear low end, punchy mids, crisp, sparkly highs and plenty of sustain. That sound doesn't change when you turn the amp up either, but playing harder produces a more compressed sound and a bit more overdrive when you crank the amp. If you're looking for more overdrive or a really distorted sound, then you'll definitely need a stompbox. Rotating the tone control, I could hear changes at each increment, but even set on 1 there was no sign of mud, and at 12, there weren't any ear-shredding highs. The Strat was a great match for The Rig, and I preferred its sound over the Les Paul's. I tend to prefer matching Fender guitars with Fender-style amps anyway. On the other hand, the Paul made the amp kick into overdrive easier.

The Rig is an amp designed with just one thing in mind—tone. If you're looking for something with switchable channels, multiple stages of gain, and as many built-in effects as can be crammed into one box, then look elsewhere. But if tone is your main concern, The Rig is definitely worthy of your attention. Its tone and intensity range from warm and sweet to punch-you-in-the-face nasty, if you ask for it. It also has the longest power cord I've ever seen on any amp, so you'll never have to worry about reaching an outlet across the stage in a club—or anywhere else in town.

POSITIVELY VINTAGE

VINTAGE GUITAR, SEPTEMBER, 1995: BY STEPHEN PATT AND
GUEST REVIEWER TIM CURTIN

This month brings us a sumptuous feast for the eyes and ears from
established luminaries and new talent, such as the new series of guitars
from Gerald Weber of Kendrick Amplifier fame. Who says vintage
ain't fascinatin'? Before getting into the goodies (and I truly mean
that — every month is a funfest at the office, wondering what new
and strange product each UPS shipment has inside), I should introduce our guest reviewer this month, multitalented guitarist Tim Curtin.

Unlike some of our previous high-profile guitarists, Tim leads a
relatively sedate life, ignored by teenyboppers and paparazzi alike.
What he does instead is spend 99% of his time playing jazz, blues,
rock, and folk music in recording studios, clubs, and some of the
more exotic locations around New York. A veteran of the Howard
Stern Show (Tim was bandleader during the 1980s for Howard's
notorious backup band Pig Vomit before progressing to the finer
things in life), Tim studied under jazz great Howard Morgen, and at-
tended Berklee College of Music ("Where I picked up all my fast
licks," he jokes). Having played with such diverse talents as Jack
Bruce, Leslie West and Corky Laing, Larry Coryell, Levon Helm, and
Phoebe Snow, Mr. Curtin now works with tremendous energy and
prodigious chops in the NY music scene doing everything from
jingles to song demos, playing in small smoky clubs to leading a full
band at prestigious society gigs. Tim's newest project is an album of
original songs co-written with rhythm and blues vocalist Mike
Nappi, due for release in early 1996.

Anyone who knows amp guru Gerald Weber will attest to the
fact (a) this guy has a practically inexhaustible supply of energy,
and (b) he eats, breathes, and sleeps guitar. The prime mover behind

Kendrick Amps, the first new amplifier company in years to be touted by studio aces and touring musicians for its unswerving dedication to tone, Gerald has built a small empire from his amplifiers, branching out into Kendrick Blackframe speakers ("The ultimate vintage style guitar speaker on the market today"), output transformers ("made just like they made 'em in the old days"), and parts of all description. If this gives you the idea that Mr. Weber speaks in hyperbole, then you're on the right track, but with his encyclopedic knowledge of vintage guitar history, novel approach to design and problem solving, and bigger-than-life dedication to making The Perfect Amplifier, you just accept him as he is. Think of Gerald as a little piece of Texas, totally dedicated to his craft.

Teaming up with master luthier Tony Nobles, Kendrick has branched out into the guitar business, and wouldn't you guess there's nothing modest about their entry level instruments, appropriately labeled "Tone Monsters from Texas." Made in limited quantities in Pflugerville, Texas are the Continental, roughly styled in the footsteps of Leo, and the Town House, derived from a very special 1959 Les Paul played by a certain Billy Gibbons. We were lucky enough to get the loan of a Town House for this month's column, and had trouble prying it out of the hands of any guitarist lucky enough to handle it. According to Gerald, the instrument is virtually blueprinted from 'Pearly Gates', the aforementioned 1959 Paul played by Billy G, and is named for the Groves, Texas supermarket-turned-club where ZZ TOP and many other Texas acts cut their teeth before climbing the ladder of fame.

"The sound of that particular instrument played through Billy's 100 watt Marshall plexi is the definitive guitar sound, as far as I'm concerned," raves Gerald. Using that tone as his benchmark, Gerald was allowed to carefully measure and weigh the 'Gates' guitar and make an initial copy for his own use. "Problem was," Mr. Weber drawls, "everyone who played or heard it being played wanted to buy it!" Due to the unexpected but tremendous response to the first instrument, Tony and Gerald began to produce the Town House in limited quantities. "We've got almost all the features of the original guitar, including the faded sunburst finish, and the feel," Weber relates enthusiastically, "but have added a few special twists..." Both

Seymour Duncan and Fralin pickups are available, and the straplocks our model was equipped with are optional. The frets are jumbo and perfectly polished, although our model's Dunlop 6105's frets extended right to the edge of the fretboard, a special nod to the owner, who bends so energetically he'd "bend the string right off the fretboard" on a regular neck. (In checking other production Town Houses, a more traditionally polished and dressed fret installation was found.) The Brazilian rosewood of the fretboard is uniform and luscious, as are the handpicked woods used for the body and neck, a combination of seasoned Honduran mahogany and select maple. "This is a niche market Tony and I are aiming for," explains Gerald, and using a true nitro finish and the high quality tonewoods certainly would prevent him from competing with Gibson in quantity. But... we're talking quality here. From the fossil ivory used on the tightly fitted nut to the to-die-for finish, this guitar is a winner.

Besides the unusual but attractive headstock, no doubt designed to throw Gibson for a loop, and the lovely little map-of-Texas inlays on the fretboard, another innovation is the exclusive offset V-shaped neck, fitting comfortably into the hand like a seasoned 35-year old guitar should. On seeing a 1930's Gretsch with an asymmetric neck, weighted towards the bass side for strength, the boys from Kendrick translated that ergonomic design to the production Town Houses, resulting in a more natural feel. In looking at research by sound engineers, confirmed by master luthier Rick Turner of California, the added mass of the neck produces more sustain and stronger fundamentals, resulting in a more toneful guitar. While a traditional neck is available, most players are requesting the offset neck, described at first glance by Tim Curtin as "a deep V-neck wearing a Wonderbra."

On sitting and playing the Town House in our studio, Tim remarked, "Acoustically, without even plugging this in, I can tell this is a really good guitar. The sustain and resonance are impressive." We put the Town House through a variety of amplifiers, from blackfaced Fender Princeton and Twin to a Rivera 120 watt monster, and produced tones ranging from jazzy—"The rhythm pickup is extremely versatile, much better than I've heard on stock Les Pauls"—to bluesy, up-to-and-past raunchy rock tones. "At first the neck was a little strange, but I'm digging it now," commented Curtin. "I can get a wide vibrato, with great control and very little effort." So, we have a perfect match of cosmetics, design, and that elusive vintage vibe in the Town House Guitar. At $3,295, it's not cheap, but since when did perfection come cheap? This instrument is a direct competitor with the Centennial and Historical Pauls, at a fair price, with perhaps more care, finer materials, and better workmanship than any of the competition. As a player's guitar or a collectible, the Town House makes an indelible impression. Contact Kendrick Amplifiers, 531 County Road 3300, Kempner, Texas 76539-5755, or call (512) 932-3130, fax (512) 932-3135 or email gerald@kendrick-amplifiers.com.

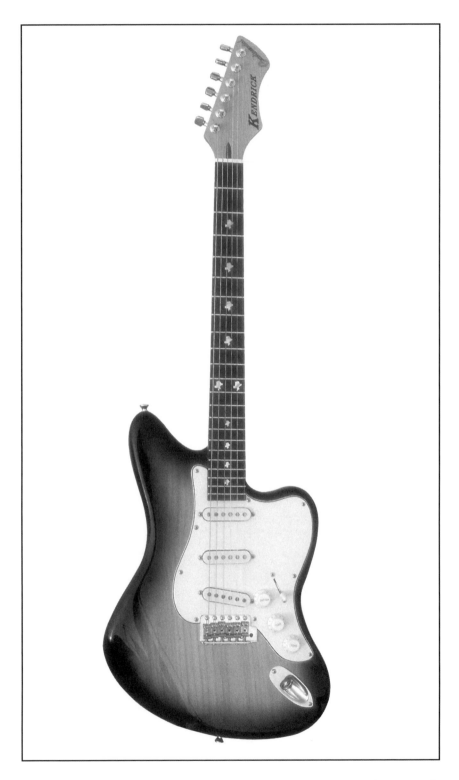

POSITIVELY VINTAGE

VINTAGE GUITAR, JANUARY, 1996: BY STEPHEN PATT AND GUEST REVIEWER ERIC WALTERS

We recently reviewed the Kendrick Town House Guitar from Pflugerville, Texas, and were mightily impressed by its high level of craftsmanship and playability. These are certainly not easy things to come by in a new guitar; lest you forget what it's like in the real world, where dollars are rationed carefully. My friend Tim Curtin recently picked up a new Fender American Strat at Sam Ash to use as an all-purpose electric for clubs, and on arriving home, found the instrument to be positively... unplayable. The main problem was lack of proper setup (something most smaller shops find time for, especially because they depend on the repeat business, but also because they care and they know better). On bringing the new guitar back to the retailer for the "free adjustment" promised by the warranty, Tim was told by the repairman, "Don't even bother. The workmanship isn't too hot to begin with, and I don't think I can make it a lot more playable than it already is." Wow. American know-how at its worst. Needless to say, the guitar was promptly returned for a full refund, but a lesson was learned.

Gerald Weber, the moving force behind Kendrick Amplifiers, has put lots of expertise and effort into his venture into the 6-string guitar market, and he's certainly doing it right. Our Kendrick Continental Guitar is another down-home inspiration drawing from the Austin blues scene and dedicated to Stevie Ray Vaughan and the Continental Club, where this vital music flourished. A true meld of function and form, the Continental is perfectly balanced in weight and aesthetics, sporting a Fender-esque body shape, sort of a cross between a Strat and a Jazzmaster, but quite ergonomic and original. The headstock is of course unique (for multiple reasons), and is at first, awkward looking... but after a bit, grew on me. As Eric put it, "We're used to seeing these headstocks that are so classic; when you see another headstock that's new and interesting, the immediate re-action is, "Ooohh!!" (I think what he meant to say is, "Yuck!") This new semi-trapezoidal headstock shape provides perfect balance (the instrument remained in perfect playing position when held on a strap, neither top- nor bottom-heavy), and may lend a little extra res-onance to the natural tone of the instrument, due to its mass. The aged one-piece body (both ash and alder are available) bore a gor-geous classic sunburst finish covered with a true nitro finish, thin and impeccably executed. The Continental literally sounded like a great guitar without ever plugging it in, due in some part to the proper aging and selection of the tonewoods, but also in part to the nature of the nitro finish, which doesn't deaden the guitar's resonance like some thicker poly-acrylic finishes do. I might add at this juncture that luthier Tony Nobles has the upper hand in these guitar-related pieces of Kendrick's, and his quality control and design capabilities are quite impressive. The trio of pickups was by Seymour Duncan, in the expected Strat configuration, and performed just as well as they looked, and controls were the familiar volume/tone/tone, with a snug five-way switch (Fralins are also an option). Our piece was equipped with a right-handed vibrato (both righty and lefty are available), al-though adjusted down to effectively function as a non-vibrato guitar. The neck... how should I describe it? A straighter, sleeker chunk of maple was ne'er seen? (Okay, so Shakespeare I'm not.) From the Brazilian rosewood fretboard to the Schaller pegs, down to the cute little mother of pearl inlays, this neck says "Play me!" Which is what we did.

Our Continental was played through a variety of amps, including the blackface Fenders, an aging-but-spunky Supro Big Star, and a Rivera 120 watt monster. It sounded consistent and had a very de-termined personality in every situation, live club or studio. The neck was fretted with tall Dunlops, and played like a dream. Sounds were classic Fender, but with the extra edge and fullness I've grown to as-sociate with the Duncan pickups. The guitar sounded wonderful through the Pennalizer, delivering a startling bite and snarl at top-gain settings full out, moving to a bright but not brittle shimmer at lower volume settings. Eric said, "The secret to the special quality of vintage guitars is that they've been played a lot. Here we have an example of a new instrument that sounds and feels great. Of course the instrument has to be well constructed to begin with. And I can imagine that with time, this will only get better." He went on to point out, "I don't really like Schaller pegs — they seem to rob the tone off of an instrument, with all that mass. Standard Klusons are good enough for me. They may have a cheesey quality to them, but there's something about them that enhances the 'ring' of a good guitar!"

Eric waxed enthusiastically, "If an instrument's too overbuilt, you know, you gotta play it for a long time to get some life into it. The overall depth of rosewood on the neck is a bit more than I'm used to. This is built like a brick s***house... really very exceptional detailing, from the quality of the electronics to the little Texas inlays on the neck. The fretwork and polish is quite nice, and I like the way Gerald has added an extra fret to the board to create a 22-fret neck. The finish and sunburst are classic vintage, really pretty. You don't see many people taking the time to do a good nitrocellulose like this these days. It allows the instrument to breathe, and sound better. I find the cut of the body reminiscent of an [Gibson] RD-Artist, more so than a Fender. But it's got a real Fender feel to it, all except the headstock, of course. The balance, by the way, is excellent. You can't go wrong with these Duncan pickups, either. They sound just right for this guitar. Overall, it's a very straight-ahead guitar with a genuine Fender vibe. I doubt that the neck will ever move — the wood quality is great, with extremely straight grain, and a very natural feel to it. This personality is 'sturdy'."

BIG GUITARS OF TEXAS

GUITAR WORLD, JUNE, 1996: BY CHRIS GILL

Kendrick has established a stellar reputation for its hand-built tube amplifiers, and now the company is seeking renown for its new line of solidbody electric guitars. With a bolt-on 22-fret neck, three single-coil pickups and a body shape similar to a Fender Jaguar's but smaller, the Continental has a distinct Fender-like vibe.

In addition to the Jag/Strat hybrid design, the Continental has a few other unusual features, including a left-handed tremolo system and an offset V-profile neck. The reversed tremolo is a tip of the Stetson to Stevie Ray via Jimi, while the neck profile, which is thick at the bass strings and thin at the treble strings, is influenced by non-truss rod Gretsch and Epiphone necks from the Thirties. For players who like to rest their left thumb along the top of the finger-board, the tip of the "v" fits comfortably in the palm, but the profile may feel uncomfortable to those who center their thumbs along the back of the neck.

To further distinguish the Continental, Kendrick has made several improvements to classic designs. The headstock is angled to increase sustain and eliminate the need for string trees, and it features a recessed truss rod that can be accessed without having to remove a cover. The custom-wound Lindy Fralin pickups are a little hotter than a normal Strat's, and the guitar is wired so the bridge and neck pickups are engaged when the 5-position switch is in the middle position, instead of just the center pickup. There are also separate tone controls for the bridge and neck pickups.

Like a competitor going for the grand prize in a chili cookoff, Kendrick didn't skimp on this guitar's ingredients. The fingerboard is a hefty slab of quarter-sawed Brazilian rosewood, the nut is cut from 40-year-old ivory and the body is crafted from a single chunk of swamp ash. Even the stock strings are wrapped with pure nickel.

Nitrocellulose lacquer covers the body and the maple neck has a warm, curry-colored glow. Our test model had a custom fingerboard inlay, but the stock model comes with tiny Texas-shaped inlays.

Plugged in, this guitar howls like a wolf in an El Paso thunderstorm. It has a fat, toothy Strat tone that's as sharp and shiny as a Bowie knife. When you roll the tone controls back, the sound gets fatter, but it doesn't get bogged down in mud. Set to the neck pickup with the tone rolled all the way off, the Continental puts out sustaining, creamy woman tone.

Since most of Kendrick's guitars are built to order, the Continental is available with a variety of options, including a traditional, rounded neck, a variety of different finishes, an alder body, a right-handed tremolo and your preference in frets.

Don't let the Continental's traditional appearance fool you into thinking that this is yet another Fender knock-off. With several thoughtful innovations and top-notch materials, this guitar stakes new territory in the vast frontier of electric solidbodies.

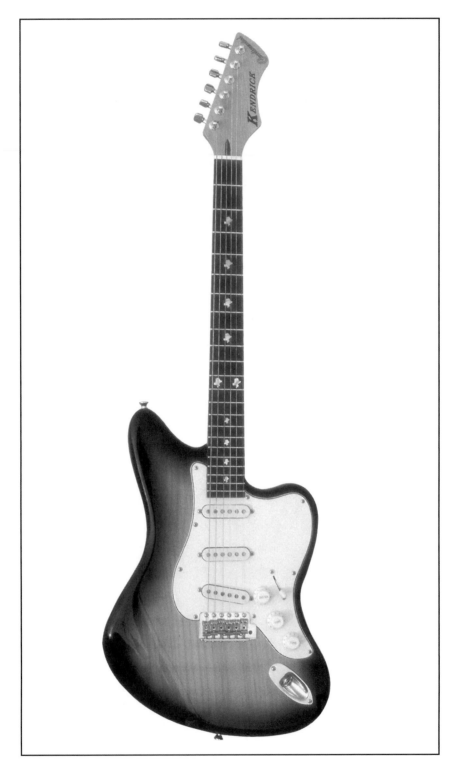

KENDRICK CONTINENTAL

GUITAR SHOP, AUGUST, 1996: BY LISA SHARKEN

Custom built-to-order guitars and amplifiers from smaller shops are becoming more and more popular, especially among discriminating players looking for something unique and a little more personal than the basic off-the-rack axe and combo. Kendrick is a small Texas-based company dedicated to building custom gear for those discerning players who recognize the difference. Gerald Weber is the owner of Kendrick and also the designer of all the guitars and amps they create.

The Continental's body shape reminds me of a Fender Jaguar. It's constructed from one piece of alder and is finished in a two-tone '50s Fender-style sunburst. Its weight is about average for a guitar of that size. The bolt-on maple neck has an offset "V" shape that's thicker on the bass side and thinner on the treble side, and squared off at the heel where it bolts into the body. This gives the neck more meat where it's sometimes needed, especially if you like to wrap your thumb over the top of the neck when you play. It has an incredibly thick Brazilian rosewood fingerboard with colorful mother-of-pearl inlays in the shape of Texas down the neck. The side marker dots are also mother-of-pearl. The headstock has its own distinctive shape that gives it its own identity. The nut was hand-carved from a generous chunk of 50 year old ivory from Weber's own private stash. Ivory is a rather scarce commodity these days, and so very few instruments come equipped with ivory anything. Although I'm impressed with the choice material, it appears to me that it may possibly be too thick and wide for the size of the slot, and I don't think that it's seated as firmly as it could be. However, I'm aware that this particular instrument was completed in a rush and then quickly sent off to me, so this detail was probably just overlooked because of the short time frame. It's not a big deal to correct, and I expect that other orders will have greater attention paid to them.

The tuners are standard chrome-plated Schallers, and the 21 frets are Dunlop 6105s, which are slightly wide and on the tall side. They've also been left longer on the sides and are not beveled inward very much, which allows for a wider finger vibrato by giving a tad more room on each side of the neck. The truss-rod is easily accessed from the top of the neck, right above the nut, and there's no truss-rod cover to mess around with or lose when you need to make a neck adjustment.

Acoustically, the wood really sings. It's got a lot of natural sustain. Electrically, it has a more traditional sound, very much like a Strat. The Continental comes with three Seymour Duncan APS-1 single-coil pickups, although other pickup options are also available, such as Lindy Fralins. Like a Strat, the guitar comes with a 5-way pickup selector switch and one volume and two tone controls. However, the wiring and controls are set up a little different from standard Strat-style 5-way wiring. With this setup, the middle position yields the two outside pickups, and there's no selection to get just the middle pickup by itself. The tone controls affect the neck and bridge pickup, whereas on a traditional Strat there's no tone control for the bridge pickup. This system gives you the best-sounding options out of the whole deal. As for the rest of the package, there's a Strat-style output jack, a traditional tremolo, and a vintage mint-green pickguard, which looks cool and definitely gives off the right vibe.

Kendrick guitars are available direct by special order. If you're ever visiting around Texas or hanging out at one of the Texas guitar shows, be sure to look them up.

INDEX

OTHER WORKS BY GERALD WEBER

Basic Tube Guitar Amplifier Servicing and Overhaul

SECOND IN A SERIES OF INSTRUCTIONAL VIDEOS

In Gerald's second and more advanced video, you will learn:

- Basics of servicing and overhauling a tube amp
- What tools are needed
- The wiring mistake made on 50% of all amps and how to correct it
- The best way to straighten preamp tube pins
- The mistake within the bias circuit of some Blackface amps and the essential correction
- Selection of rectifier tubes for tone and performance
- How to perform a cap job correctly
- Soldering and de-soldering
- Proper servicing of potentiometers, jacks, sockets, tubes and more

One hour and thirty minutes with Gerald Weber, diagnosing, inspecting, troubleshooting, servicing, repairing and even playing the tube amp! It's an hour and a half of essential tube amp knowledge.

Price is only $34.95 plus $3 shipping and handling

Check your local dealer or contact Kendrick Books (512) 932 3130